QUANTUM CONCEPTS IN THE SOCIAL, ECOLOGICAL AND BIOLOGICAL SCIENCES

Quantum mechanics is traditionally associated with microscopic systems; however, quantum concepts have also been successfully applied to a diverse range of macroscopic systems both within and outside of physics. This book describes how complex systems from a variety of fields can be modeled using quantum mechanical tools, from biology and ecology to sociology and decision-making. The mathematical basis of these models is covered in detail, furnishing a self-contained and consistent approach. This book provides unique insight into the dynamics of these macroscopic systems and opens new interdisciplinary research frontiers. It will be an essential resource for students and researchers in applied mathematics or theoretical physics who are interested in applying quantum mechanics to dynamical systems in the social, biological or ecological sciences.

FABIO BAGARELLO is Professor of Mathematical Physics and Mathematical Methods at the University of Palermo. His research interests include the application of quantum mechanics to macroscopic systems and the application of functional analysis and operator theory to quantum mechanics. He is the author of numerous scientific articles on these topics in addition to four books and several edited volumes.

QUANTUM CONCEPTS IN THE SOCIAL, ECOLOGICAL AND BIOLOGICAL SCIENCES

FABIO BAGARELLO

University of Palermo

CAMBRIDGE UNIVERSITY PRESS

CAMBRIDGE
UNIVERSITY PRESS

University Printing House, Cambridge CB2 8BS, United Kingdom

One Liberty Plaza, 20th Floor, New York, NY 10006, USA

477 Williamstown Road, Port Melbourne, VIC 3207, Australia

314-321, 3rd Floor, Plot 3, Splendor Forum, Jasola District Centre, New Delhi - 110025, India

79 Anson Road, #06-04/06, Singapore 079906

Cambridge University Press is part of the University of Cambridge.

It furthers the University's mission by disseminating knowledge in the pursuit of
education, learning, and research at the highest international levels of excellence.

www.cambridge.org
Information on this title: www.cambridge.org/9781108492126
DOI: 10.1017/9781108684781

First published 2019

Printed in the United Kingdom by TJ International Ltd, Padstow Cornwall

A catalog record for this publication is available from the British Library.

ISBN 978-1-108-49212-6 Hardback

As always, I dedicate this book to my beloved parents, Giovanna and Benedetto, and to Federico, Giovanna and Grazyna, with pure and diverging (to $+\infty$!) love. I wish for them to enjoy their lives much more than I do (and please consider that I am already quite satisfied).

Contents

Preface

I like to do research, and I like to move from one topic to another. Otherwise, I easily get bored! But from time to time, I feel the necessity to stop and think of what I have produced up to that moment, alone or with a group of colleagues, who quite often are also friends. And this is exactly the right moment to stop and summarize what has been done during the past few years after adopting my operatorial approach to mathematical modeling – which is exactly what this book is about. Writing a book also gives you a nice opportunity: one needs to reflect again and again on things which, when you write an article, may appear simple and clear. But sometimes you realize that what appeared to be simple years ago is not simple at all, and it possibly deserves a deeper analysis and a better understanding. And this is something I like: my old (or recent) results live again, give new suggestions and drive me toward unexpected directions. I am forced to read material on topics which are, in principle, far away from my usual *know-how*: ecology, economy, biology and anthropology are just a few such topics, but aren't the only ones. And this is a nice way to learn many amazing things.... Hence, I consider this book as the conclusion of a long period of study and the beginning of a new one. At least, this is what I hope. And I also do hope that you will enjoy reading this book, and that it will spark your curiosity, encourage you to start forming ideas of your own and then see if those ideas can be transformed into formulas according to the framework described in the chapters to follow. In fact, this is also what a book is for: to attract people! So, please, come and join me: everyone is welcome!

Acknowledgments

It is really a pleasure to thank all the people who significantly contributed to my research in this field along the years, starting with the old ones, Franco Oliveri and Francesco Gargano, going to the recent friends, Emmanuel Haven and Rosa Di Salvo, and ending with the brand-new ones, Andrei Khrennikov, Irina Basieva, Emmanuel Pothos, Lucia Tamburino and Giangiacomo Bravo. Their help has been precious in ∞^∞ situations! And they will soon discover I still need them!

Some of the figures you will see appeared (as they are, or in some slightly modified form) in a few articles of mine. The editors kindly gave me the permission to use them, and for this reason I thank all of them. In particular, I wish to thank SIAM for some figures in Chapters 3 and 4, Elsevier for others in Chapters 3, 5, 6, 8 and 12 and Springer for more figures in Chapter 3. I also wish to thank Entropy, Plos One and Phylosophical Transaction A for allowing me to use the figures in Chapters 7, 9 and 11.

1

Why a Quantum Tool in Classical Contexts? (Part II)

The reader may wonder where *Part I* is. In fact, there is no *Part I* here. Part I is in [1]. When I wrote that book I felt strongly the responsibility to justify my approach, since it was, in fact, rather unusual, and the reaction of most referees, when submitting a research paper of mine, was quite often the same: "Why are you adopting this technique? What is wrong with a *classical approach*?" However, since 2012, I realized that this approach was not so crazy, and I discovered that many people, in many different fields of research, were adopting similar strategies, using quantum ideas and, in particular, mathematical tools deeply connected with quantum mechanics, to deal with problems that are not necessarily related to the microscopic world. For this reason I do not really feel anymore the necessity of justifying myself. However, I think that giving some words of explanation can still be useful for readers, and this is what the next few sentences are about.

The driving idea behind my approach is that the lowering and raising operators related to the canonical (anti-)commutation relations (CCR or CAR) can be used in the description of processes where some relevant quantities change discontinuously. To cite a few examples, stock markets, migration processes or some biological systems: multiples of share are exchanged, in a market; one, two or more people (and not just *half of a person*) move from one place to another; and one cell duplicates producing two cells. This suggests that objects labeled by natural numbers are important, in some situations. People with a background in quantum mechanics know that nonnegative integers can be seen as eigenvalues of some suitable number operator constructed, in a natural way, using ladder operators. Also, ladder operators can be efficiently used to describe systems where discrete quantities of some kind are exchanged between different agents. Hence ladder operators, and some combinations of them, can be used in the description of particular systems. We refer to [1] for some results and models in this direction. But, due to the fact that the *observables* of a system S are now operators, one of the main questions to be answered is the following: *How should we assign a time evolution to these observables?* The answer which I have adopted so far is, in my opinion, completely natural: I use a Hamiltonian-like operator H, which describes the interactions occurring between the various elements of S and their *free energy*, and then I adopt the Heisenberg or, equivalently, the Schrödinger equation of motion borrowed from quantum mechanics to deduce the time evolution for S. More specifically, I use the first when I am interested to describe the time evolution of the observables of S, while I adopt the second to find the time evolution of the state of the system. Both choices are useful, and we know that they are equivalent [2,3], at least under suitable conditions for H.[1] Quite often, in my previous applications, I have used the Heisenberg representation, so that I have dealt with differential equations with unknowns that are time-dependent operators. However, and this has proved to be particularly useful from a technical point of view, I have also used the Schrödinger representation, in particular when some explicit time dependence was to be included in the Hamiltonian of S. This was the case, for instance, in [4], where my operator strategy was adopted in connection of some recently available demographic data for prehistoric South America, and proved to work well in reproducing some aspects of these data, providing some possible scenarios for the agents considered by the model: humans and the natural resources.

Despite of their nature, ladder operators are not necessarily restricted to the analysis of cases in which the observables can only have discrete variations: the same general settings was adopted in other situations, to describe the time evolution of densities in some biological or sociological contexts [5–8], or the *degrees of affection* in love affairs [9, 10], or yet in the description of some decision-making processes [11–16], some of which will be described in this book. We will see how, in many explicit situations, the dynamics of a physical system can be deduced not from a set of ordinary or partial differential equations, but from a single operator living in a non-commuting world. Of course, this approach makes particular sense *if it works*, i.e., if the dynamics deduced in this way describes what is observed in the real world, or, even better, if we can make predictions out of the model. However, I should admit that, in my opinion, a similar approach is also interesting in itself, since it suggests the possibility of using an unusual way to look at a certain system. And it is a way that is quite promising. In fact, a huge literature has been produced, and is still being produced, all having as its focus *quantum ideas outside quantum mechanics*. Just to cite few recent monographs, we refer to the following contributions by several authors: [17–22], in which these quantum ideas are applied in quite different contexts.

The reader should be aware of the fact that, within the book, more so than mathematical rigor, we will be interested in the possibility of getting concrete (exact, or at

---

[1] In particular, this is true if $H = H^{\dagger}$, but not when H and its adjoint $H^{\dagger}$ are different, which is the case in some of the applications discussed in these notes.

least approximated) results. However, we will try to stress the various approximations done all along the book, and to state explicitly whether these approximations and assumptions are *under control*, or if they are only useful to take some steps toward the solution of the problem. In other words, we will do our best to clarify to what extent, within the context of the model we are considering, our results can be trusted or should be refined further.

1.1 Organization of the Book

The book is divided into two main sections. In Part I, I will discuss useful facts from quantum dynamics that will be used later. Some of these facts are well-known, and can be also found in many textbooks, but some others are not, and they are needed for a better comprehension of some of the applications considered in Part II of the book. In particular, in Section 2.8 I will introduce and describe a sort of algebraic dynamics that is not only driven by a given Hamiltonian H, but that is also linked to other features of the system not (easily) included in any such H.

Looking at the table of contents the reader can see that the applications considered in Part II are really different. I will discuss applications to politics, to biology, to economics, to social sciences and to decision-making. This should not be surprising, since all the models proposed in all these areas will be considered as different *dynamical systems*. And we all know that dynamical systems cover quite different areas in research and in the real world. I will also apply quantum ideas to two different (non-dynamical) problems, typical in decision-making: the description of compatible and incompatible questions and the analysis of order effects. In this case, the natural non-commutativity of the observables in quantum mechanics will be the relevant tool adopted to describe these effects.

Going more into detail, in Chapter 3 I will describe a problem of alliances in politics, as a sort of decision-making procedure driven by the interactions of the political parties with a base of electors. In Chapter 4, an application to ecology will be discussed: desertification in a two-dimensional region, and a possible way to control it. A dynamical approach to escape strategies is the content of Chapter 5. In Chapter 6, I will describe two closed ecological systems, with one or two different garbages, and with several *kinds of organism*, discussing, in particular, the existence of some equilibrium, corresponding to a long-time survival of the system. Another biologically oriented application is the topic of Chapter 7, where a simple model of tumor cell proliferation is proposed. Chapter 8 contains a possible extension of the Game of Life, where quantum rules and quantum evolution are adopted. A somewhat *unexpected* application of quantum techniques to prehistoric data miming is given in Chapter 9, while in Chapter 10, I discuss a way to introduce the information in a simple model of stock market. Information is also the core of my interest in Chapter 11, in connection with decision-making. In Chapter 12, a different application of quantum tools to a problem of order and compatibility of questions is discussed, again in the context of decision-making. My final remarks are given in Chapter 13. The variety of topics discussed in Part II is evident. To me, this is a nice indication that the framework I am going to describe here is really promising. Also, it is not hard to imagine that more, and more refined, applications can be considered. So, there is still a lot of work to do.

But now: let's start!

Part I

The General Framework

2

Some Preliminaries

In this chapter we briefly review some basic facts in quantum mechanics, together with a few new concepts and results used in some applications later. In particular, we will first recall some aspects of what physicists usually call *second quantization*, which will be one of the essential tools all throughout this book. For us, second quantization is nothing but the functional settings associated with CCR and with CAR. For completeness, we will also briefly present some deformed versions of these relations, which have been considered in the literature in recent years, because of their mathematical relevance, or for their appearance in pseudo-Hermitian quantum mechanics [23–25], or even for their use in recent models [26]. This will be done in Section 2.3. Then, we will discuss some definitions and results on quantum dynamics, focusing on those aspects which are relevant for us, and discussing several aspects of the Schrödinger and Heisenberg representations, and of their dynamical contents. Next we will recall a few facts on the Heisenberg uncertainty principle and on states of a quantum system, and then, in view of their relevance in Part II of this book, we will discuss and compare two different strategies which produce a stable asymptotic behavior in the time evolution of some given observables, see Section 2.7. We will finally introduce and analyze what has been called (H, ρ)-induced dynamics [5], which can be thought, roughly speaking, as a modified version of the Heisenberg dynamics in presence of what is called a *rule*, whose effect cannot be described in terms of any Hamiltonian operator. This is the content of Sections 2.8 and 2.9, where we will also discuss how a suitable (H, ρ)-induced dynamics can produce some stabilization in the large time limit.

2.1 The Bosonic Number Operator

Let $\mathcal{H}$ be a Hilbert space and $B(\mathcal{H})$ the set of all the bounded operators on $\mathcal{H}$. $B(\mathcal{H})$ is a so-called C^*-*algebra*, that is an algebra with involution which is complete under a norm, $\|\cdot\|$, satisfying the C^*-*property*: $\|A^*A\| = \|A\|^2$, for all

$A \in B(\mathcal{H})$. As a matter of fact, $B(\mathcal{H})$ is usually seen as a *concrete realization* (or, in mathematical terms, a *representation*) of an abstract C^*-algebra. Let $\mathcal{S}$ be our physical system and $\mathcal{A}$ the set of all the operators useful for a complete description of $\mathcal{S}$, which includes the observables of $\mathcal{S}$, i.e., those quantities which are measured in a concrete experiment. Let $\mathcal{H}$ be the Hilbert space where $\mathcal{S}$ is defined. For simplicity, it would be convenient to assume that $\mathcal{A}$ is a C^*-algebra by itself, possibly coinciding with the original set $B(\mathcal{H})$, or, at least, with some closed subset of $B(\mathcal{H})$. However, this is not always possible in our concrete applications. This is because of the crucial role of some unbounded operators within our scheme: unbounded operators do not belong to any C^*-algebra. However, if X is such an operator, and if it is self-adjoint (as the observables of a quantum system are usually assumed to be), then $U(t) = e^{iXt}$ is unitary and, therefore, bounded. In particular we have $\|e^{iXt}\| = 1$, for all $t \in \mathbb{R}$, and X can be recovered from $U(t)$ by taking its time derivative in $t = 0$ and multiplying the result by $-i$. For this reason, C^*-algebras and their subsets are relevant even when unbounded operators appear. However, if we want to define X properly, the first problem we meet is the definition of the domain of X, $D(X)$, i.e., the set of vectors of $\mathcal{H}$ on which X can act, returning vectors which are again in $\mathcal{H}$. If X is bounded, then $D(X)$ coincides with $\mathcal{H}$. Otherwise $D(X)$ is a proper subset of $\mathcal{H}$. It is crucial that $D(X)$ is not too small. In fact, X is well defined only if $D(X)$ is dense in $\mathcal{H}$ [27].

Remarks: (1) In this book the mathematical aspects related to unbounded operators will not play an essential role, even when they appear in the game. The reason is the following: In some models, like in [9], even if the observables of the system $\mathcal{S}$ are unbounded, an integral of motion I exists for the model. This means, as we will show in Section 2.4, that a (self-adjoint) operator I exists which commutes with the Hamiltonian of $\mathcal{S}$. When this happens, not all the (infinite-dimensional) Hilbert space $\mathcal{H}$ attached to $\mathcal{S}$ is available: the only vectors of the orthonormal (o.n.) basis $\mathcal{F}_\varphi = \{\varphi_{\mathbf{n}}\}$, $\mathbf{n} = (n_1, n_2, \ldots)$, $n_j \geq 0$, of $\mathcal{H}$ which are needed in the description of $\mathcal{S}$ are those labeled by quantum numbers which are compatible with the existence of I, and these are only a finite number. This implies that, even if $\dim(\mathcal{H}) = \infty$, the dynamics of $\mathcal{S}$ is restricted to an effective Hilbert space $\mathcal{H}_{eff}$, finite-dimensional, which is the span of those $\varphi_{\mathbf{n}}$ whose $\mathbf{n}$ satisfies the selection rule induced by I. Therefore, the observables of $\mathcal{S}$ become finite matrices in $\mathcal{H}_{eff}$, so that they are all bounded. We refer the reader interested in mathematical aspects of unbounded operators and of unbounded operator algebras to [27–30], where it is discussed how an algebraic structure can be constructed also for unbounded operators, at least under some suitable mathematical assumptions.

(2) It is also convenient to notice that, in most of our computations, we will be interested in computing mean values of the form $\langle f, ABf \rangle$, with A and B possibly unbounded, and with $f \in \mathcal{H}$ with a very special property: f belongs (always!) to the

domain of B, $D(B)$, and is such that $Bf \in D(A)$, the domain of A. Hence, even if AB does not in general make sense, ABf is perfectly defined in $\mathcal{H}$. Hence, $|\langle f, ABf \rangle| < \infty$, for all such operators A and B, and for all these vectors f.

A special role in our analysis is played by the CCR: we say that a set of operators $\{a_l, a_l^\dagger, l = 1, 2, \dots, L\}$, acting on the Hilbert space $\mathcal{H}$, satisfies the CCR, if the following hold:

$$[a_l, a_n^\dagger] = \delta_{ln} \mathbb{1}, \qquad [a_l, a_n] = [a_l^\dagger, a_n^\dagger] = 0, \qquad (2.1)$$

for all $l, n = 1, 2, \dots, L$, $\mathbb{1}$ being the identity operator on $\mathcal{H}$. These operators, which are widely analyzed in any textbook in quantum mechanics, see [2,31] for instance, are those which are used to describe L different *modes* of bosons. From these operators we can construct $\hat{n}_l = a_l^\dagger a_l$ and $\hat{N} = \sum_{l=1}^{L} \hat{n}_l$ which are both self-adjoint: $\hat{n}_l = \hat{n}_l^\dagger$ and $\hat{N} = \hat{N}^\dagger$. In particular $\hat{n}_l$ is the *number operator* for the l-th mode, while $\hat{N}$ is the *number operator* for the system $\mathcal{S}$ described by our operators.

> *Remark:* The equality $\hat{n}_l = \hat{n}_l^\dagger$ above can be seen as the result of a simple, but formal, computation: $\hat{n}_l^\dagger = (a_l^\dagger a_l)^\dagger = a_l^\dagger (a_l^\dagger)^\dagger = a_l^\dagger a_l = \hat{n}_l$. To be more rigorous, we should do much more than this. In particular, we should also consider the domains of $\hat{n}_l$ and of $\hat{n}_l^\dagger$, since these could be different.[1] This is particularly relevant here, since each $\hat{n}_l$ is unbounded. However, we will be satisfied with this formal result, since these mathematical aspects are not particular relevant for what we will discuss in the following.

An o.n. basis of $\mathcal{H}$ can be constructed as follows: we introduce the *vacuum* of the theory, that is, a vector φ_0 which is annihilated by all the operators a_l: $a_l \varphi_0 = 0$ for all $l = 1, 2, \dots, L$. Then we act on φ_0 with the operators $a_l^\dagger$ and with their powers,

$$\varphi_{n_1, n_2, \dots, n_L} := \frac{1}{\sqrt{n_1! \, n_2! \dots n_L!}} \left(a_1^\dagger\right)^{n_1} \left(a_2^\dagger\right)^{n_2} \dots \left(a_L^\dagger\right)^{n_L} \varphi_0, \qquad (2.2)$$

$n_l = 0, 1, 2, \dots$, for all l, and we normalize the vectors obtained in this way. The set of the $\varphi_{n_1, n_2, \dots, n_L}$'s in (2.2) forms a complete and o.n. set in $\mathcal{H}$, and they are eigenstates of both $\hat{n}_l$ and $\hat{N}$:

$$\hat{n}_l \varphi_{n_1, n_2, \dots, n_L} = n_l \varphi_{n_1, n_2, \dots, n_L}$$

---

[1] Given an operator X acting on $\mathcal{H}$ and with domain $D(X)$, its adjoint $X^\dagger$ is defined by introducing first its domain $D(X^\dagger)$ as follows:

$$D(X^\dagger) = \{g \in \mathcal{H} : \exists g_X \in \mathcal{H} : \langle f, g_X \rangle = \langle Xf, g \rangle, \ \forall f \in D(X)\}.$$

Then we put $X^\dagger g = g_X$, for all $g \in D(X^\dagger)$. X is self-adjoint if $\langle Xf, g \rangle = \langle f, Xg \rangle$, for all $f, g \in D(X)$, and if $D(X) = D(X^\dagger)$.

and

$$\hat{N}\varphi_{n_1,n_2,\ldots,n_L} = N\varphi_{n_1,n_2,\ldots,n_L},$$

where $N = \sum_{l=1}^{L} n_l$. Hence, n_l and N are eigenvalues of $\hat{n}_l$ and $\hat{N}$ respectively. Moreover, using the CCR we deduce that

$$\hat{n}_l \left(a_l \varphi_{n_1,n_2,\ldots,n_L}\right) = (n_l - 1) \left(a_l \varphi_{n_1,n_2,\ldots,n_L}\right),$$

for $n_l \geq 1$ while, if $n_l = 0$, a_l annihilates the vector, and

$$\hat{n}_l \left(a_l^{\dagger} \varphi_{n_1,n_2,\ldots,n_L}\right) = (n_l + 1) \left(a_l^{\dagger} \varphi_{n_1,n_2,\ldots,n_L}\right),$$

for all l and for all n_l. For these reasons the following interpretation is given in the literature: if the L different modes of bosons of $\mathcal{S}$ are described by the vector $\varphi_{n_1,n_2,\ldots,n_L}$, this means that n_1 bosons are in the first mode, n_2 in the second mode and so on. The operator $\hat{n}_l$ acts on $\varphi_{n_1,n_2,\ldots,n_L}$ and returns n_l, which is exactly the number of bosons in the l-th mode. The operator $\hat{N}$ counts the total number of bosons. Moreover, the operator a_l destroys a boson in the l-th mode, if there is at least one. Otherwise a_l simply destroys the state. Its adjoint, $a_l^{\dagger}$, creates a boson in the same mode. This is why in the physical literature a_l and $a_l^{\dagger}$ are usually called the *annihilation* and the *creation* operators.

The vector $\varphi_{n_1,n_2,\ldots,n_L}$ in Equation (2.2) defines a *vector (or number) state* over the set $\mathcal{A}$ as

$$\omega_{n_1,n_2,\ldots,n_L}(X) = \langle \varphi_{n_1,n_2,\ldots,n_L}, X\varphi_{n_1,n_2,\ldots,n_L} \rangle, \tag{2.3}$$

where $\langle \, , \, \rangle$ is the scalar product in the Hilbert space $\mathcal{H}$, and $X \in \mathcal{A}$. These states will be used to *project* from quantum to classical dynamics and to fix the initial conditions of the system under consideration, in a way which will be clarified later on. Something more concerning states will be discussed later in this chapter.

As observed previously, the operators a_l, $a_l^{\dagger}$, $\hat{n}_l$ and $\hat{N}$ are all unbounded. This can be easily understood since, for instance,

$$\|\hat{n}_l\| = \sup_{0 \neq \varphi \in \mathcal{H}} \frac{\|\hat{n}_l \varphi\|}{\|\varphi\|} \geq \sup_{\{n_j \geq 0, \, j=1,2,\ldots,L\}} \|\hat{n}_l \varphi_{n_1,\ldots,n_l,\ldots,n_L}\|$$

$$= \sup_{\{n_j \geq 0, \, j=1,2,\ldots,L\}} n_l = \infty,$$

and it is clearly related, as already observed, to the fact that $\mathcal{H}$ is infinite-dimensional. We have already pointed out that unbounded operators have severe domain problems, since they cannot be defined in all of $\mathcal{H}$ [27], but only on a dense subset of $\mathcal{H}$. In fact, an operator X acting on $\mathcal{H}$ is bounded if, and only if, $D(X) = \mathcal{H}$. We want to stress once more that this will not be a major problem for us: in fact, first of all, each vector $\varphi_{n_1,n_2,\ldots,n_L}$ belongs to the domains of all the operators which are relevant for us, and the linear span of the set $\mathcal{F}_\varphi = \{\varphi_{n_1,n_2,\dots,n_L}, n_j \geq 0\}$ is left stable under the action of all these operators. Secondly, as already discussed, it may happen that the infinite-dimensional Hilbert space $\mathcal{H}$ is replaced by an effective Hilbert space, $\mathcal{H}_{eff}$, which becomes *dynamically finite-dimensional* because of the existence of some conserved quantity and because of the initial conditions, which impose some constraints on the accessible levels to the agents of the system [9].

2.2 The Fermionic Number Operator

Given a set of operators $\{b_l, b_l^\dagger, \ell = 1, 2, \dots, L\}$ acting on a certain Hilbert space $\mathcal{H}_F$, we say that they satisfy the CAR if the conditions

$$\left\{b_l, b_n^\dagger\right\} = \delta_{l,n} \mathbb{1}, \qquad \{b_l, b_n\} = \left\{b_l^\dagger, b_n^\dagger\right\} = 0 \tag{2.4}$$

hold true for all $l, n = 1, 2, \dots, L$. Here, $\{x, y\} := xy + yx$ is the *anticommutator* of x and y and $\mathbb{1}$ is now the identity operator on $\mathcal{H}_F$. These operators, which are considered in many textbooks on quantum mechanics, see, for instance [2, 31], are those which are used to describe L different modes of fermions. As for bosons, from these operators we can construct $\hat{n}_l = b_l^\dagger b_l$ and $\hat{N} = \sum_{l=1}^L \hat{n}_l$, which are both self-adjoint. In particular, $\hat{n}_\ell$ is the *number operator* for the ℓ-th mode, while $\hat{N}$ is the *global number operator* for $\mathcal{S}$. Compared with bosonic operators, the operators introduced here satisfy a very important feature: if we try to square them (or to rise to higher powers), we simply get zero: for instance, (2.4) implies that $b_l^2 = 0$. This is of course related to the fact that fermions satisfy the Pauli exclusion principle [31], while bosons do not.

The Hilbert space of our system is constructed as for bosons: we introduce the *vacuum* of the theory, that is, a vector Φ_0 which is annihilated by all the operators b_l: $b_l \Phi_0 = 0$ for all $l = 1, 2, \dots, L$. Then we act on Φ_0 with the operators $(b_l^\dagger)^{n_l}$:

$$\Phi_{n_1,n_2,\dots,n_L} := \left(b_1^\dagger\right)^{n_1} \left(b_2^\dagger\right)^{n_2} \cdots \left(b_L^\dagger\right)^{n_L} \Phi_0, \tag{2.5}$$

$n_l = 0, 1$, for all l. Of course, we do not consider higher powers of the $b_j^\dagger$'s, since these powers would simply destroy the vector. This explains why no normalization appears. In fact, for all allowed values of the n_l's, the normalization constant $\sqrt{n_1! \, n_2! \dots n_L!}$ in (2.2) is equal to one. These vectors form an o.n. set which spans all of $\mathcal{H}_F$, and they are eigenstates of both $\hat{n}_l$ and $\hat{N}$, similar to what we have seen for CCR:

$$\hat{n}_l \Phi_{n_1,n_2,\dots,n_L} = n_l \Phi_{n_1,n_2,\dots,n_L}$$

and

$$\hat{N} \Phi_{n_1,n_2,\dots,n_L} = N \Phi_{n_1,n_2,\dots,n_L},$$

where $N = \sum_{l=1}^{L} n_l$. A major difference with respect to what happens for bosons is that the eigenvalues of $\hat{n}_l$ are simply zero and one, and consequently N can take any nonnegative integer value (as for bosons), but smaller or equal to L. Moreover, using the CAR, we deduce that

$$\hat{n}_l \left(b_l \Phi_{n_1, n_2, \dots, n_L} \right) = \begin{cases} (n_l - 1)(b_l \Phi_{n_1, n_2, \dots, n_L}), & n_l = 1 \\ 0, & n_l = 0, \end{cases}$$

and

$$\hat{n}_l \left(b_l^\dagger \Phi_{n_1, n_2, \dots, n_L} \right) = \begin{cases} (n_l + 1)(b_l^\dagger \Phi_{n_1, n_2, \dots, n_L}), & n_l = 0 \\ 0, & n_l = 1, \end{cases}$$

for all l. The interpretation does not differ much from that for bosons, and then b_l and $b_l^\dagger$ are again respectively called the *annihilation* and the *creation* operators. However, in some sense, $b_l^\dagger$ is also an annihilation operator since, acting on a state with $n_l = 1$, it destroys that state: we are trying to put together two identical fermions, and this operation is forbidden by the Pauli exclusion principle.

Of course, $\mathcal{H}_F$ has a finite dimension. In particular, for just one mode of fermions, $dim(\mathcal{H}_F) = 2$, while $dim(\mathcal{H}_F) = 4$ if $L = 2$. This also implies that, contrary to what happens for bosons, all the fermionic operators are bounded and can be represented by finite-dimensional matrices, independently of the existence of any integral of motion.

As for bosons, the vector $\Phi_{n_1, n_2, \dots, n_L}$ in (2.5) defines a *vector (or number) state* over the algebra $\mathcal{A}$ of the operators over $\mathcal{H}_F$ as

$$\omega_{n_1, n_2, \dots, n_L}(X) = \langle \Phi_{n_1, n_2, \dots, n_L}, X \Phi_{n_1, n_2, \dots, n_L} \rangle, \tag{2.6}$$

where $\langle \, , \, \rangle$ is the scalar product in $\mathcal{H}_F$, and $A \in \mathcal{A}$. Again, these states will be used to project from quantum to classical dynamics and to fix the initial conditions of the considered system. This will be clarified in the next chapters, when discussing concrete applications of both CCR and CAR.

2.3 Other Possibilities

The ones considered so far are not the only ladder operators arising when dealing with quantum mechanical systems. Many other possibilities have been introduced along the years in the literature, in many different contexts and for many different situations. Each one of these alternatives has some particularly interesting mathematical features and turns out to be useful in certain concrete applications. In the rest of this section we will briefly review some of them.

2.3.1 Quons

One such alternative is based on the following (one-mode) q-mutation relation:

$$\left[A, A^{\dagger}\right]_q := AA^{\dagger} + qA^{\dagger}A = \mathbb{1}, \tag{2.7}$$

where q is a real number between plus and minus one, A is a given operator acting on some Hilbert space $\mathcal{H}$, and $A^{\dagger}$ is its adjoint. When $q = 1$ the q-mutation relation produces CAR (when supplemented with the condition $A^2 = 0$), while CCR are recovered when $q = -1$. If $q \in]-1, 1[$, Equation (2.7) describes particles which are neither bosons nor fermions. Many details on *quons* are discussed in [32–37]. Particularly relevant for us is the fact that A behaves as a lowering operator, while $A^{\dagger}$ is a raising operator, as discussed in the remainder of this section.

In [37] it is proved that the eigenstates of $N_0 = A^{\dagger}A$ are analogous to the bosonic ones, except that for the normalization. A simple concrete realization of Equation (2.7) can be deduced as follows: let $\mathcal{F}_e = \{e_k, k = 0, 1, 2, \ldots\}$ be the canonical o.n. basis in $\mathcal{H} = l^2(\mathbb{N}_0)$, the set of all the square summable sequences, with all zero entries except in the $(k+1)$-th position, which is equal to one: $\langle e_k, e_m \rangle = \delta_{k,m}$. If we take

$$A = \begin{pmatrix} 0 & \beta_0 & 0 & 0 & 0 & 0 & \cdots \\ 0 & 0 & \beta_1 & 0 & 0 & 0 & \cdots \\ 0 & 0 & 0 & \beta_2 & 0 & 0 & \cdots \\ 0 & 0 & 0 & 0 & \beta_3 & 0 & \cdots \\ 0 & 0 & 0 & 0 & 0 & \beta_4 & \cdots \\ \cdots & \cdots & \cdots & \cdots & \cdots & \cdots & \cdots \\ \cdots & \cdots & \cdots & \cdots & \cdots & \cdots & \cdots \end{pmatrix}, \tag{2.8}$$

it follows that (2.7) is satisfied if $\beta_0^2 = 1$ and $\beta_n^2 = 1 + q\beta_{n-1}^2$, $n \geq 1$. Then β_n^2 coincides with $n + 1$ if $q = 1$, and with $\frac{1-q^{n+1}}{1-q}$ if $q \in]-1, 1[$. It is convenient to fix $\beta_n > 0$ for all $n \geq 0$. Moreover, it is clear that $A\, e_0 = 0$, and $A^{\dagger}$ behaves, as stated, as a raising operator since from Equation (2.8) we deduce

$$e_{n+1} = \frac{1}{\beta_n} A^{\dagger}e_n = \frac{1}{\beta_n!} \left(A^{\dagger}\right)^{n+1} e_0, \tag{2.9}$$

for all $n \geq 0$. Here we have introduced the notation $\beta_n! := \beta_n\beta_{n-1} \cdots \beta_2\beta_1$. In the literature this quantity is sometimes called the *q-factorial*. Of course, from (2.9) it follows that $A^{\dagger}e_n = \beta_n e_{n+1}$. Using the matricial form for A in (2.8) it is also easy to check that A acts as a lowering operator on $\mathcal{F}_e$: $A\, e_m = \beta_{m-1}e_{m-1}$, for all $m \in \mathbb{N}_0$, where we have also introduced $\beta_{-1} = 0$, to ensure that $A\, e_0 = 0$.

Then, calling $\hat{N}_0 = A^{\dagger}A$, we have

$$N_0 e_m = \beta_{m-1}^2 e_m, \tag{2.10}$$

for all $m \in \mathbb{N}_0$. The operator $\hat{N}$, formally defined in [37] as $\hat{N} = \frac{1}{\log(q)}$ $\log(\mathbb{1} - \hat{N}_0(1 - q))$ for $q > 0$, satisfies the eigenvalue equation $\hat{N} e_m = m e_m$, for all $m \in \mathbb{N}_0$. For this reason $\hat{N}$ is called the *number operator* for the quons, while $\hat{N}_0$ is not. Notice that, however, they share the same eigenvectors.

It should be stressed that the one in Equation (2.8) is not the only possible way to represent an operator A satisfying Equation (2.7). For instance, in [34], the authors adopt the following representation of A and $A^\dagger$ in $\mathcal{L}^2(\mathbb{R})$:

$$A = \frac{e^{-2i\alpha x} - e^{i\alpha \frac{d}{dx}} e^{-i\alpha x}}{-i\sqrt{1 - e^{-2\alpha^2}}}, \qquad A^\dagger = \frac{e^{2i\alpha x} - e^{i\alpha x} e^{i\alpha \frac{d}{dx}}}{i\sqrt{1 - e^{-2\alpha^2}}}, \qquad (2.11)$$

where $\alpha = \sqrt{-\frac{\log(q)}{2}}$ or, which is the same, $q = e^{-2\alpha^2}$. In this case, since α is assumed to belong to the set $[0, \infty)$, q ranges in the interval $]0, 1]$.

For completeness we also remark that other possible q-mutation relations have also been proposed along the years. Another such possibility is

$$\left[A, A^\dagger\right]_q = q^{-2\hat{N}},$$

where $\hat{N}$ is the number operator introduced before. This rule, again, gives rise to an interesting functional structure, and is applied to some physical situations, when particles do not obey the more common CCR and CAR.

2.3.2 Truncated Bosons

As we have seen above, in a sense quons interpolate between fermions and bosons, depending on the value of q in Equation (2.7). This is not the only way to perform a similar interpolation. A completely different method makes use of the so-called *truncated bosons*, discussed, for instance, in [38, 39].

We consider an operator B which obeys the following rule:

$$\left[B, B^\dagger\right] = \mathbb{1} - LK, \qquad (2.12)$$

in which $L = 2, 3, 4, \ldots$ is a fixed natural number, while K is a self-adjoint projection operator, $K = K^2 = K^\dagger$, satisfying the equality $KB = 0$. The presence of the term LK in Equation (2.12) makes it possible to find a representation of K and B in terms of $L \times L$ matrices, which would not be possible in absence of such a correction. In fact, in this case, we would recover the CCR, which does not admit any finite-dimensional representation. Here, on the other hand, K, B and $B^\dagger$ act on an L-th-dimensional Hilbert space, which we call $\mathcal{H}_L$.

> *Remark:* Assume, contrary to what was just stated, that the CCR $[a, a^\dagger] = \mathbb{1}$ can
> be represented in a finite-dimensional Hilbert space $\mathcal{H}_0$. This implies that a, $a^\dagger$ and

$\mathbb{1}$ are finite matrices, with finite trace tr. But any trace satisfies, in particular, the following properties: $tr(A + B) = tr(A) + tr(B)$ and $tr(AB) = tr(BA)$, for all operators A and B on which the trace is defined.[2] Hence we get $tr([a, a^\dagger]) = tr(aa^\dagger) - tr(a^\dagger a) = 0$, while $tr(\mathbb{1}) = M$, with $M = dim(\mathcal{H}_0)$. This is clearly impossible, and shows that CCR cannot live in any finite-dimensional Hilbert space. On the other hand, this contradiction does not appear if we use (2.12). In fact, again we have $tr([B, B^\dagger]) = 0$, but we also find that $tr(\mathbb{1} - LK) = tr(\mathbb{1}) - L\,tr(K) = L - L = 0$. This is because, see [38, 39], $tr(K) = 1$.

In [39] it is shown that the matrices for B and $B^\dagger$ are essentially the truncated versions of the analogous, infinite-dimensional matrices for the bosonic annihilation and creation operators. In [39] it is also discussed how to construct an o.n. basis of eigenvectors of the self-adjoint operator $H_0 = \frac{1}{2}(Q_0^2 + P_0^2)$, where $Q_0 = \frac{B+B^\dagger}{\sqrt{2}}$ and $P_0 = \frac{B-B^\dagger}{\sqrt{2}\,i}$ are the truncated position and momentum operators. These vectors turn out to be eigenvectors of both H_0 and K, and their explicit construction is strongly based on the fact that H_0 is a positive operator, other than being self-adjoint. Hence they are labeled by two quantum numbers. That's why the operators B and $B^\dagger$ still behave as ladder operators, but in a slightly more elaborated way [38, 39]. We refer to these papers for more details on this particular aspect. Here we just want to stress that, if $L = 2$, B and $B^\dagger$ are two-by-two matrices, as for the fermionic case, while the size of the matrices representing B and $B^\dagger$ increases with L, and in particular B becomes an infinite matrix when L diverges. In this sense, we can say that, roughly speaking, the truncated bosons in Equation (2.12) interpolate between fermions and bosons.

2.3.3 Pseudo-Versions of These Rules

All the commutation rules considered so far, from the CCR in Equation (2.1) to the CAR in Equation (2.4), passing for the q-mutator in Equation (2.7) and for the rule in Equation (2.12), can be deformed further in such a way that the raising operator, which is usually the adjoint of the lowering operator, is something different. For instance, $[a, a^\dagger] = \mathbb{1}$ can be replaced by $[A, B] = \mathbb{1}$, where $B \neq A^\dagger$. This, on some particular occasions, proved to be useful, since it turns out that many models originally proposed in the so-called *PT* or *pseudo-Hermitian quantum mechanics* [24, 25] can be rewritten in terms of these operators. We refer to [40] for a rather rich discussion on the so-called $\mathcal{D}$-*pseudo-bosons*, whose mathematics is somehow complicated due to the essential appearances of unbounded operators.

---

[2] In this case, for all matrices A and B on $\mathcal{H}_0$.

Deformed versions of quons and of truncated bosons have been proposed and analyzed quite recently [33, 41]. The relevant rules in these cases are, respectively,

$$[A,B]_q = \mathbb{1}, \qquad [C,D] = \mathbb{1} - LK,$$

where, in particular, $B \neq A^\dagger$ and $D \neq C^\dagger$, while q, L are as before, and K is again an
orthogonal projector, $K = K^\dagger = K^2$, with $KC = DK = 0$. Our interest in these kind of
operators in the context of this book is related to the fact that they can be used, in
principle, in connection with Hamiltonians which are manifestly not self-adjoint.
This is important since we can produce, in this way, some sort of *irreversible dynamics* which cannot be easily deduced with the original (i.e., canonical) forms of
the commutation relations proposed before.

In the rest of this section we just focus on a few aspects of the simplest deformation, among those cited here. In particular, we will discuss what *pseudo-fermions*
are, just to give the flavor of the kind of mathematics needed in their description.

The starting point in this section is a modification of the CAR (for a single mode)
in (2.4), which is replaced here by the following rules:

$$\{a,b\} = \mathbb{1}, \quad \{a,a\} = 0, \quad \{b,b\} = 0, \tag{2.13}$$

where the interesting situation is when $b \neq a^\dagger$. Since $\det(a) = \det(b) = 0$, a and b
are not invertible. Therefore two nonzero vectors, φ_0 and Ψ_0, exist in $\mathcal{H} = \mathbb{C}^2$, the
Hilbert space of our system, such that $a \varphi_0 = 0$ and $b^\dagger \Psi_0 = 0$. In general, $\varphi_0 \neq \Psi_0$.

We introduce further the following nonzero vectors

$$\varphi_1 := b\,\varphi_0, \quad \Psi_1 = a^\dagger \Psi_0, \tag{2.14}$$

as well as the non-self-adjoint operators

$$N = ba, \quad N^\dagger = a^\dagger b^\dagger. \tag{2.15}$$

The reason why φ_1 and Ψ_1 are surely nonzero vectors is because they satisfy,
among other properties, the following equality: $\langle \varphi_1, \Psi_1 \rangle = 1$, see Equation (2.19).
This would not be possible if one of these two vectors were zero. Of course, it
makes no sense to consider $b^n \varphi_0$ or $a^{\dagger n} \Psi_0$ for $n \geq 2$, since all these vectors are
automatically zero, due to Equation (2.13). This is analogous to what happens for
ordinary fermions. Let now introduce the self-adjoint operators S_φ and S_Ψ via their
action on a generic $f \in \mathcal{H}$:

$$S_\varphi f = \sum_{n=0}^{1} \langle \varphi_n, f \rangle\, \varphi_n, \quad S_\Psi f = \sum_{n=0}^{1} \langle \Psi_n, f \rangle\, \Psi_n. \tag{2.16}$$

It is very easy to get the following results:

1.

$$a\varphi_1 = \varphi_0, \quad b^\dagger \Psi_1 = \Psi_0. \tag{2.17}$$

2.

$$N\varphi_n = n\varphi_n, \quad N^\dagger \Psi_n = n\Psi_n, \tag{2.18}$$

for $n = 0, 1$.

3. If the normalizations of φ_0 and Ψ_0 are chosen in such a way that $\langle \varphi_0, \Psi_0 \rangle = 1$, then

$$\langle \varphi_k, \Psi_n \rangle = \delta_{k,n}, \tag{2.19}$$

for $k, n = 0, 1$.

4. S_φ and S_Ψ are bounded, strictly positive, self-adjoint and invertible. They satisfy

$$\|S_\varphi\| \le \|\varphi_0\|^2 + \|\varphi_1\|^2, \quad \|S_\Psi\| \le \|\Psi_0\|^2 + \|\Psi_1\|^2, \tag{2.20}$$

$$S_\varphi \Psi_n = \varphi_n, \quad S_\Psi \varphi_n = \Psi_n, \tag{2.21}$$

for $n = 0, 1$, as well as $S_\varphi = S_\Psi^{-1}$ and the following intertwining relations:

$$S_\Psi N = N^\dagger S_\Psi, \quad S_\varphi N^\dagger = N S_\varphi. \tag{2.22}$$

The above formulas show that (i) N and $N^\dagger$ behave as (non-Hermitian[3]) fermionic number operators, having (real) eigenvalues 0 and 1; (ii) their related eigenvectors are respectively the vectors in $\mathcal{F}_\varphi = \{\varphi_0, \varphi_1\}$ and $\mathcal{F}_\Psi = \{\Psi_0, \Psi_1\}$; (iii) a and $b^\dagger$ are lowering operators for $\mathcal{F}_\varphi$ and $\mathcal{F}_\Psi$ respectively; (iv) b and $a^\dagger$ are raising operators for $\mathcal{F}_\varphi$ and $\mathcal{F}_\Psi$ respectively; (v) the two sets $\mathcal{F}_\varphi$ and $\mathcal{F}_\Psi$ are biorthonormal; (vi) the very well-behaved operators S_φ and S_Ψ map $\mathcal{F}_\varphi$ into $\mathcal{F}_\Psi$ and vice versa; (vii) S_φ and S_Ψ intertwine between the operators N and $N^\dagger$, which are manifestly not self-adjoint.

As already claimed, a and b, as well as similar ladder operators, can be used, quite efficiently, in the analysis of quantum systems driven by Hamiltonians which are not self-adjoint, as those appearing in connection with gain and loss systems. We refer to [40] for more details.

Remark: A rather general way to construct ladder operators uses, as its main ingredient, an o.n. basis $\mathcal{F}_e = \{e_n, n \ge 0\}$ of the Hilbert space $\tilde{\mathcal{H}}$ where the system is defined, and a sequence of nonnegative numbers: $0 \le \epsilon_0 < \epsilon_1 < \epsilon_2 < \dots$. Then we can define an operator c as follows:

$$cf = \sum_{k=0}^{\infty} \epsilon_k \langle e_{k+1}, f \rangle e_k,$$

---

[3] In this book *Hermitian* and *self-adjoint* will be used as synonyms.

for all f in the domain of c, $D(c)$, i.e., for all those $f \in \tilde{\mathcal{H}}$ for which this series is convergent. It is clear that c is a lowering operator and that $e_k \in D(c)$ for all k: $c\, e_k = \epsilon_{k-1} e_{k-1}$ for $k \geq 1$ with $c\, e_0 = 0$, while its adjoint, $c^\dagger$, turns out to be a raising operator. Moreover, if $\epsilon_k = k$, c and $c^\dagger$ are just the bosonic annihilation and creation operators considered in Section 2.1, and a simple computation shows that $[c, c^\dagger] = \mathbb{1}$. However, if $\epsilon_k \neq k$, c and $c^\dagger$ are operators of some different nature. They are introduced in some applications of quantum optics, for instance in connection with the so-called *nonlinear coherent states*.

2.3.4 Simple Ladder Operators for Finite-Dimensional Hilbert Spaces

Ladder operators for finite-dimensional Hilbert spaces already appeared before, for fermions, in Section 2.2, for truncated bosons, in Section 2.3.2 and for their pseudo-versions, in Section 2.3.3. Fermions and pseudo-fermions live in two-dimensional Hilbert spaces, while truncated bosons and truncated pseudo-bosons can be defined in any finite-dimensional Hilbert space, with dimension equal or greater than two. The way in which these latter are introduced is not so simple, and one may wonder if other kind of ladder operators for finite-dimensional Hilbert spaces can be constructed in some simpler way. The answer is affirmative and this alternative strategy will be relevant, in particular in Chapter 7, where similar operators will be used in the construction of a model for tumor cell proliferation.

For concreteness, to show how our construction works, we will restrict to a five-dimensional Hilbert space $\mathcal{H}_5 = \mathbb{C}^5$. The same approach can be easily extended to $\mathcal{H}_N$, for all finite $N = 2, 3, 4, \ldots$.

Let $\mathcal{E}_5 = \{e_j, j = 0, 1, 2, 3, 4\}$ be the canonical o.n. basis of $\mathcal{H}_5$. Each e_j is a five-dimensional vector with all zero entries, except the $(j + 1)$-th entry, which is one. We define an operator $b^\dagger$ via its action on the e_j's:

$$b^\dagger e_0 = e_1, \quad b^\dagger e_1 = \sqrt{2}\, e_2, \quad b^\dagger e_2 = \sqrt{3}\, e_3, \quad b^\dagger e_3 = \sqrt{4}\, e_4, \quad b^\dagger e_4 = 0. \quad (2.23)$$

It is clear that $b^\dagger$ behaves as a sort of fermionic raising operator, destroying the upper level. The matrix expression for $b^\dagger$ in the basis $\mathcal{E}_5$ is the following:

$$b^\dagger = \begin{pmatrix} 0 & 0 & 0 & 0 & 0 \\ 1 & 0 & 0 & 0 & 0 \\ 0 & \sqrt{2} & 0 & 0 & 0 \\ 0 & 0 & \sqrt{3} & 0 & 0 \\ 0 & 0 & 0 & \sqrt{4} & 0 \end{pmatrix},$$

and the adjoint is

$$
b = \begin{pmatrix}
0 & 1 & 0 & 0 & 0 \\
0 & 0 & \sqrt{2} & 0 & 0 \\
0 & 0 & 0 & \sqrt{3} & 0 \\
0 & 0 & 0 & 0 & \sqrt{4} \\
0 & 0 & 0 & 0 & 0
\end{pmatrix}.
$$

These operators cannot satisfy the canonical commutation relation $[b, b^\dagger] = \mathbb{1}_5$, $\mathbb{1}_5$ being the identity operator in $\mathcal{H}_5$, since this would be possible only in an infinite-dimensional Hilbert space. In fact, straightforward computations show that

$$
\left[b, b^\dagger \right] = \mathbb{1}_5 - 5P_4, \tag{2.24}
$$

where P_4 is the projection operator on e_4: $P_4 f = \langle e_4, f \rangle \, e_4$, for all $f \in \mathcal{H}_5$. Notice that $\mathbb{1}_5 - 5P_4$ is the following diagonal matrix: $\mathbb{1}_5 - 5P_4 = diag\{1, 1, 1, 1, -4\}$, which differs from the identity matrix on $\mathcal{H}_5$ only for the last component in its main diagonal.[4] It is easy to check that b behaves as a lowering operator for $\mathcal{E}_5$:

$$
be_0 = 0, \quad be_1 = e_0, \quad be_2 = \sqrt{2} \, e_1, \quad be_3 = \sqrt{3} \, e_2, \quad be_4 = \sqrt{4} \, e_3. \tag{2.25}
$$

$\hat{N} = b^\dagger b = diag\{0, 1, 2, 3, 4\}$ is the number operator, and the following eigenvalue equation holds:

$$
\hat{N} e_k = k e_k, \tag{2.26}
$$

$k = 0, 1, 2, 3, 4$. It is further easy to see that $b^5 = (b^\dagger)^5 = 0$, which is similar to the analogous equation for fermionic ladder operators (with 5 replaced by 2).

> *Remarks:* (1) To extend the construction to $N \neq 5$ it is sufficient to consider the canonical o.n. basis for $\mathcal{H}_N$, $\mathcal{E}_N$, and then use its vectors to define $b^\dagger$ in analogy with Equation (2.23). Then b is just the adjoint of $b^\dagger$. These are ladder operators such that $b \, e_0 = 0$ and $b^\dagger e_N = 0$. Together they produce $\hat{N}$ as above, $\hat{N} = b^\dagger b$.
>
> (2) If $N = 2$, (2.24) should be replaced by $[b, b^\dagger] = \mathbb{1}_2 - 2P_1 = diag\{1, -1\}$. Also $b = \begin{pmatrix} 0 & 1 \\ 0 & 0 \end{pmatrix}$ and $b^\dagger = \begin{pmatrix} 0 & 0 \\ 1 & 0 \end{pmatrix}$, which is in agreement with the well known expressions for the fermionic ladder operators.

In view of the application we are interested in, see Chapter 7, we need to construct three different families of ladder operators of this kind, one for each agent of the biological model we will describe there, and then put all these ingredients together, by working on a suitable tensor product Hilbert space. More in details, and adopting the same notation which will be used in Chapter 7, the agents of the system $\mathcal{S}$ we want to describe are the healthy cells, *attached* to the ladder operators h and $h^\dagger$, living in a Hilbert space $\mathcal{H}_h$. Then we have the sick cells, described in

---

[4] Notice that (2.24) is a particular realization of (2.12).

terms of the operators s and $s^\dagger$, which are defined on a second Hilbert space $\mathcal{H}_s$, and the medical treatment, related to m and $m^\dagger$, acting on a third Hilbert space $\mathcal{H}_m$. We call $N_\alpha = dim(\mathcal{H}_\alpha)$, where $\alpha = h, s, m$, and $\mathcal{E}_\alpha = \{e_j^{(\alpha)}, j = 0, 1, 2, \dots, N_\alpha - 1\}$ the o.n. basis of $\mathcal{H}_\alpha$. The operators h, s and m satisfy relations which extend those in Equations (2.23) and (2.25). First of all we have

$$h\, e_0^{(h)} = 0, \qquad s\, e_0^{(s)} = 0, \qquad m\, e_0^{(m)} = 0,$$

and then

$$e_1^{(h)} = h^\dagger e_0^{(h)}, \quad e_2^{(h)} = \frac{1}{\sqrt{2}} h^\dagger e_1^{(h)}, \dots, e_{N_h-1}^{(h)} = \frac{1}{\sqrt{N_h - 1}} h^\dagger e_{N_h-2}^{(h)},$$

$$e_1^{(s)} = s^\dagger e_0^{(s)}, \quad e_2^{(s)} = \frac{1}{\sqrt{2}} s^\dagger e_1^{(s)}, \dots, e_{N_s-1}^{(s)} = \frac{1}{\sqrt{N_s - 1}} s^\dagger e_{N_s-2}^{(s)},$$

$$e_1^{(m)} = m^\dagger e_0^{(m)}, \quad e_2^{(m)} = \frac{1}{\sqrt{2}} m^\dagger e_1^{(m)}, \dots, e_{N_m-1}^{(m)} = \frac{1}{\sqrt{N_m - 1}} m^\dagger e_{N_m-2}^{(m)}.$$

Finally, we have that

$$h^\dagger e_{N_h-1}^{(h)} = 0, \quad s^\dagger e_{N_s-1}^{(s)} = 0, \quad m^\dagger e_{N_m-1}^{(m)} = 0.$$

The Hilbert space of our full system is now the tensor product $\mathcal{H} = \mathcal{H}_h \otimes \mathcal{H}_s \otimes \mathcal{H}_m$, whose dimension is clearly $N = N_h \times N_s \times N_m$. Calling $\mathbb{1}_\alpha$ the identity operator on $\mathcal{H}_\alpha$, each operator X_h on $\mathcal{H}_h$ is identified with the following tensor product on $\mathcal{H}$: $X_h \otimes \mathbb{1}_s \otimes \mathbb{1}_m$. Analogously, the operators X_s and X_m on $\mathcal{H}_s$ and $\mathcal{H}_m$ should be identified respectively with $\mathbb{1}_h \otimes X_s \otimes \mathbb{1}_m$ and with $\mathbb{1}_h \otimes \mathbb{1}_s \otimes X_m$, both acting on $\mathcal{H}$. Furthermore,

$$(X_h \otimes X_s \otimes X_m)\,(f_h \otimes f_s \otimes f_m) = (X_h f_h) \otimes (X_s f_s) \otimes (X_m f_m),$$

for all $f_h \in \mathcal{H}_h, f_s \in \mathcal{H}_s$ and $f_m \in \mathcal{H}_m$. From now on, when no confusion arises, we will just write X_h, X_s, X_m instead of $X_h \otimes \mathbb{1}_s \otimes \mathbb{1}_m, \mathbb{1}_h \otimes X_s \otimes \mathbb{1}_m$ and $\mathbb{1}_h \otimes \mathbb{1}_s \otimes X_m$, and their action is obviously meant on the whole $\mathcal{H}$.

An o.n. basis for $\mathcal{H}$ is the following:

$$\mathcal{E} = \left\{ \varphi_{n_h, n_s, n_m} := e_{n_h}^{(h)} \otimes e_{n_s}^{(s)} \otimes e_{n_m}^{(m)}, \ n_\alpha = 0, 1, \dots, N_\alpha - 1, \ \alpha = h, s, m \right\}, \qquad (2.27)$$

so that any vector Ψ of $\mathcal{H}$ can be expressed as a combination of these vectors:

$$\Psi = \sum_{n_h, n_s, n_m} c_{n_h, n_s, n_m} \varphi_{n_h, n_s, n_m}. \qquad (2.28)$$

Here the sum is extended to all the possible values of n_h, n_s and n_m, which of course depend on our choice of N_h, N_s and N_m, and c_{n_h, n_s, n_m} are complex scalars which can be deduced from Ψ as follows: $c_{n_h,n_s,n_m} = \langle \varphi_{n_h,n_s,n_m}, \Psi \rangle$. Here $\langle .,. \rangle$ is the scalar product in $\mathcal{H}$.

> *Remark:* In our particular application, see Chapter 7, each vector in $\mathcal{E}$ has a clear meaning: $\varphi_{k,0,0}$, with $k > 0$, describes a situation in which the system consists only of k healthy cells, with no sick cell and with no active medical treatment, whereas $\varphi_{n,3n,1}$, with $n > 0$, represents a state in which the sick cells are three times the number of the healthy ones, and a medical treatment is acting on the system.

A simple computation shows that all the ladder operators of the different agents commute. For instance, using the properties of the tensor product, we see that

$$[h, m]\varphi_{n_h,n_s,n_m} = \left(h\, e_{n_h}^{(h)} \right) \otimes e_{n_s}^{(s)} \otimes \left(m\, e_{n_m}^{(m)} \right) - \left(h\, e_{n_h}^{(h)} \right) \otimes e_{n_s}^{(s)} \otimes \left(m\, e_{n_m}^{(m)} \right) = 0.$$

Of course, this implies that all the operators on $\mathcal{H}_\alpha$ commute with all the operators acting on $\mathcal{H}_\beta$, if $\alpha \neq \beta$: $[X_\alpha, Y_\beta] = 0$, $\alpha, \beta = h, s, m$ and $\alpha \neq \beta$.

The number operators can be introduced as usual:

$$\hat{N}_h = h^\dagger h, \qquad \hat{N}_s = s^\dagger s, \qquad \hat{N}_m = m^\dagger m, \tag{2.29}$$

and they act on the elements of $\mathcal{E}$ as follows:

$$\hat{N}_h \varphi_{n_h,n_s,n_m} = n_h \varphi_{n_h,n_s,n_m}, \quad \hat{N}_s \varphi_{n_h,n_s,n_m} = n_s \varphi_{n_h,n_s,n_m},$$

$$\hat{N}_m \varphi_{n_h,n_s,n_m} = n_m \varphi_{n_h,n_s,n_m}. \tag{2.30}$$

Then we conclude that the same general settings discussed for bosons and fermions are also recovered here in each $\mathcal{H}_N$, $N = 2, 3, 4, ..., N < \infty$, except, at most, for the commutation relations between ladder operators.[5] We also want to stress that the procedure proposed in this section is by far simpler than that in Section 2.3.2, where we have also worked in $\mathcal{H}_N$. The operators h, s and m will be the main ingredient of our analysis in Section 7.2.

The conclusion of this long section is that bosons and fermions are not the only classes of operators which can be used in the analysis of those physical systems for which ladder operators are relevant (which are many!); not surprisingly, the use of a particular class, rather than another, is usually connected with the system we want to describe, and with the explicit interpretation we can give to these operators.

2.4 Dynamics for a Quantum System

So far, we have discussed some algebraic and functional properties of quantities (e.g., operators, vectors, bases) which can be useful in the analysis of quantum,

---

[5] Incidentally we observe that, while the CCR and CAR were the starting points for the construction in Sections 2.1 and 2.2, here the starting point is the o.n. basis $\mathcal{E}_N$, and the result of the commutators are consequences.

or quantum-like, systems. Now we briefly discuss how the time evolution of these systems are usually described in the literature. We refer to [2, 3] for more details and other possibilities.

Let S be a closed quantum system. This means that S does not interact with any external reservoir (this is why, it is *closed*) and that its size is comparable with that of, say, the hydrogen atom: S is a *microscopic* system (so, it is a *quantum system*).

> *Remark:*   It is useful to observe that, if S also interacts with another system $S_\mathcal{E}$, of the same size, then we can always consider the union $S_{full} = S \cup S_\mathcal{E}$ as a larger closed system, again microscopic, for which all the general ideas and results we are going to describe in the next few pages apply. If the size of $S_\mathcal{E}$ is much larger than that of S, then we will say that S is an *open quantum system*. We will discuss the dynamics for systems like these in Section 2.7.

2.4.1 Schrödinger Representation

In this section we will briefly describe how to find the time evolution of S, starting from a simple (and natural) assumption: in a closed system, the total probability is preserved. The physical interpretation of this is clear: suppose, to be concrete, that S is a particle. It is clear that the particle, at $t = t_0$, should be somewhere in the space. Then, recalling the probabilistic interpretation of the wave-function in quantum mechanics, we must have $\|\Psi(t_0)\|^2 = \int_{\mathbb{R}^3} |\Psi(\underline{r}, t_0)|^2 d^3\underline{r} = 1$. If the particle is not destroyed, during its time evolution (i.e., for $t \geq t_0$) it can change position in space, but, again, it should be found somewhere. Therefore, the total probability should stay constantly equal to one: $\|\Psi(t)\|^2 = \int_{\mathbb{R}^3} |\Psi(\underline{r}, t)|^2 d^3\underline{r} = 1$, for all $t \geq t_0$. This is guaranteed if the time evolution of the particle is described by a unitary operator, which maps $\Psi(\underline{r}, t_0)$ into $\Psi(\underline{r}, t)$:

$$\Psi(\underline{r}, t) = U(t, t_0)\Psi(\underline{r}, t_0), \tag{2.31}$$

where $U^{-1}(t, t_0) = U^\dagger(t, t_0)$. In this way, in fact, the $\mathcal{L}^2(\mathbb{R}^3)$-norm[6] of Ψ is preserved for all t: $\|\Psi(t)\| = \|\Psi(t_0)\|$. From now on, when not explicitly needed, we will not write explicitly the dependence of Ψ from other variables, since this is not relevant for what we are discussing here. In particular, we will not write explicitly the dependence of Ψ on the space variable $\underline{r}$.

Going back to the operator $U(t, t_0)$, it is clear that $U(t_0, t_0) = \mathbb{1}$, and that $U(t_2, t_1)U(t_1, t_0) = U(t_2, t_0)$, for all $t_0 \leq t_1 \leq t_2$. Then, if we take in particular $t_0 = t_2$,

---

[6]   The relevant Hilbert space, in this context, is $\mathcal{L}^2(\mathbb{R}^3)$, with its natural scalar product $\langle f, g \rangle = \int_{\mathbb{R}^3} \overline{f(\vec{r})} g(\vec{r}) d^3\vec{r}$, for all $f(\vec{r}), g(\vec{r}) \in \mathcal{L}^2(\mathbb{R}^3)$ and with the norm induced by this: $\|f\|^2 = \langle f, f \rangle$. The wave function $\Psi(\vec{r}, t)$ is assumed, of course, to depend on the spatial variables $\vec{r} = (x, y, z)$ and on time, t. Hence the norm of Ψ depends, in general, on t.

we deduce that $U^{-1}(t_1, t_0) = U(t_0, t_1)$, for all t_0 and t_1: hence the inverse of $U(t_0, t_1)$ coincides with $U(t_0, t_1)$ itself, but with t_0 and t_1 exchanged.

It is now easy to deduce the differential equation for $\Psi(t)$. This is the well-known Schrödinger equation

$$i\,\frac{\partial \Psi(t)}{\partial t} = H(t)\Psi(t), \tag{2.32}$$

where $H(t)$ is a self-adjoint operator, the Hamiltonian of $\mathcal{S}$, $H(t) = H^\dagger(t)$, which can be, in general, explicitly dependent on time. This point of view is known as *Schrödinger representation*: the wave function depends on time, while the observables, in general, do not.[7] To show how Equation (2.32) follows from Equation (2.31) we compute first

$$\frac{\partial \Psi(t)}{\partial t} = \frac{\partial U(t,t_0)}{\partial t}\,\Psi(t_0) = \frac{\partial U(t,t_0)}{\partial t}\,U^{-1}(t,t_0)\Psi(t) = \left[\frac{\partial U(t,t_0)}{\partial t}\,U^\dagger(t,t_0)\right]\Psi(t),$$

which can be written as in Equation (2.32) defining

$$H(t) = i\,\frac{\partial U(t,t_0)}{\partial t}\,U^\dagger(t,t_0). \tag{2.33}$$

The fact that $H(t)$ is self-adjoint and that it does not depend on t_0 is not evident. However, both these statements are easy to prove. First we observe that

$$H^\dagger(t) = -i\,U(t,t_0)\,\frac{\partial U^\dagger(t,t_0)}{\partial t}.$$

Now, since $U(t,t_0)U^\dagger(t,t_0) = \mathbb{1}$, we have $\frac{\partial U(t,t_0)U^\dagger(t,t_0)}{\partial t} = 0$. Hence

$$\frac{\partial U(t,t_0)}{\partial t}\,U^\dagger(t,t_0) + U(t,t_0)\,\frac{\partial U^\dagger(t,t_0)}{\partial t} = 0.$$

Therefore

$$H^\dagger(t) = -i\,U(t,t_0)\,\frac{\partial U^\dagger(t,t_0)}{\partial t} = (-1)^2 i\,\frac{\partial U(t,t_0)}{\partial t}\,U^\dagger(t,t_0) = H(t),$$

which proves our first claim. To check now that $H(t)$ is independent of t_0 we observe that, since $U(t_0, t_1)U(t_1, t_0) = \mathbb{1}$ for all t_0 and t_1,

$$\frac{\partial U(t,t_0)}{\partial t}\,U^\dagger(t,t_0) = \frac{\partial U(t,t_0)}{\partial t}\,(U(t_0,t_1)U(t_1,t_0))\,U^\dagger(t,t_0)$$

$$= \frac{\partial\,(U(t,t_0)U(t_0,t_1))}{\partial t}\,(U(t_1,t_0)U(t_0,t)) = \frac{\partial U(t,t_1)}{\partial t}\,U^\dagger(t,t_1).$$

---

[7] The time dependence which may be present in this case is, for instance, the one arising because some part of $\mathcal{S}$ is driven by external forces depending explicitly on t.

Hence we have

$$H(t) = i \frac{\partial U(t, t_0)}{\partial t} U^{\dagger}(t, t_0) = i \frac{\partial U(t, t_1)}{\partial t} U^{\dagger}(t, t_1),$$

so that our second assertion follows. Incidentally, this is the reason why in Equation (2.33) we have used $H(t)$ rather than $H(t, t_0)$.

> *Remark:* It is interesting to show that the above result can be somehow inverted: we have shown that (2.32) follows from (2.31) and by the unitarity of $U(t, t_0)$. It is now easy to show that Equation (2.32) implies the equality $\|\Psi(t)\| = \|\Psi(t_0)\|$, for all $t \geq t_0$. In fact,
>
> $$\frac{d \|\Psi(t)\|^2}{dt} = \frac{d}{dt} \langle \Psi(t), \Psi(t) \rangle = \langle \dot{\Psi}(t), \Psi(t) \rangle + \langle \Psi(t), \dot{\Psi}(t) \rangle = \langle -i H(t) \Psi(t), \Psi(t) \rangle$$
>
> $$+ \langle \Psi(t), -i H(t) \Psi(t) \rangle = i \langle H(t) \Psi(t), \Psi(t) \rangle - i \langle \Psi(t), H(t) \Psi(t) \rangle = 0,$$
>
> since $H(t) = H^{\dagger}(t)$. It is useful to stress that this result holds independently of the fact that H depends or not explicitly on time.

We have seen that $\Psi(t)$ and $H(t)$ can both be derived, in principle, from the operators $U(t, t_0)$ as shown in Equations (2.31) and (2.33). Reversing the procedure, $U(t, t_0)$ can be deduced from $H(t)$. The explicit result, however, is deeply linked to the time dependence of H, as we show now. First we rewrite Equation (2.32) as

$$i \frac{\partial U(t, t_0)}{\partial t} \Psi(t_0) = H(t) U(t, t_0) \Psi(t_0),$$

which should be satisfied for all possible initial states $\Psi(t_0)$. This means that the equation for $U(t, t_0)$ is

$$i \frac{\partial U(t, t_0)}{\partial t} = H(t) U(t, t_0). \tag{2.34}$$

This equation can be solved easily if $H(t) = H$, i.e., if the Hamiltonian of $\mathcal{S}$ does not depend explicitly on time. In this case, in fact, the solution of (2.34) is

$$U(t, t_0) = e^{-iH(t-t_0)}. \tag{2.35}$$

> *Remark:* When we say that the operator $e^{-iH(t-t_0)}$ is a solution of Equation (2.34) we are, in fact, a bit optimistic. The reason is the following: from one side, this operator, when replaced in (2.34), really produces an identity. Hence it is really, in mathematical terms, a solution of the differential equation. However, the technical computation of $e^{-iH(t-t_0)}$ can be quite complicated, since we are trying to compute the exponential of some operator. Of course, if $H = H^{\dagger}$, it is enough to use the spectral theorem [27], to compute this exponential, by means of the spectral family of H. So, again, from a mathematical side we have no problem at all to prove the existence of $U(t, t_0)$, which automatically turns out to be unitary, as it should. However, finding a useful expression (for computing things!) for $e^{-iH(t-t_0)}$ can be a different story. And

this is particularly hard for those physical systems whose dynamics can be deduced out of some (physically motivated) non-self-adjoint Hamiltonian, $H \neq H^{\dagger}$, as it happens for some systems in Quantum Optics, for instance. This problem is particularly complicated if H is an operator on an infinite-dimensional Hilbert space. But it can also be a hard problem when H is a finite matrix, if its dimensionality is not small enough. This, of course, depends on the system we are analyzing.

Equation (2.34) can also be solved when $H(t)$ depends explicitly on time, but $H(t_1)$ commutes with $H(t_2)$: $[H(t_1), H(t_2)] = 0$. This is the case, for instance, of a driven harmonic oscillator with Hamiltonian

$$H = \omega a^{\dagger} a + \frac{1}{2} \mathbb{1} + \lambda (a + a^{\dagger}) V(t),$$

where $[a, a^{\dagger}] = \mathbb{1}$, ω and λ are real constants, and $V(t)$ is a real valued force. In this case Equation (2.35) must be replaced by

$$U(t, t_0) = \exp\left\{-i \int_{t_0}^{t} H(t') dt'\right\}, \tag{2.36}$$

which, in particular, returns Equation (2.35) if $H(t) = H$. To prove that (2.36) solves (2.34) we first notice that, calling $X(t, t_0) = -i \int_{t_0}^{t} H(t') dt'$, then $\dot{X}(t, t_0) = -iH(t)$. Then we have

$$[\dot{X}(t, t_0), X(t, t_0)] = (-i)^2 \left[H(t), \int_{t_0}^{t} H(t') dt'\right] = (-i)^2 \int_{t_0}^{t} [H(t), H(t')] dt' = 0.$$

Using this commutativity it is a matter of simple computations to show that

$$\frac{d}{dt} e^{X(t, t_0)} = \dot{X}(t, t_0) e^{X(t, t_0)} = e^{X(t, t_0)} \dot{X}(t, t_0), \tag{2.37}$$

which can be written as

$$\frac{d}{dt} e^{X(t, t_0)} = -iH(t) e^{X(t, t_0)}.$$

Also, since $X(t_0, t_0) = -i \int_{t_0}^{t_0} H(t') dt' = 0$, $e^{X(t_0, t_0)} = \mathbb{1}$, which coincides with $U(t_0, t_0)$. Therefore $U(t, t_0)$ satisfies the same differential equation as $e^{X(t, t_0)}$, with the same initial condition. Hence (assuming that the Cauchy's theorem for ordinary differential equations can be applied in our context),

$$U(t, t_0) = e^{X(t, t_0)} = e^{-i \int_{t_0}^{t} H(t') dt'},$$

which is exactly what we had to show, see Equation (2.36).

The situation is by far more complicated if $[H(t_1), H(t_2)] \neq 0$, for $t_1 \neq t_2$. In this case we should introduce the operator of *time ordering* $\mathcal{T}$. It is probably more convenient to rewrite Equation (2.34) in its integral form, recalling that $U(t_0, t_0) = \mathbb{1}$:

$$U(t, t_0) = \mathbb{1} - i \int_{t_0}^{t} H(t_1)\, U(t_1, t_0)\, dt_1 \qquad (2.38)$$

This equation gives rise to a perturbative series:

$$U(t, t_0) = \mathbb{1} - i \int_{t_0}^{t} H(t_1)\, dt_1 + (-i)^2 \int_{t_0}^{t} \int_{t_0}^{t_1} H(t_1)H(t_2)\, dt_2\, dt_1 + \cdots, \qquad (2.39)$$

where the order in which the Hamiltonian $H(t_j)$ appears is essential, and where $t_0 < t_1 < t_2 < \ldots < t$. It is known, see [42] for instance, that $U(t, t_0)$ can be rewritten in the following form:

$$U(t, t_0) = \mathcal{T} \exp \left\{ -i \int_{t_0}^{t} H(t_1)\, dt_1 \right\}, \qquad (2.40)$$

where $\mathcal{T}$ is the *Dyson time-ordering operator* satisfying the following:

$$\mathcal{T}(A(t_1)B(t_2)) = \begin{cases} A(t_1)B(t_2), & \text{if } t_1 < t_2 \\ B(t_2)A(t_1), & \text{if } t_2 < t_1 \end{cases}$$

Of course, $\mathcal{T}$ does not modify the order of the operators $A(t_1)$ and $B(t_2)$ when they commute, as it was assumed for $H(t)$ in the derivation of Equation (2.36). In fact, when this happens, Equation (2.40) coincides with Equation (2.36). In general, however, they are different and Equation (2.40) is only a formal formula which should be taken as the starting point for a perturbative expansion like the one in Equation (2.39). But this goes (much) beyond the scope of this book.

2.4.2 Heisenberg Representation

The possibility of changing representation in quantum mechanics is based on the fact that what is usually relevant for us is not really the time evolution of the state of $\mathcal{S}$, or the time evolution of its observables, but only *the mean values of the observables in the state of $\mathcal{S}$*. This is because what is usually measured in experiments are exactly these mean values, and not the operators themselves.

In what follows it is convenient to introduce a suffix to distinguish between the Schrödinger and Heisenberg representations[8]: in particular we use Ψ_S and X_S to indicate the state Ψ and the observable X in the Schrödinger representation, while

---

[8] This suffix will not be used in the rest of the book to avoid useless complications in the notation.

we adopt Ψ_H and X_H for the same objects in the Heisenberg representation. The link between the two representations is provided by the following equality:

$$\langle \Psi_S(t), A_S \Psi_S(t) \rangle = \langle \Psi_H, A_H(t) \Psi_H \rangle, \qquad (2.41)$$

where it is explicitly shown that the wave function depends on time in the Schrödinger but not in the Heisenberg representation, and that, vice versa, the observables depend on time in the Heisenberg but not in the Schrödinger representation. From Equation (2.41) and from Equation (2.31), with Ψ identified with Ψ_S, it follows that (putting for simplicity $t_0 = 0$),

$$A_H(t) = U^\dagger(t,0)\, A_S\, U(t,0). \qquad (2.42)$$

It is clear that, in particular, $A_H(0) = A_S$ and that $\mathbb{1}_H = \mathbb{1}_S$. Moreover, $A_H(t) = A_S$ if A_S commutes with $U(t,0)$. Also, if A and B are two observables, calling $C_S = A_S B_S$ their product in the Schrödinger representation, then $C_H(t) = A_H(t)B_H(t)$.

> *Remark:* This last result is true since $U(t,0)$ is unitary, which is the case when H (or $H(t)$) is self-adjoint. However, as already stated in Section 2.3.3, this is not always the case: in some applications it can be convenient to adopt an effective, non-self-adjoint, Hamiltonian. In this case, it is not granted that $C_H(t) = A_H(t)B_H(t)$. This is because $A_H(t)B_H(t) = U^\dagger(t,0)\, A_S\, U(t,0)U^\dagger(t,0)B_S\, U(t,0) \neq U^\dagger(t,0)\, A_S B_S\, U(t,0)$, since $U(t,0)U^\dagger(t,0)$ does not need to be equal to the identity operator. In other words, $U(t,0)$ is unitary if H is self-adjoint, but not in general. This is particularly evident from, say, Equation (2.35) or from Equation (2.36). However, this kind of problems will not appear often in this book, except in a few points, where we will say more on these aspects.

Going back to the relation between the Schrödinger and the Heisenberg representation, we observe that, if $C_S = [A_S, B_S]$, then $C_H(t) = [A_H(t), B_H(t)]$.

It is interesting to notice that, if $[H(t_1), H(t_2)] = 0$, then the Hamiltonian H is the same in both representations, even in presence of an explicit time dependence:

$$H_H(t) = U^\dagger(t,0)\, H_S(t)\, U(t,0) = U^\dagger(t,0)\, U(t,0)\, H_S(t) = H_S(t),$$

using Equation (2.37). The reason why we are using here $H_S(t)$ is because we are also admitting the possibility that H_S has an explicit time dependence, as it happens when the system $\mathcal{S}$ is driven also by some external, time-dependent, classical field.

The next step consists in finding the differential equation for $A_H(t)$. For that we observe that, taking the adjoint of Equation (2.34), we get

$$-i\, \frac{\partial U^\dagger(t,t_0)}{\partial t} = U^\dagger(t,t_0)\, H_S(t),$$

so that, after a few simple computations,

$$i\dot{A}_H(t) = [A_H(t), H_H(t)] + i\left(\frac{\partial A_S(t)}{\partial t}\right)_H, \qquad (2.43)$$

where

$$\left(\frac{\partial A_S(t)}{\partial t}\right)_H = U^\dagger(t, t_0)\left(\frac{\partial A_S(t)}{\partial t}\right)U(t, t_0).$$

Of course, this term disappears if $A_S(t)$ does not depend explicitly on time, i.e., if $A_S(t) = A_S$. In this case, in particular, if A_S commutes with H_S, $[A_S, H_S] = 0$, then, for what we have already seen, $[A_H(t), H_H(t)] = 0$ as well, and Equation (2.43) implies that $A_H(t) = A_H(0) = A_S$. When this happens, the operator A_S is called a *constant* or *integral of motion*. The existence of integrals of motion for a given system $\mathcal{S}$ can be useful in the analysis of $\mathcal{S}$ and its dynamics. A particularly illuminating example of this relevance is discussed in [9].

> *Remark:* It is well known that other quantum mechanical representations also exist, as, for instance, the *interaction representation*. Since this representation will not be useful for us here, we refer to specialized books in quantum mechanics for its detailed description, see [2, 3, 31] for instance.

2.5 Heisenberg Uncertainty Principle

Due to its relevance in ordinary quantum mechanics and in decision-making, we give here the mathematical details of the derivation of the Heisenberg uncertainty principle. This is useful in connection with what discussed in Chapter 12, where we will consider its role in the analysis of compatible and incompatible questions, a very hot topic in decision-making.

Consider two self-adjoint, possibly non-commuting, operators, A and B, acting on the Hilbert space $\mathcal{H}$: $A = A^\dagger$, $B = B^\dagger$, and let us assume that iC is the commutator between A and B: $[A, B] = iC$. It is easy to show that, since $A = A^\dagger$ and $B = B^\dagger$, the operator C must necessarily be self-adjoint as well. In fact, taking the adjoint of both sides of the equality $[A, B] = iC$, we easily find that $[A, B] = iC^\dagger$ as well, and the two equalities are compatible only if $C = C^\dagger$.

> *Remark:* Sometimes the operators A and B are unbounded. This is the case, for instance, for the position and the momentum operators. In this case the rule $[A, B] = iC$ must be supplemented with information on the domains of the operators A, B and C, but not only. In fact, in order to give a rigorous meaning to this equality we should work, for instance, on some $\mathcal{D}$, dense in $\mathcal{H}$, such that $\mathcal{D} \subseteq D(AB) \cap D(BA) \cap D(C)$. If such a $\mathcal{D}$ exists, then the commutation rule should be understood as follows: $ABf - BAf = iCf$, for all $f \in \mathcal{D}$.

Let now $\varphi \in \mathcal{H}$ be a fixed vector and $<X>:=\langle \varphi, X\varphi \rangle$ the expectation value of the operator X on the vector φ. If X is a bounded operator, $<X>$ is well defined for all $\varphi \in \mathcal{H}$. But, if X is unbounded, φ must be taken in the domain of X, $D(X)$. In particular we assume here that $\varphi \in D(A^2) \cap D(B^2)$, since this guarantees that $\varphi \in D(A) \cap D(B)$ as well.[9] Let us further define $(\Delta A)^2 := \langle \varphi, (A-<A>)^2\varphi \rangle = \langle A^2 \rangle - \langle A \rangle^2$. Analogously we have $(\Delta B)^2 = \langle B^2 \rangle - \langle B \rangle^2$. Then, since each self-adjoint operator $X = X^\dagger$ has an expectation value $<X>$ which is real, we can rewrite

$$(\Delta X)^2 = \left\langle \varphi, (X-<X>)^2 \varphi \right\rangle = \|(X-<X>)\varphi\|^2. \tag{2.44}$$

Notice that both $\langle X \rangle$ and ΔX depend on φ, but we are not making explicit this dependence here, to simplify the notation. ΔX is called the *uncertainty* on X [2], and measures how well we really know the value of the observable X. In fact, from (2.44) we conclude that $\Delta X = 0$ if and only if $(X - \langle X \rangle)\varphi = 0$, which means that φ must be an eigenstate of X with eigenvalue $\langle X \rangle$. On the other hand, if $\Delta X > 0$, then φ cannot be an eigenstate of X. If we are measuring several times the observable X, then $\langle X \rangle$ gives the mean value of these results, while ΔX is proportional to the width of the distribution of these results, see Figure 2.1. Of course, the smaller ΔX, the better our knowledge of X.

Now, using the Schwartz inequality,[10] we get

$$|\langle (A-<A>)\varphi, (B-<B>)\varphi \rangle| \leq \|(A-<A>)\varphi\| \|(B-<B>)\varphi\| = (\Delta A)(\Delta B).$$

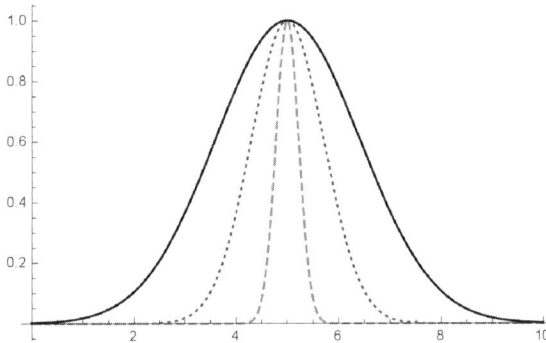

Figure 2.1 Black line: large ΔX; Dotted line: medium ΔX; Dashed line: small ΔX.

---

[9] This is because, for instance, $D(A^2) = \{\phi \in D(A) : A\phi \in D(A)\}$. Of course, if A is bounded, $A \in B(\mathcal{H})$, $D(A) = D(A^2) = \mathcal{H}$.

[10] For all $f, g \in \mathcal{H}$, the Schwartz (or Cauchy-Schwarz) inequality states that $|<f, g>| \leq \|f\| \|g\|$.

Moreover, using the properties of A and $<A>$, and in particular the fact that $A = A^\dagger$ and that $<A>$ is real,

$$\langle (A - <A>)\varphi, (B - <B>)\varphi \rangle = \langle \varphi, (A - <A>)(B - <B>)\varphi \rangle .$$

For our purposes it is now useful to write

$$(A - <A>)(B - <B>) = \frac{(A - <A>)(B - <B>) + (B - <B>)(A - <A>)}{2}$$
$$+ \frac{(A - <A>)(B - <B>) - (B - <B>)(A - <A>)}{2} = F + \frac{i}{2}C,$$

where we have defined a new self-adjoint operator

$$F := \frac{1}{2}\left[(A - <A>)(B - <B>) + (B - <B>)(A - <A>) \right],$$

$F = F^\dagger$, and we have observed that $(A - <A>)(B - <B>) - (B - <B>)(A - <A>) = AB - BA = iC$. Summarizing, we have

$$(\Delta A)^2(\Delta B)^2 \geq |\langle (A - <A>)\varphi, (B - <B>)\varphi \rangle|^2 = \left| \left\langle \varphi, \left(F + \frac{i}{2}C \right)\varphi \right\rangle \right|^2$$
$$= \left| <F> + \frac{i}{2}<C> \right|^2 .$$

Taking now into account the fact that both $<F>$ and $<C>$ are real quantities, we conclude that

$$(\Delta A)^2(\Delta B)^2 \geq \left| <F> + \frac{i}{2}<C> \right|^2 = <F>^2 + \frac{<C>^2}{4} \geq \frac{<C>^2}{4},$$

so that

$$\Delta A \, \Delta B \geq \frac{|<C>|}{2}, \tag{2.45}$$

which gives a lower bound on the product of the uncertainties on A and B: whenever A and B do not commute, they cannot be known together with arbitrary precision! For instance, if we consider the position and the self-adjoint momentum operators x and p, and we take $A = x$, $B = p$ and $C = \mathbb{1}$, we go back to the well-known Heisenberg inequality: the commutator becomes $[x, p] = i\,\mathbb{1}$, while the inequality in (2.45) takes the following well-known form: $\Delta x \, \Delta p \geq \frac{1}{2}$. Notice that we are working here in suitable units, taking $\hbar = 1$. Incidentally we observe that, from Equation (2.45), we get the obvious inequality $0 \geq 0$ if φ is an eigenstate of A or of B. Notice also that φ can be simultaneously an eigenstate of both A and B if $[A, B] = 0$, i.e., if $C = 0$, but not in general. When this happens we have $\Delta A = \Delta B = 0$, which means that both A and B can be measured exactly. Otherwise, i.e., if φ is an eigenstate of, say, A, but not of B, only A can be measured with no error ($\Delta A = 0$), while B cannot. Still, as we have already noticed, Equation (2.45) holds in the trivial form $0 \geq 0$.

We have recently used commutativity between observables as an efficient way to describe compatible or incompatible questions. For instance, if A and B are two self-adjoint operators connected to two different questions (e.g., A can be related to the question "*is Alice happy?*" and B to the other question "*is Alice employed?*," both with just two answers, yes or not; therefore A and B can be represented as two spin operators). Then, $[A, B] = 0$ can be naturally understood as if A and B describe two compatible questions (they do not affect each other), while these are necessarily incompatible if $[A, B] \neq 0$. Hence the commutator between A and B reflects the nature of the questions described by these operators. We will discuss this particular application in detail in Chapter 12.

2.6 A few Words on States

In our approach, states play an essential role, since they are used to project down the time evolution of the system S we are dealing with from an operatorial to a classical level. The strategy is the following: we use operators to describe certain macroscopic systems. Let S be one such system, and X be an observable relevant in the analysis of S. Its time evolution $X(t)$ is deduced from the Hamiltonian of S, as discussed before, adopting, for instance, the Heisenberg representation. The Hamiltonian H is written by considering all the interactions existing between the various agents of S, following a few minimal (and natural) rules listed in [1]. Hence, if Ψ is the vector describing S at time zero (and therefore also at time t, see Section 2.4), $x(t) = \langle \Psi, X(t)\Psi \rangle$ is interpreted as the time evolution of the classical dynamical variable represented, in our approach, by the operator X, when S is described, at $t = 0$, by Ψ. In this sense $x(t)$ is the projection of the quantum-like dynamics of X, $X(t)$ to its original classical level.

A state of this kind, $\langle \varphi, Y\varphi \rangle$, for some normalized vector $\varphi \in \mathcal{H}$ and some operator Y on $\mathcal{H}$ is called *a vector state*. More in general, a state ω over a certain *-algebra $\mathcal{A}$ is a positive linear functional which is normalized. In other words, for each element $A \in \mathcal{A}$, ω satisfies the following:

$$\omega(A^\dagger A) \geq 0, \quad \text{and} \quad \omega(\mathbb{1}) = 1,$$

where $\mathbb{1}$ is the identity of $\mathcal{A}$. It is known, [43], that not all the *-algebras necessarily possess an identity. But it is also known that $\mathbb{1}$ can be added to $\mathcal{A}$, keeping the mathematical structure essentially unchanged, see again [43]. Moreover, $\omega(A^\dagger) = \overline{\omega(A)}$. These properties are clearly satisfied by any vector state. In fact, independently of the choice of normalized φ, we have $\langle \varphi, 1\!\!1\, \varphi \rangle = \|\varphi\|^2 = 1$. Moreover, $\langle \varphi, A^\dagger A \varphi \rangle = \|A\varphi\|^2 \geq 0$. Finally, $\langle \varphi, A^\dagger \varphi \rangle = \langle A\varphi, \varphi \rangle = \overline{\langle \varphi, A\varphi \rangle}$.

States over bounded operators share many nice properties: they are automatically continuous [43]; if A_n is a sequence of bounded operators converging to A in the uniform topology, then $\omega(A_n)$ converges to $\omega(A)$ in the topology of $\mathbb{C}$:

$$\|A_n - A\| \to 0, \qquad \Longrightarrow \qquad |\omega(A_n) - \omega(A)| \to 0,$$

when $n \to \infty$. A very useful inequality which each state satisfies is the *Cauchy-Schwarz inequality* already used, in a slightly different and simpler form, in Section 2.5: for each $A, B \in \mathcal{A}$, the following holds:

$$|\omega(A^\dagger B)|^2 \leq \omega(A^\dagger A)\, \omega(B^\dagger B).$$

The vector states already introduced in this chapter, see Equations (2.3) and (2.6), describe a situation in which the numbers of all the different modes of bosons or fermions are known, since the vectors we use are eigenstates of the single-mode number operators. This kind of states will be used several times in this book, but they are not the only useful states in quantum mechanics. In fact, other types of states also exist and are relevant in many physical applications. For instance, the so-called KMS states, i.e., the equilibrium states for systems with infinite degrees of freedom, are usually used to prove the existence of phase transitions or to find conditions for the existence of some thermodynamical equilibrium. Without going into the mathematical rigorous definition, see [43], a KMS state ω with inverse temperature β satisfies the following equality, known as *the KMS condition*:

$$\omega(A\, B(i\beta)) = \omega(B\, A), \tag{2.46}$$

where A and B are general elements of $\mathcal{A}$, and $B(i\beta)$ is the time evolution of the operator B computed at the complex value $i\beta$ of the time.[11] It is well known [44] that, when restricted to a finite size system, a KMS state is nothing but a Gibbs state which can be expressed in terms of a trace with a suitable weight involving the Hamiltonian of the system:

$$\omega(A) = \frac{tr\left(e^{-\beta H} A\right)}{tr\left(e^{-\beta H}\right)}.$$

In general, a KMS state is used to describe a thermal bath interacting with a physical system $\mathcal{S}$, if the bath has a nonzero temperature.

It is interesting to observe that, as each normalized vector defines a state, each state can be used to define a normalized vector in a certain Hilbert space. The way in which this is achieved is known as the *GNS construction*, where GNS stands

---

[11] More precisely, $B(i\beta)$ is an analytic extension of $B(t)$ to the complex domain [44].

for Gelfand-Naimark-Segal [43]. This is quite important in physical applications, since different states can give rise to different, and inequivalent, representations of the same system S. These inequivalent representations can be related, for instance, to different thermodynamical phases of S.

2.7 More on Dynamics

For what we will discuss in Part II, we need to know something more on the dynamical aspects of quantum systems. In fact, for instance, when trying to describe a procedure of decision-making, we expect that the time evolution of the system produces, at least for those dynamical variables associated with the mechanism of the decision, some reasonable asymptotic value. In other words, if $d(t)$ is the function which describes the time evolution of the decision process of, say, Alice, it is quite natural to expect that, for time large enough, $d(t)$ converges to some value, d_∞. This is what should be understood as Alice's final decision. If $d(t)$ reaches no asymptotic value, as it is the case if Alice always oscillates between different moods, no decision is taken. Of course, one natural question (but not the only one!) is *how large the time should be*, i.e., what is a reasonable time needed to take a decision. There is no sharp answer to this question: in some situations the decision should be extremely fast, while in other cases there is no real need to hurry up. Some authors just fix the *decision time* t_d from the very beginning. Hence, even if $d(t)$ is oscillating, when $t = t_d$, Alice just takes $d(t_d)$ as her final decision. Of course, this approach implies that fixing a different value for t_d Alice takes a (possibly) completely different decision. Hence the result, and the whole procedure, is extremely sensitive to this choice. For this reason, we prefer to adopt here (and in many applications discussed later in the book) a different approach, based on what in quantum mechanics are called *open quantum systems*, in which the decision is driven by the interaction between Alice and her environment. In particular, we will see how different systems give rise to different interpretations for this environment, which could be seen as a group of people, or a set of information, or even a group of cells (or neurons).

If S is a closed quantum system with time-independent, self-adjoint, Hamiltonian H, it is natural to suspect that only periodic or quasi-periodic effects can take place, during the time evolution of S. This is because the energy of S is preserved, and this seems to prevent to have any damping effect. For instance, if we work in the Schrödinger representation, the time evolution $\Psi(t)$ of the wave-function of the system is simply $\Psi(t) = e^{-iHt}\Psi(0)$ and, since the operator e^{-iHt} is unitary, we do not expect that $\Psi(t)$ decreases (in any reasonable sense) to zero when t diverges. On the other hand, we will show that a similar decay feature is possible if

$\mathcal{S}$ is coupled to a reservoir $\mathcal{R}$, but only if $\mathcal{R}$ is rather large when compared to $\mathcal{S}$, or, more explicitly, if $\mathcal{R}$ has an infinite number of degrees of freedom. This aspect is essential, indeed.

We start considering a first system, $\mathcal{S}$, interacting with a second system, $\tilde{\mathcal{S}}$, and we assume for the time being that both $\mathcal{S}$ and $\tilde{\mathcal{S}}$ are *of the same size*: to be concrete, we imagine here that $\mathcal{S}$ describes a single particle whose related operators are a, $a^\dagger$ and $\hat{n}_a = a^\dagger a$. Analogously, $\tilde{\mathcal{S}}$ describes a second particle whose related operators are b, $b^\dagger$ and $\hat{n}_b = b^\dagger b$. These operators obey the following CCR: $[a, a^\dagger] = [b, b^\dagger] = \mathbb{1}$, while all the other commutators are assumed to be zero. In particular, a and b commute: $[a, b] = 0$. Hence we are dealing with two originally independent bosonic modes. A natural choice for the Hamiltonian of the interacting system $\mathcal{S} \cup \tilde{\mathcal{S}}$, adopting a very simple type of interaction between the particles, is the following: $h = \omega_a \hat{n}_a + \omega_b \hat{n}_b + \mu(a^\dagger b + b^\dagger a)$, where ω_a, ω_b and μ must be real quantities if we are interested to have $h = h^\dagger$. We should recall that losing self-adjointness of h would produce, among other consequences, a non-unitary time evolution, and this is out of the scheme usually considered in ordinary quantum mechanics.[12] The Hamiltonian h contains a free part plus an interaction which is such that, if the eigenvalue of $\hat{n}_a$ increases of one unit during the time evolution, the eigenvalue of $\hat{n}_b$ must decrease, again by one unit, and vice versa. This can be easily understood since because $[h, \hat{n}_a + \hat{n}_b] = 0$, so that $\hat{n}_a + \hat{n}_b$ is an integral of motion. The equations of motion for $a(t)$ and $b(t)$ can be deduced as in Equation (2.43) and turn out to be

$$\dot{a}(t) = i[h, a(t)] = -i\omega_a a(t) - i\mu b(t), \qquad \dot{b}(t) = i[h, b(t)] = -i\omega_b b(t) - i\mu a(t),$$

$$(2.47)$$

whose solution can be written as $a(t) = \alpha_a(t)\, a + \alpha_b(t)\, b$ and $b(t) = \beta_a(t)\, a + \beta_b(t)\, b$. The functions $\alpha_j(t)$ and $\beta_j(t), j = a, b$, are linear combinations of $e^{\lambda_\pm t}$, with $\lambda_\pm = \frac{-i}{2}(\omega_a + \omega_b - \sqrt{(\omega_a - \omega_b)^2 + 4\mu^2})$. Moreover, $\alpha_a(0) = \beta_b(0) = 1$ and $\alpha_b(0) = \beta_a(0) = 0$, in order to have $a(0) = a$ and $b(0) = b$. Hence we see that both $a(t)$ and $b(t)$, and $\hat{n}_a(t) = a^\dagger(t) a(t)$ and $\hat{n}_b(t) = b^\dagger(t) b(t)$ as a consequence, are linear combinations of oscillating functions, so that no damping, and no stable asymptotic limit, is possible for this simple model. This simple remark is important for us, because of the role that number operators like $\hat{n}_a(t)$ and $\hat{n}_b(t)$ will have in the rest of the book.

Of course the reader could be not really satisfied with this simple model because, in particular, of our specific choice of h. In fact, this is useful just because it produces an analytical solution for $a(t)$ and $b(t)$. However, a similar conclusion

---

[12] We should also stress that in recent years a larger and larger group of physicists and mathematicians started to be interested in what happens when the dynamics is defined in terms of some Hamiltonian which is not self-adjoint but satisfies certain symmetry properties which are physically motivated. This is what, in the literature, it is usually called *pseudo-Hermitian quantum mechanics*, see also Sections 2.3.3 and 2.7.2.

can still be deduced also if h is defined differently, even if h not necessarily constructed in terms of ladder operators. Suppose, to be concrete, that h is a self-adjoint $n \times n$ matrix. This means that the system we are interested in, $\hat{S}$, lives in a finite-dimensional Hilbert space $\mathcal{H}_n$. This excludes, in principle, the possibility that bosonic operators play some role in the analysis of this particular system, while other ladder operators (coming from the CAR, or from truncated bosonic operators, for instance) fit into this situation. Moreover, as it was widely discussed in [9], some bosonic systems can still be efficiently described in a finite-dimensional Hilbert space, $\mathcal{H}_{eff}$. This is what happens when some integral of motion exists which relates, and constrains, the various number operators of the model. When this happens, not all the *energetic levels* of $\mathcal{H}$ can be reached by the system, and those that can give rise to $\mathcal{H}_{eff}$ [1,9].

Now, the time evolution of a certain observable D of $\hat{S}$ deduced by the Hamiltonian h is, of course, $D(t) = e^{iht} D e^{-iht}$. But, since $h = h^\dagger$, a unitary matrix U exists such that $UhU^{-1} = h_d$, with diagonal h_d. Of course, the diagonal elements of h_d are the real eigenvalues of h. Then

$$D(t) = U^{-1} e^{ih_d t} U D U^{-1} e^{-ih_d t} U,$$

where, of course, $e^{\pm ih_d t}$ are diagonal matrices. Hence the mean value $\langle D(t) \rangle$ of $D(t)$ on any vector can only be periodic or quasi-periodic, depending on the relation between the eigenvalues of h: no stable asymptotic value is possible for $\langle D(t) \rangle$, which keeps on oscillating for all t. Therefore, if we imagine that $\langle D(t) \rangle$ describes how Alice is processing her decision, the only natural way to stop the procedure, and concretely decide something, is to introduce, from outside, an ad hoc decision time t_d.

Suppose now that the system $\tilde{S}$ is replaced by an (infinitely extended) reservoir $\mathcal{R}$, whose particles are described by an infinite set of bosonic operators $b(k), b^\dagger(k)$ and let $\hat{n}(k) = b^\dagger(k) b(k), k \in \mathbb{R}$, be the corresponding number operators. We assume now the following Hamiltonian for $\mathcal{S}_{full} = \mathcal{S} \cup \mathcal{R}$, which extends the operator h for the two bosons proposed above:

$$H = H_0 + \lambda H_I, \quad H_0 = \omega \hat{n}_a + \int_\mathbb{R} \omega(k) \hat{n}(k) \, dk, \quad H_I = \int_\mathbb{R} \left(a b^\dagger(k) + a^\dagger b(k) \right) f(k) dk, \tag{2.48}$$

where $[a, a^\dagger] = \mathbb{1}$, $[b(k), b^\dagger(q)] = \mathbb{1} \delta(k - q)$, while all the other commutators are zero. All the constant appearing in (2.48), as well as the regularizing function $f(k)$, are real, so that $H = H^\dagger$. In our simple system, H models Alice and her interactions with some environment, which can be understood as the set of her friends, parents and relatives, for instance.

Notice that an integral of motion exists also for $\mathcal{S}_{full}$, $\hat{n}_a + \int_{\mathbb{R}} \hat{n}(k)\, dk$, which extends the one for $\mathcal{S} \cup \tilde{\mathcal{S}}$, $\hat{n}_a + \hat{n}_b$. With this choice of H, the Heisenberg equations of motion are

$$\begin{cases} \dot{a}(t) = i[H, a(t)] = -i\omega a(t) - i\lambda \int_{\mathbb{R}} f(k)\, b(k,t)\, dk, \\ \dot{b}(k,t) = i[H, b(k,t)] = -i\omega(k)b(k,t) - i\lambda f(k)\, a(t), \end{cases} \tag{2.49}$$

which are supplemented by the initial conditions $a(0) = a$ and $b(k,0) = b(k)$. The last equation in (2.49) can be rewritten in integral form as $b(k,t) = b(k)e^{-i\omega(k)t} - i\lambda f(k) \int_0^t a(t_1)e^{-i\omega(k)(t-t_1)}\, dt_1$. We fix now, as is often done in the literature, $f(k) = 1$, and we take $\omega(k) = k$ and we replace $b(k,t)$ in the first equation in (2.49). Then, with a change of the order of integration, and recalling that $\int_{\mathbb{R}} e^{-ik(t-t_1)}\, dk = 2\pi\delta(t-t_1)$ and that $\int_0^t g(t_1)\delta(t-t_1)\, dt_1 = \frac{1}{2}g(t)$ for any test function $g(t)$, we conclude that

$$\dot{a}(t) = -(i\omega + \pi\lambda^2)a(t) - i\lambda \int_{\mathbb{R}} b(k)\, e^{-ikt}\, dk. \tag{2.50}$$

This equation can be solved, and the solution is

$$a(t) = \left(a - i\lambda \int_{\mathbb{R}} dk\, \eta(k,t) b(k) \right) e^{-(i\omega + \pi\lambda^2)t}, \tag{2.51}$$

where $\eta(k,t) = \frac{1}{\rho(k)}(e^{\rho(k)t} - 1)$ and $\rho(k) = i(\omega - k) + \pi\lambda^2$. Using complex contour integration, it is possible to check that $[a(t), a^\dagger(t)] = \mathbb{1}$ for all t: this means that the apparent decay of $a(t)$, described in Equation (2.51), is balanced by a contribution of the reservoir.

> *Remark:*   The choices $f(k) = 1$ and $\omega(k) = k$ are convenient since they allow us to get analytical solutions, see Equation (2.51). However, other choices could be adopted, and some of them will be considered in Section 2.7.1.

Let us now consider a state over $\mathcal{S}_{full}$, $\langle X_S \otimes X_R \rangle = \langle \varphi_{n_a}, X_S \varphi_{n_a} \rangle \langle X_R \rangle_R$, in which φ_{n_a} is the eigenstate of the number operator $\hat{n}_a$ and $<>_R$ is a state of the reservoir, see Section 2.6. Here $X_S \otimes X_R$ is the tensor product of an operator of the system, X_S, and an operator of the reservoir, X_R. Stated differently, $\langle . \rangle_R$ is a normalized, positive linear functional over the algebra of the operators of R, while $= \langle \varphi_{n_a}, \cdot \varphi_{n_a} \rangle$ is a vector state over the algebra of S. The state $\langle . \rangle_R$ is assumed to satisfy some properties which extend to R similar properties of the vector states over S. In particular, we require that $\langle b^\dagger(k)b(q) \rangle_R = n_b(k)\delta(k-q)$ and that $\langle b(k)b(q) \rangle_R = 0$. These are standard choices, see for instance [45], at least in the context of open quantum systems and of quantum optics in particular. Then, if for simplicity we take the function $n_b(k)$ to be constant in k, we get

$$n_a(t) := <\hat{n}_a(t)> = <a^\dagger(t)a(t)> = n_a\, e^{-2\lambda^2\pi t} + n_b\left(1 - e^{-2\lambda^2\pi t}\right), \tag{2.52}$$

which goes to n_b as $t \to \infty$. Hence, if $0 \le n_b < n_a$, the value of $n_a(t)$ decreases with time. If, on the other hand, $n_b > n_a$, then the value of $n_a(t)$ increases for large t. In both cases, $n_a(t)$ admits an asymptotic value which, identifying $n_a(t)$ with the decision function $\langle D(t) \rangle$ introduced before, can be interpreted as Alice's final decision. Notice that it is essential to assume here that $\mathcal{R}$ has an infinite number of degrees of freedom. Notice also that, in particular, if the reservoir is originally empty, $n_b = 0$, then $n_a(t) = n_a\, e^{-2\lambda^2 \pi t}$ decreases exponentially to zero: the system becomes empty (and Alice's decision is zero). On the other hand, since $\hat{n}_a + \int_{\mathbb{R}} \hat{n}(k)\, dk$ is a constant of motion, the reservoir starts to be filled up.

It might be interesting to remark that the continuous reservoir considered here could be replaced by a discrete one, describing again an infinite number of particles, but labeled by a discrete index. This could be useful to provide a more natural interpretation for the reservoir in terms of agents of some kind interacting with the *core* of $\mathcal{S}_{full}$, the system $\mathcal{S}$, which in our case is just Alice. Hence, to obtain a Dirac delta distribution, which is the crucial mathematical ingredient in the derivation above, we have to replace the integral formula $\int_{\mathbb{R}} e^{-ik(t-t_1)}\, dk = 2\pi \delta(t - t_1)$ with the *Poisson summation formula*, which we write here as $\sum_{n \in \mathbb{Z}} e^{inxc} = \frac{2\pi}{|c|} \sum_{n \in \mathbb{Z}} \delta$ $(x - n\frac{2\pi}{c})$, for all nonzero $c \in \mathbb{R}$. Not many other differences arise when continuing the analysis of $\mathcal{S}_{full}$.

2.7.1 Inhomogeneous Reservoir

In the explicit computations above it is technically important to work with a sort of homogeneous reservoir, i.e., to assume that $n_b(k) = n_b$, for all $k \in \mathbb{R}$. However, it is also natural to imagine that other reasonable (and well-motivated) choices for $n_b(k)$ also exist. This aspect was considered, for instance, in [11], where the inhomogeneity of the reservoir was used to model different sources of information reaching some agents during a procedure of decision-making. We have also used an explicitly k-dependent function $n_b(k)$ in [10], in connection with love affairs. Of course, due to formula $\langle b^\dagger(k)b(q) \rangle_{\mathcal{R}} = n_b(k)\delta(k - q)$, $n_b(k)$ must be real and nonnegative. This does not fix (at all!) the analytic expression of $n_b(k)$. Still, if we are willing to derive analytic results, it is convenient to restrict our choice of this function to some particular mathematical expressions. This is important if we want to use complex integration to perform the computations, as it is usually done in quantum optics [45]. For instance, in [11] we have taken a decreasing function of k. More in details, it is useful to assume that $n_b(k)$ is analytic in k, and that $|n_b(k)|$ does not diverge when $|k|$ diverges. In this way we can use the Jordan's lemma and compute the integrals involved in our computations using complex methods. We fix, just as an example,

$$n_b(k) = \frac{n_b}{k^2 + \alpha^2}, \tag{2.53}$$

for some positive n_b and for some $\alpha > 0$: this describes a situation in which not all the parts of the bath interact with Alice (the decision maker) in the same way. We also take $\omega_b(k) = \omega_b k$. In the computation of $\int_\mathbb{R} n_b(k)|\eta(k,t)|^2 dk$, which appears when checking that $[a(t), a^\dagger(t)] = \mathbb{1}$ for all t, we have to consider two singularities of the integrating functions both in the upper and lower complex semi-planes, so the result of the integral is the sum of two residues.

The computation could be given in detail, and the analytic form of $n_a(t)$ could be deduced for all t. However, if we are interested only in deducing what the final decision is, what is relevant for us is just the large time limit of $n_a(t)$. This turns out to be

$$n_a(\infty) := \lim_{t,\infty} n_a(t) = \frac{n_b(\alpha\omega_b^2 + \pi\lambda^2)}{\alpha\lambda^2\omega_b^2\left(\frac{\omega^2}{\omega_b^2} + \left(\alpha + \frac{\pi\lambda^2}{\omega_b^2}\right)^2\right)}. \tag{2.54}$$

Notice that this result makes sense only if α and λ are not zero. This suggests that the role of the complex pole in $n_b(k)$, and of the interaction between Alice and her bath, is really essential to get some stabilization within the scheme proposed here. We observe also that the asymptotic value of the decision function $n_a(t)$ strongly depends on the various parameters of the model, which could be adjusted to fit experimental data, if needed. From this point of view the choice of $n_b(k)$ in Equation (2.53) is more interesting than the one considered before, of a uniform reservoir with $n_b(k) = n_b$. In that case, in fact, n_b turned out to be exactly Alice's final decision: this means that Alice does not really have a constructing role in her process of decision-making, since she always decides n_b independently of what was her original *mood*. On the other hand, see Equation (2.54), if we enrich the form of the reservoir, $n_a(\infty)$ is also related to ω and to λ, which are both parameters of H directly related to Alice. In particular, ω is a measure of Alice's inertia, [1], while λ measures the strength of the interaction between Alice and her reservoir.

> *Remark:* Our results suggest that the asymptotic limit of $n_a(t)$ does not really depend on the fact that, at $t = 0$, Alice is described by a pure state (some φ_{n_a}) or by some linear combination of these states. It is further clear that our approach does not give directly information on the output state, i.e., on the state after the decision is taken. This is because we are working here in the Heisenberg representation.

2.7.1.1 Other Choices for f(k) and ω(k)

So far we have fixed $f(k) = 1$ and $\omega(k) = \omega_b k$. As we have already claimed, this is not the only possible choice, in particular if we want to model possible differences between different parts of the reservoir. Technically, what is particularly useful for us is that Dirac's delta distribution appears when deducing the differential equation for $a(t)$ in Equation (2.50). And this is possible also with other choices: suppose that the two functions $f(k)$ and $\omega(k)$ are such that

$$\omega(k) \to \pm\infty \text{ for } k \to \pm\infty, \qquad \text{and} \qquad \frac{d\omega(k)}{dk} = \frac{1}{\beta} f^2(k),$$

for some real $\beta \neq 0$. These assumptions are satisfied if $f(k) = 1$ and $\omega(k) = \omega_b k$, but not only. Hence

$$\int_{\mathbb{R}} f^2(k) e^{-i\omega(k)(t-t_1)} \, dk = \beta \int_{\mathbb{R}} e^{-i\omega(k)(t-t_1)} \, d\omega(k) = 2\pi\beta\delta(t-t_1),$$

and we can again significantly simplify the equation for $a(t)$, which now becomes

$$\dot{a}(t) = - \left(i\omega + \beta\pi\lambda^2 \right) a(t) - i\lambda \int_{\mathbb{R}} f(k)b(k) \, e^{-i\omega(k)t} \, dk. \qquad (2.55)$$

This is the equation which replaces (2.50) in this new situation. The solution can be found as before, and we get

$$a(t) = \left(a - i\lambda \int_{\mathbb{R}} dk \, \tilde{\eta}(k,t)b(k)f(k) \right) e^{-\left(i\omega+\beta\pi\lambda^2\right)t}, \qquad (2.56)$$

where $\tilde{\eta}(k,t) = \frac{1}{\tilde{\rho}(k)}(e^{\tilde{\rho}(k)t} - 1)$ and $\tilde{\rho}(k) = i(\omega - \omega(k)) + \beta\pi\lambda^2$. At a first sight, the situation is not particularly different from what was deduced before. However, to find $[a(t), a^\dagger(t)]$ and $n_a(t)$, we need to compute integrals, and the poles of the integrating functions are strongly dependent on the analytic expression of $\omega(k)$: in particular, if $\omega(k)$ is not linear, the computations become rather hard. For this reason, this strategy will not be considered further in this book, even if it could be possibly relevant for other applications.

2.7.1.2 What If We Use the CAR?

In the Hamiltonian (Equation 2.48) the operators a and $b(k)$ are assumed to satisfy the CCR. However, from the point of view of decision-making, it might be more interesting to consider the case in which the agent's decisions are described by fermionic operators, so that the CCR assumed for the operators in Equation (2.48) are now replaced by the following CAR: $\{a, a^\dagger\} = \mathbb{1}$, with $\{a, a\} = 0$, and

$$\{b(k), b^\dagger(q)\} = \delta(k-q) \, \mathbb{1}, \qquad \{b(k), b(q)\} = 0.$$

Moreover, $\{a^\sharp, b^\sharp(k)\} = 0$, where $x^\sharp$ indicates either x or $x^\dagger$. Assuming the same Hamiltonian (Equation 2.48), the differential equations of motion for the annihilation operators $a(t)$ and $b(k,t)$ can be deduced. They turn out to coincide with those in Equation (2.49), except that λ must be replaced by $-\lambda$. In other words, we get

$$\begin{cases} \dot{a}(t) = i[H, a(t)] = -i\omega a(t) + i\lambda \int_{\mathbb{R}} f(k) \, b(k,t) \, dk, \\ \dot{b}(k,t) = i[H, b(k,t)] = -i\omega(k)b(k,t) + i\lambda f(k) \, a(t). \end{cases} \qquad (2.57)$$

As in the simplest bosonic case, we assume here that the function $n_b(k)$ for the state of the reservoir is constant, $n_b(k) \equiv n_b \geq 0$. The CAR implies that this constant cannot exceed one, $0 \leq n_b \leq 1$.

Our previous results deduced with the bosonic settings show that $n_a(t)$ depends on λ^2, see (2.52). This is true even with the choice (2.53) of $n_b(k)$, see (2.54). Therefore we do not expect any difference in the results when replacing the CCR with the CAR, since, as we have just seen, this only causes λ to be replaced by $-\lambda$ in the differential equations of motion. Of course, choosing the CCR or CAR should be related to the nature of the decision we want to model: in case of a binary question (yes or not), the natural setting is probably the fermionic one. But if we imagine that infinite possible answers are possible, then we should use a bosonic version of the model. When a finite number of answers are possible, but this number is larger than two, then we could try to adopt the truncated version of bosonic operators, see Section 2.3.2. But these are just details which do not change the general strategy: decisions (of any possible kind) are driven by interactions with some (homogeneous or not) environment. We will discuss several consequences of this idea in the second part of the book.

2.7.2 Non-Self-Adjoint Hamiltonians for Damping

An alternative simple way to describe damping effects in a quantum system makes use of non-self-adjoint Hamiltonians to deduce the dynamics. In this way there is no need to consider any reservoir interacting with the system S. This is, indeed, the point of view of many papers in quantum optics. Just to cite a single simple example, in [46, 47] the authors considered an effective non-self-adjoint Hamiltonian describing a two-level atom interacting with an electromagnetic field. This was done in connection with pseudo-Hermitian systems, as those naturally connected with the deformed commutation rules considered in Section 2.3.3. In fact, pseudo-fermions were used later [48], in the analysis of the same system previously considered in [46, 47]. More explicitly, the starting point in the analysis of the system is the Schrödinger equation:

$$i\dot{\Psi}(t) = H_{eff}\Psi(t), \qquad H_{eff} = \begin{pmatrix} -i\gamma_a & v \\ \bar{v} & -i\gamma_b \end{pmatrix}. \tag{2.58}$$

Here γ_a and γ_b are real quantities, related to the decay rates for the two levels, while the complex parameter v characterizes the radiation-atom interaction. We refer to [46,47], and to the references therein, for further details. Here we just want to stress that $H_{eff} \neq H_{eff}^{\dagger}$ and that the analytical expression of $\Psi(t)$ can be explicitly deduced. The resulting time evolution describes a decay of the wave-function, essentially because $e^{-iH_{eff}t}$ is not a unitary operator, so that $\|e^{-iH_{eff}t}\Psi(0)\| \neq \|\Psi(0)\|$, in general.

Having this kind of examples in mind, it seems a natural choice to consider, for instance, the following operator:

$$\hat{h} = \omega_a \hat{n}_a + \omega_b \hat{n}_b + \mu\, a\, b^\dagger, \tag{2.59}$$

where a, $a^\dagger$ and $\hat{n}_a$ are the operators of a first system $\mathcal{S}_a$, while b, $b^\dagger$ and $\hat{n}_b$ are those of a second system, $\mathcal{S}_b$, which in our idea should play here the role of a *hyper-simplified reservoir*. Here we take ω_a, ω_b and μ real, but this is not sufficient, of course, to make of h a self-adjoint operator, if $\mu \neq 0$. To make it self-adjoint, we should add a contribution $\mu\, a^\dagger b$ in its definition. However, it is exactly the lack of this term which makes $\hat{h}$ interesting for us. In fact, the interaction term $\mu\, a\, b^\dagger$ in $\hat{h}$ causes the eigenvalue of the operator $\hat{n}_a$ to decrease of one unit (because of the presence of the annihilation operator a) and that of $\hat{n}_b$ to increase simultaneously (because of $b^\dagger$), again of one unit. In other words, $\hat{h}$ seems to describe a situation in which some quanta of the system flow from $\mathcal{S}_a$ to $\mathcal{S}_b$. If a and b describe the cash of two different traders, trader τ_a is giving money to τ_b. After some time, and several interactions with τ_b, it is clear that τ_a ends up with no cash at all. This seems great, since it would suggest an effective procedure to have damping without making use of reservoirs with infinite degrees of freedom. Unfortunately, this very simple and apparently efficient method doesn't work! This can be explicitly seen if we assume that, also in this case, the time evolution of any fixed operator, x, is given by $x(t) = e^{i\hat{h}t}xe^{-i\hat{h}t}$. This is just the same choice we have discussed for self-adjoint Hamiltonians, but the main difference here is that $e^{\pm i\hat{h}t}$ are no longer unitary operators. However, rather than a problem, this could be considered as a positive feature of this approach, since we really want to deduce damping, and damping does not really go well with unitary operators (even if they do not necessarily exclude each other, see Equation (2.52), for instance). A second technical (but not only) problem arises because, with this definition of $x(t)$, it is not true in general that $(x(t))^\dagger = (x^\dagger)(t)$, so that $x(t)$ and $x^\dagger(t)$ should be considered as independent dynamical variables. For our simple system we find

$$a(t) = a\, e^{-i\omega_a t}, \qquad a^\dagger(t) = a^\dagger\, e^{i\omega_a t} + \frac{\mu}{\omega_b - \omega_a}\, b\left(e^{-i\omega_b t} - 1\right).$$

Hence, contrarily to what one could expect, $a^\dagger(t)$ is not simply the Hermitian conjugate of $a(t)$. Moreover, and quite relevant for us, despite our original interpretation of $\hat{h}$, it is clear from the above formulas for $a(t)$ and $a^\dagger(t)$ that the mean value of the number operator $\hat{n}_a(t) = a^\dagger(t)\, a(t)$ presents no damping at all.

Then we conclude that our simple minded point of view is simply wrong. One possible source of mistakes could be the fact that we have assumed here, following Equation (2.42), that the time evolution of the operator x is $x(t) = e^{i\hat{h}t} x e^{-i\hat{h}t}$, even if $\hat{h} \neq \hat{h}^{\dagger}$. This is, in fact, not so reasonable, as the following rather general argument shows: let us suppose, as in (2.58), that the Schrödinger equation which describes the evolution of our system is the usual one: $i\dot{\Phi}(t) = \hat{h}\,\Phi(t)$, even if $\hat{h}$ is not self-adjoint. This is quite often the starting point in many papers in quantum optics, but not only. Here $\hat{h}$ needs not being the one in Equation (2.59). Hence, if $\hat{h}$ does not depend on time, the formal solution of this equation is $\Phi(t) = e^{-i\hat{h}t}\Phi(0)$. As we have already pointed out before, what we do measure in a concrete experiment is really the mean value of some relevant observable related to the system in a state of the system. Hence we expect that the time evolution of such a mean value cannot be dependent on the representation chosen. In other words, if x is such an observable, we have

$$\langle \Phi(t), x\,\Phi(t)\rangle = \left\langle e^{-i\hat{h}t}\Phi(0), x\, e^{-i\hat{h}t}\Phi(0)\right\rangle = \left\langle \Phi(0), e^{i\hat{h}^{\dagger}t} x\, e^{-i\hat{h}t}\Phi(0)\right\rangle,$$

and this suggests to call $x(t)$, rather than $e^{i\hat{h}t} x e^{-i\hat{h}t}$ as we have done before, the following operator: $x(t) := e^{i\hat{h}^{\dagger}t} x\, e^{-i\hat{h}t}$. It is interesting to notice that this definition cures the anomaly we have seen above, i.e., the fact that, with our previous choice, $(x(t))^{\dagger} \neq (x^{\dagger})(t)$. In fact, with this different definition, we easily check that $(x(t))^{\dagger} = (x^{\dagger})(t)$ for all possible x and for all t. On the other hand, in general, it is more difficult to deduce the explicit form of the differential equations for $x(t)$: let us write $\hat{h} = H_r + iH_i$, with $H_r = \frac{1}{2}(\hat{h} + \hat{h}^{\dagger}) = H_r^{\dagger}$ and $H_i = \frac{1}{2i}(\hat{h} - \hat{h}^{\dagger}) = H_i^{\dagger}$. Then, the differential equation of motion for $x(t) = e^{i\hat{h}^{\dagger}t} x e^{-i\hat{h}t}$ can be written as

$$\frac{dx(t)}{dt} = e^{i\hat{h}^{\dagger}t}(\hat{h}^{\dagger}x - x\hat{h})e^{-i\hat{h}t} = e^{i\hat{h}^{\dagger}t}([H_r,x] - i\{H_i,x\})e^{-i\hat{h}t},$$

which involves both a commutator and an anti commutator. This makes the situation rather more difficult, of course. There is still another reason to prefer, when possible, working with self-adjoint rather than with non-self-adjoint Hamiltonians: the reason is that, using formula $x(t) = e^{i\hat{h}^{\dagger}t} x\, e^{-i\hat{h}t}$, we lose the group property of the time evolution: $(xy)(t) \neq x(t)y(t)$, for two generic operators x and y. This has unpleasant drawbacks in the deduction of the equations of motion for the system, since this property is essential in order to get some closed system of differential equations, which is often the starting point to look for the solution of the dynamical problem.

   Of course, most of these problems are just hidden if we work in the Schrödinger representation, since the validity of the Schrödinger equation is exactly the main working assumption. With this in mind, despite all these problems, sometimes we will use some non-self-adjoint Hamiltonians of a special type. For instance, this will be done in Chapters 4 and 6, where we will introduce first self-adjoint Hamiltonians depending on real parameters, and then we will replace some of these with complex quantities, keeping unchanged the formal expression of the Hamiltonians.[13] These complex parameters will be shown to model quite well the existence of stress factors and positive effects acting on the system, depending on the sign of their imaginary parts. A different choice of non-self-adjoint Hamiltonian will be discussed in Chapter 7, more on the line of the Hamiltonian in (2.59), in connection with the analysis of cellular tumor growth.

2.8 The (*H*, *ρ*)-Induced Dynamics

As we have already discussed before, one of the technical limitations of the approach considered in this book is that the Hamiltonian driving the time evolution of the system $\mathcal{S}$ is (almost) always assumed to be self-adjoint. Losing self-adjointness usually produces non-unitary time evolution. However, it may not be so for those non-self-adjoint Hamiltonians which are, for instance, similar (but not unitarily equivalent) to self-adjoint ones. In fact, in these cases, it is possible to replace the original scalar product $\langle .,. \rangle$ of the Hilbert space $\mathcal{H}$ with a different one, $\langle .,. \rangle_{new}$, which makes of these non-self-adjoint operators new, and self-adjoint, operators. Let us show how this is possible. For that, let us consider a Hamiltonian H such that $H \neq H^\dagger$, and let us assume that a self-adjoint, positive and invertible (but not unitary) operator S exists such that $H = SH_0 S^{-1}$, where $H_0 = H_0^\dagger$. Then we can prove that H is self-adjoint, but with respect to the scalar product $\langle .,. \rangle_S$ defined as follows:

$$\langle f, g \rangle_S := \left\langle S^{-2} f, g \right\rangle,$$

for all $f, g \in \mathcal{H}$. Here, to avoid useless complications, we are assuming that S, S^{-1} and H_0 are bounded.[14] Hence, H is bounded as well.

First of all, we need to check that $\langle .,. \rangle_S$ is really a scalar product. We just prove here that $\langle f, f \rangle_S \geq 0$ for all $f \in \mathcal{H}$, and that it is zero only if $f = 0$. Since S^{-1} is self-adjoint, we have

$$\langle f, f \rangle_S = \left\langle S^{-2} f, f \right\rangle = \left\langle S^{-1} f, S^{-1} f \right\rangle = \| S^{-1} f \|^2 \geq 0,$$

of course. In particular, this formula shows that $\langle f, f \rangle_S = 0$ if, and only if, $S^{-1} f = 0$, which implies that $f = 0$ as well.

---

[13] Of course, Hamiltonians like the one in (2.59) are not of this kind.

[14] This in particular implies that the two norms arising from $\langle .,. \rangle$ and $\langle .,. \rangle_S$, $\|f\| = \sqrt{\langle f, f \rangle}$ and $\|f\|_S = \sqrt{\langle f, f \rangle_S}$, are equivalent: a sequence $\{f_n \in \mathcal{H}\}$ converges with respect to $\|.\|$ if and only if it converges with respect to $\|.\|_S$. So the two norms induce the same topology on $\mathcal{H}$, but the conjugation is different: calling $\langle Xf, g \rangle = \langle f, X^\dagger g \rangle$ and $\langle Xf, g \rangle_S = \langle f, X^\sharp g \rangle_S$, for all $f, g \in \mathcal{H}$, then $X^\dagger \neq X^\sharp$. Of course, we are implicitly assuming that all the mean values introduced here are well defined.

Now, to prove that H is self-adjoint with respect to $\langle .,. \rangle_S$, it is enough to check that $\langle Hf, g \rangle_S = \langle f, H g \rangle_S$, for all $f, g \in \mathcal{H}$. But, since $H = S H_0 S^{-1}$, we first observe that $S^{-1} H = H_0 S^{-1}$ and that $H^\dagger S^{-2} = S^{-2} H$. Then, the following holds:

$$\langle Hf, g \rangle_S = \left\langle S^{-2} Hf, g \right\rangle = \left\langle H^\dagger S^{-2} f, g \right\rangle = \left\langle S^{-2} f, Hg \right\rangle = \langle f, Hg \rangle_S,$$

which is what we had to prove. This shows that, in some cases, losing self-adjointness of H is not really dramatic and does not produce results which are far away from what one could get using self-adjoint operators: it is enough to change the topology of the Hilbert space! But this can be done, of course, if H satisfies the condition above, i.e., if H is similar to a self-adjoint operator H_0. This is possible if all the eigenvalues of H are real. Otherwise, the situation is much more complicated but can still be interesting for us. In fact, this will be the relevant case in Chapters 4 and 6.

From what we have discussed so far, it is clear that working with a non-self-adjoint Hamiltonian H is a possible way to extend the standard dynamical approach to quantum mechanics, see Sections 2.4 and 2.7, where H is always taken to be self-adjoint. A different possibility consists in introducing what we will call a *rule* to the time evolution, i.e., some effect which cannot be easily modeled adding some further term in the Hamiltonian H of the system, or admitting that its eigenvalues could be complex. This gives rise to a series of interesting results and considerations, which will be discussed, at a rather general level, in this section, together with some preliminary applications. More, and more interesting, applications will be considered in the second part of this book. As it will appear clear from our treatment, the idea of considering the dynamics as driven by some $H = H^\dagger$ and some further rule not included in H links the general framework of quantum dynamics with the possibility that the dynamics may be periodically disturbed because of some external (or internal) action on the system, whose effect cannot be (easily) described by any extra term in the self-adjoint Hamiltonian. This is what we call (H, ρ)-induced dynamics.

Let $\mathcal{S}$ be our physical system, and Q_j ($j = 1, \dots, M$) a set of M commuting self-adjoint operators with eigenvectors $\varphi_{\alpha_n}^{(j)}$ and eigenvalues $\alpha_n^{(j)}$ needed for a full description of $\mathcal{S}$. Then

$$[Q_j, Q_k] = Q_j Q_k - Q_k Q_j = 0, \qquad Q_j = Q_j^\dagger, \qquad Q_j \varphi_{n_j}^{(j)} = \alpha_{n_j}^{(j)} \varphi_{n_j}^{(j)}, \tag{2.60}$$

$j, k = 1, 2, \dots, M$, $n_j = 1, 2, 3, \dots, N_j$. Setting $\mathbf{n} = (n_1, n_2, \dots, n_M)$, the vector

$$\varphi_{\mathbf{n}} = \varphi_{n_1}^{(1)} \otimes \varphi_{n_2}^{(2)} \otimes \cdots \varphi_{n_M}^{(M)} \tag{2.61}$$

represents an eigenstate of all the operators Q_j, say

$$Q_j \, \varphi_{\mathbf{n}} = \alpha^{(j)}_{n_j} \, \varphi_{\mathbf{n}}. \tag{2.62}$$

The existence of a common eigenstate for all the operators Q_j is guaranteed by the fact that they all mutually commute. It is convenient to assume that these vectors are mutually orthogonal and normalized:

$$\langle \varphi_{\mathbf{n}}, \varphi_{\mathbf{m}} \rangle = \delta_{\mathbf{n},\mathbf{m}} = \prod_{j=1}^{M} \delta_{n_j, m_j}. \tag{2.63}$$

This is the case, of course, if the eigenvalues are all nondegenerate, which we will assume here to simplify the treatment. However, it is not hard to extend our results to the case in which this is not true. The Hilbert space $\mathcal{H}$ where $\mathcal{S}$ is defined as the closure of the linear span of all the vectors $\varphi_{\mathbf{n}}$, which in this way turn out to be, by construction, an o.n. basis for $\mathcal{H}$. Now, let H be a (quadratic, time-independent) self-adjoint Hamiltonian, describing the kinetic term and the interactions occurring in $\mathcal{S}$. Notice that H, in general, does not commute with the Q_j's. Therefore, these operators are not, in principle, integrals of motion. In absence of any other information, the wave function $\Psi(t)$, describing $\mathcal{S}$ at time t, evolves according to the Schrödinger equation $i\dot{\Psi}(t) = H\Psi(t)$, where $\Psi(0) = \Psi_0$ describes the initial status of $\mathcal{S}$. As we know, the formal solution of the Schrödinger equation is, since H does not depend explicitly on t, $\Psi(t) = e^{-iHt}\Psi(0) = e^{-iHt}\Psi_0$. We can now compute the mean value of each operator Q_j in the state $\Psi(t)$,

$$q_j(t) = \langle \Psi(t), Q_j \Psi(t) \rangle, \tag{2.64}$$

and we use the various functions $q_j(t)$ to define an M-dimensional time-dependent vector $\mathbf{q}(t) = (q_1(t), q_2(t), \dots, q_M(t))$. Of course, there is a dual possibility in which Ψ_0 stays constant and Q_j evolves in time according to the Heisenberg recipe[15]: $Q_j(t) = e^{iHt}Q_j e^{-iHt}$. And in a similar way we can construct again the same vector $\mathbf{q}(t)$, with $q_j(t) = \langle \Psi_0, Q_j(t)\Psi_0 \rangle$.

We are now ready to introduce two different ways of extending the dynamics through the introduction of a specific *rule*.

In the first approach, the rule ρ is a map from $\mathcal{H}$ to $\mathcal{H}$. Its explicit action depends on the expression of $\mathbf{q}(t)$ at particular instants $k\tau$, where $k \in \mathbb{N}$ and $\tau \in \mathbb{R}^+$ is some fixed time which is relevant (for some reason) for $\mathcal{S}$. For instance, some external action could take place on $\mathcal{S}$ for $t = \tau, 2\tau, 3\tau, \dots$. In other words, according to what $\mathbf{q}(\tau)$ looks like, ρ maps the input vector $\Psi(\tau)$ (here $k = 1$) into a (possibly different) output vector Ψ_{new}, and we write $\rho(\Psi(\tau)) = \Psi_{new}$. This is not very different from what happens in scattering theory, where an incoming state, after the occurrence

---

[15] We recall that $H = H^\dagger$, here.

of the scattering, is transformed into a different outgoing state [31]. However, a big difference exists, which has to do with the fact that Ψ_{new} is connected, in some way still to be specified, to the vector $\mathbf{q}(\tau)$, while in a scattering process the only essential fact is that the scattering must take place. Now, the new vector Ψ_{new} can be considered as the initial state of the system $\mathcal{S}$ and its time evolution is driven again, for another time interval of length τ, by the Hamiltonian H. At $t = 2\tau$ the rule is applied again, and a (second) new vector is introduced, which also evolves according to H. This procedure can be iterated several times. As an example, in [5], the rule ρ is used to map an incoming state into one of the elements of a fixed o.n. basis of $\mathcal{H}$. We have used this rule to extend the game of life, see also Chapter 8. Our approach has some aspects in common with the *repeated quantum measurements*, according to which the state of a system is perturbed by a nontrivial quantum measurement, and subsequent measurements of the same system reveal the presence of disturbances if compared to the situation in which no previous measurements were performed. In our case, the disturbance manifests with the creation of the new state Ψ_{new}, which is prepared as the result of the choice of the rules. We also like to mention that the repeated action of the rule ρ in our approach reminds of what is done in the context of the quantum Zeno effect (see [49] and references therein), where some measures on a given quantum system are repeated again and again, projecting the system on some particular state connected to the result of the measure. We will say more on this connection later.

Another aspect of the (H, ρ)-induced dynamics, which makes this idea interesting and useful for applications, has to do with the fact that the rule does not necessarily act to modify the state of the system, as discussed so far. A second possibility exists, in which the rule ρ works on the space of the parameters of the Hamiltonian H, rather than on the wave function. In this case, if our self-adjoint Hamiltonian involves p real parameters, the rule ρ is a map from $\mathbb{R}^p$ to $\mathbb{R}^p$ that, again at particular instants $k\tau$, depending on some check performed on the system, modifies some of the values of these parameters. In this way the model adjusts itself, by slightly modifying some details of H, and is able to describe more complex (and possibly more realistic) behaviors. As one can easily understand, this approach introduces a sort of time dependence in H, but this time dependence is, in a sense, rather specific: H is, in fact, piecewise time-independent. This is, of course, technically much simpler than when H presents a more complicated time dependence, since in each interval $[k\tau, (k + 1)\tau[$ one can solve the Schrödinger (or the Heisenberg) equation of motion as if H were time-independent, and then glue the solutions at the various instants $k\tau$, requiring some suitable regularity conditions. We will discuss more on the two different kind of rules in Sections 2.8.1 and 2.8.2, while some concrete applications of (H, ρ)-induced dynamics will be considered in Chapters 3 and 8.

2.8.1 The Rule ρ *as a Map from* $\mathcal{H}$ *to* $\mathcal{H}$

We begin our analysis by discussing the case in which the rule ρ is a map from $\mathcal{H}$ to $\mathcal{H}$. First, we observe that there exists a one-to-one correspondence between the vector $\varphi_{\mathbf{n}}$ in Equation (2.61) and its label $\mathbf{n}$: once we know $\mathbf{n}$, $\varphi_{\mathbf{n}}$ is clearly identified, and vice versa. Suppose now that, at time $t = 0$, the system S is in a state $\mathbf{n}^0$ or, which is the same, S is described by the vector $\varphi_{\mathbf{n}^0}$. Then, once fixed a positive value of τ, this vector evolves in the time interval $[0, \tau[$ according to the Schrödinger recipe: $e^{-iHt}\varphi_{\mathbf{n}^0}$. Here $H = H^\dagger$ is the Hamiltonian of S. Let us set

$$\Psi(\tau^-) = \lim_{t \to \tau^-} e^{-iHt}\varphi_{\mathbf{n}^0}, \tag{2.65}$$

where t converges to τ from below.[16] Now, at time $t = \tau$, ρ is applied to $\Psi(\tau^-)$, and the output of this action is a new vector which we assume here to be again an eigenstate of each operator Q_j, but with different eigenvalues, $\varphi_{\mathbf{n}^1}$. In other words, ρ looks at the explicit expression of $\Psi(\tau^-)$ and, according to its form, returns a new vector $\mathbf{n}^1 = (n_1^1, n_2^1, \ldots, n_M^1)$ and, as a consequence, a new vector $\varphi_{\mathbf{n}^1}$ of $\mathcal{H}$: $\varphi_{\mathbf{n}^1} = \rho(\Psi(\tau^-))$. Some preliminary examples of how ρ explicitly acts will be briefly presented in Section 2.9.

> *Remark:* As we have already said, this is not the only possibility to set up a rule. In fact, other possibilities can also be considered. In particular, it is not necessary to stay inside the set of eigenstates of the Q_j's: the rule could produce a vector Φ which has nothing to do with any vector $\varphi_{\mathbf{k}}$ in Equation (2.61). The common aspect for all possible choices of ρ is that it behaves as a control over the system S and modifies some of its ingredients according to the result of this check. This aspect will be discussed further in Section 2.8.2.

Now, the procedure is iterated, taking $\varphi_{\mathbf{n}^1}$ as the initial vector and letting it evolve with H for another time interval of length τ. Hence we compute

$$\Psi(2\tau^-) = \lim_{t \to \tau^-} e^{-iHt}\varphi_{\mathbf{n}^1}, \tag{2.66}$$

and the new vector $\varphi_{\mathbf{n}^2}$ is deduced as the result of the action of rule ρ on $\Psi(2\tau^-)$: $\varphi_{\mathbf{n}^2} = \rho(\Psi(2\tau^-))$. In general, for all $k \geq 1$, we have

$$\Psi(k\tau^-) = \lim_{t \to \tau^-} e^{-iHt}\varphi_{\mathbf{n}^{k-1}}, \quad \text{and then} \quad \varphi_{\mathbf{n}^k} = \rho\left(\Psi(k\tau^-)\right). \tag{2.67}$$

Now, let X be a generic operator on $\mathcal{H}$, bounded or unbounded. In this latter case, we need to check that the various $\varphi_{\mathbf{n}^k}$ belong to the domain of $X(t) = e^{iHt}Xe^{-iHt}$ for all $t \in [0, \tau]$. In the applications we have considered so far, this condition is always satisfied.

---

[16] We use here τ^-, $2\tau^-$, …, as argument of Ψ to emphasize that, for instance, before τ the time evolution is due only to H. In fact, as we will see now, ρ really acts at $t = \tau$.

Definition 2.8.1   The sequence of functions

$$x_{k+1}(t) := \langle \varphi_{\mathbf{n}^k}, X(t)\varphi_{\mathbf{n}^k} \rangle, \qquad (2.68)$$

for $t \in [0, \tau]$ and $k \in \mathbb{N}_0$, is called the (H, ρ)–induced dynamics of X.

Some consequences of Definition 2.8.1 and some properties of the sequence $\underline{X}(\tau) = (x_1(\tau), x_2(\tau), ...)$ have been discussed in [5]. Moreover, from $\underline{X}(t) = (x_1(t), x_2(t), ...)$ it is possible to define a new function of time, which can be understood as the evolution of the observable X of $\mathcal{S}$ under the action of H and ρ, in the following way:

$$\tilde{X}(t) = \begin{cases} x_1(t) & t \in [0, \tau[, \\ x_2(t - \tau) & t \in [\tau, 2\tau[, \\ x_3(t - 2\tau) & t \in [2\tau, 3\tau[, \\ ... \end{cases} \qquad (2.69)$$

It is clear that $\tilde{X}(t)$ may have discontinuities in $k\tau$, for positive integers k. The meaning of $\tilde{X}(t)$ is the following: For $t \in [0, \tau[$ $\tilde{X}(t)$ is just the mean value of the operator $X(t)$ in the state defined by $\varphi_{\mathbf{n}^0}$. For larger values of t, and in particular if $t \in [\tau, 2\tau[$, then $\tilde{X}(t) = \langle \varphi_{\mathbf{n}^1}, X(t - \tau)\varphi_{\mathbf{n}^1} \rangle$, which is again the mean value of $X(t)$ shifted back in time, but in a state labeled by $\varphi_{\mathbf{n}^1}$, and so on. In other words, $\tilde{X}(t)$ is constructed by considering a set of mean values of $X(s)$ with $s \in [0, \tau[$, but on different states, $\varphi_{\mathbf{n}^0}$, $\varphi_{\mathbf{n}^1}$, $\varphi_{\mathbf{n}^2}$ and so on, which are the vectors identified by ρ by its successive actions. This is because, in our interpretation of the rule, we assume that the time evolution of any observable starts again and again any time the rule acts on the system.

2.8.2 The Rule ρ as a Map in the Space of the Parameters of H

Formula (2.67) shows that the action of the rule considered in the previous section produces a change in the state of the system, from an input to an output vector – and, we would add, a very special change: from an eigenstate of the Q_j's to another eigenstate of the same compatible operators. The other ingredients of $\mathcal{S}$, and in particular its Hamiltonian H, are not modified by the rule and stay unchanged. As we have already mentioned, this is not the only possibility. We show now how a rule can also act on $\mathcal{S}$, changing some details of the Hamiltonian of $\mathcal{S}$, and in particular adjusting the values of some of the parameters of H.

To be concrete, we assume $\mathcal{S}$ is a system described in terms of M fermionic (or bosonic) modes, and we suppose that its evolution is ruled by the following quadratic time-independent self-adjoint Hamiltonian:

$$H = \sum_{j=1}^{M} \omega_j a_j^\dagger a_j + \sum_{j=1}^{M-1} \sum_{k=j+1}^{M} \lambda_{j,k}(a_j a_k^\dagger + a_k a_j^\dagger), \qquad (2.70)$$

involving the $p = M(M+1)/2$ real parameters (not necessarily all different from zero) ω_j and $\lambda_{j,k}$, and where a_j and $a_j^\dagger$, $j = 1, \ldots M$, are annihilation and creation operators, respectively.

The time evolution of the lowering operators a_j's can be deduced from the following linear system of ordinary differential equations, resulting from Equation (2.43) applied to the present situation:

$$\dot{a}_j(t) = i\left(-\omega_j a_j(t) + \sum_{k=1, k\neq j}^{M} \lambda_{j,k} a_k(t)\right), \qquad j = 1, \ldots, M. \qquad (2.71)$$

Restricting ourselves to the fermionic case, these are $M \times 2^{2M}$ linear differential equations which should be solved, considering the initial conditions for the matrices representing the operators a_j: $a_j(0) = a_j$.

Remark: It is worth observing that, if we replace the CAR with the CCR, then the set Equation (2.71) produces an infinite number of linear differential equations, for which finding an analytic solution is, in principle, much harder (if possible). However, cases when this can be done do exist, see [9] as an example, but in the analysis presented here we prefer to avoid difficulties related to working in infinite-dimensional Hilbert spaces. In fact, Equation (2.71) is already a set of linear differential equations. Hence, an analytic solution can be found. But this solution, easy to be formally deduced, could be very difficult to manage to get *numbers*.

Going back to (2.71), and noticing that the system is linear, we may write it in the following compact form:

$$\dot{A}(t) = UA(t), \qquad (2.72)$$

where $A(t) = (a_1(t), a_2(t), \ldots, a_M(t))^T$, and U is an $M \times M$ constant matrix such that $U_{j,j} = -i\omega_j$, $U_{j,k} = i\lambda_{j,k}$, and each component of A is a $2^M \times 2^M$ matrix. The formal solution is immediately deduced, namely

$$A(t) = e^{Ut}A(0) = V(t)A(0). \qquad (2.73)$$

Thus, if $V_{\ell,m}(t)$ is the generic entry of matrix $V(t)$, we have

$$a_\ell(t) = \sum_{k=1}^{M} V_{\ell,k}(t)a_k(0), \qquad \ell = 1, \ldots, M. \qquad (2.74)$$

Now, we need to compute the mean value of the number operator for the ℓ-th mode (which is intended to represent a physical quantity which is relevant for the description of $\mathcal{S}$)

$$\hat{n}_\ell(t) = a_\ell^\dagger(t) a_\ell(t) \tag{2.75}$$

on an eigenvector $\varphi_{n_1, n_2, \ldots, n_M}$ of all the $\hat{n}_\ell(0)$,

$$\hat{n}_\ell(0)\varphi_{n_1, n_2, \ldots, n_M} = n_\ell \varphi_{n_1, n_2, \ldots, n_M}, \qquad \ell = 1, 2, \ldots, M. \tag{2.76}$$

Simple computations show that

$$n_\ell(t) = \langle \varphi_{n_1, n_2, \ldots, n_M}, \hat{n}_\ell(t)\varphi_{n_1, n_2, \ldots, n_M}\rangle = \sum_{k=1}^{M} |V_{\ell,k}(t)|^2 \, n_k, \qquad \ell = 1, \ldots, M. \tag{2.77}$$

This is what we get from a quadratic Hamiltonian such as the one in (2.70). Of course, even without entering into the detailed meaning of the model, which is not particularly relevant here, we see that the dynamics is related to the details of the matrix $V(t)$. But, since $V(t) = e^{Ut}$, these are related to the matrix elements of U which, in turn, are fixed by the parameters of the Hamiltonian H. Hence, we may enrich the dynamics of $\mathcal{S}$ by introducing a rule, acting several times at specific instants, and accounting for a sort of dependence of the parameters ω_j and $\lambda_{j,k}$ in Equation (2.70) on the state of the system at the instants in which the rule acts. In some sense, as we have already observed, the model adjusts itself as a consequence of its evolution. However, the way in which $\mathcal{S}$ is adjusted is quite special: the rule modifies the value of the parameters of H, while keeping unchanged its analytical expression. In other words, once ρ is applied, the values of ω_j and $\lambda_{j,k}$ in H can be modified, but the expression of H in Equation (2.70) cannot. This is not the only choice we can adopt. In fact, also the interacting part of H could be modified after the action of the rule. However, we will not consider this possibility here, since going from a quadratic to a, say, cubic Hamiltonian would produce equations of motion for which we cannot possibly find any exact analytical solution.

Let us now go more in the details of how our approach works concretely. We start considering a self-adjoint, time-independent, quadratic Hamiltonian operator $H^{(0)}$, obtained by choosing a first set of values of the parameters in (2.70). Then, the time evolution of a certain observable X driven by $H^{(0)}$ is

$$X(t) = e^{iH^{(0)}t} X e^{-iH^{(0)}t}, \tag{2.78}$$

and its mean value is

$$x(t) = \langle \varphi_{n_1, n_2, \ldots, n_M}, X(t)\varphi_{n_1, n_2, \ldots, n_M}\rangle \tag{2.79}$$

in a time interval of length $\tau > 0$ on a vector $\varphi_{n_1, n_2, \ldots, n_M}$, which can again be assumed to be a common eigenstate of the operators Q_j's. Then, according to the particular value of $x(\tau)$, we modify some of the parameters appearing in the definition of $H^{(0)}$. In this way, we get a new Hamiltonian operator, $H^{(1)}$, having the same functional form as $H^{(0)}$, but (in general) different values of (some of) its parameters, and the system now evolves under the action of this new Hamiltonian for the next time interval of length τ. We are not restarting the time evolution of the system from a new initial condition, but simply continuing to follow the evolution with the only difference that for $t \in]\tau, 2\tau]$ the Hamiltonian $H^{(1)}$ rules the process, rather than $H^{(0)}$ and so on. Therefore, the rule now has to be thought of as a map from $\mathbb{R}^p$ into $\mathbb{R}^p$ acting on the space of the parameters involved in the Hamiltonian, and the global evolution is governed by a sequence of analogous Hamiltonian operators, with the same analytic expression but with different value of the parameters. In a few words, these parameters of H become nothing but piecewise (in time) constant functions, whose values are not fixed a priori, but decided by the evolution itself, and by the rule.

In general, in the k-th subinterval $[k\tau, (k+1)\tau[$, the dynamics will be driven by the Hamiltonian $H^{(k)}$. Hence, the global dynamics arises from the sequence of Hamiltonians

$$H^{(0)} \xrightarrow{\tau} H^{(1)} \xrightarrow{\tau} H^{(2)} \xrightarrow{\tau} \ldots . \tag{2.80}$$

In every subinterval we therefore have a system like

$$\dot{A}(t) = U^{(k)} A(t), \qquad t \in [k\tau, (k+1)\tau[, \tag{2.81}$$

$k \geq 0$, where $U^{(k)}$ is constructed out of the parameters of $H^{(k)}$. The analytical solution of this system is not difficult, due to the fact that each equation is linear. The complete evolution is obtained by gluing the local evolutions.

Some numerical aspects of our procedure are discussed in [6], which we refer to for more details on this particular aspect of the strategy proposed here.

2.9 A Two-Mode System

In Section 2.7 we have discussed how some observable of a system $\mathcal{S}$ can reach an equilibrium. In particular, we have shown that this is not possible, within the approach discussed in Section 2.4, if $\mathcal{S}$ lives in a finite-dimensional Hilbert space and if its dynamics is driven by a self-adjoint Hamiltonian. On the other hand, in presence of an (infinitely extended) environment, decay processes can be modeled and stable asymptotic values can be found. Also, as discussed in Section 2.7, it is possible to use very specific non-self-adjoint Hamiltonians with the same purpose, but this approach should be pursued *cum grano salis*, since several complications arise. To avoid these difficulties, we are going to show how the (H, ρ)-induced dynamics can be useful to produce, after some time, a nontrivial equilibrium for the time evolution of the mean value of some observables of the system. This will now be shown by considering a very simple toy model considered first in [7].

Let us consider a system $\mathcal{S}$, having two (fermionic) degrees of freedom, and described by the Hamiltonian

$$H = H_0 + \lambda H_I, \qquad H_0 = \omega_1 a_1^\dagger a_1 + \omega_2 a_2^\dagger a_2, \qquad H_I = a_1^\dagger a_2 + a_2^\dagger a_1, \qquad (2.82)$$

where ω_j and λ are real (and positive) quantities in order to ensure that H is self-adjoint. The operators a_j and $a_j^\dagger$ are assumed to satisfy the following CAR:

$$\{a_i, a_j^\dagger\} = \delta_{i,j} \, \mathbb{1}, \qquad \{a_i, a_j\} = \{a_i^\dagger, a_j^\dagger\} = 0, \qquad (2.83)$$

$i, j = 1, 2$, where, as usual, $\mathbb{1}$ is the identity operator. Of course, when $\lambda = 0$, the two agents of $\mathcal{S}$ are not interacting. Notice that H in (2.82) is a fermionic version of the Hamiltonian for $\mathcal{S} \cup \tilde{\mathcal{S}}$ introduced in Section 2.7, for which we proved that no equilibrium exists at all.

The eigenstates of the number operators $\hat{n}_j := a_j^\dagger a_j$ are easily obtained: if $\varphi_{0,0}$ is the *ground vector* of $\mathcal{S}$, $a_1 \varphi_{0,0} = a_2 \varphi_{0,0} = 0$, an o.n. basis of the four-dimensional Hilbert space $\mathcal{H}$ of $\mathcal{S}$ is given by the following vectors:

$$\varphi_{0,0}, \qquad \varphi_{1,0} := a_1^\dagger \varphi_{0,0}, \qquad \varphi_{0,1} := a_2^\dagger \varphi_{0,0}, \qquad \varphi_{1,1} := a_1^\dagger a_2^\dagger \varphi_{0,0}. \qquad (2.84)$$

We have

$$\hat{n}_1 \varphi_{n_1,n_2} = n_1 \varphi_{n_1,n_2}, \qquad \hat{n}_2 \varphi_{n_1,n_2} = n_2 \varphi_{n_1,n_2}. \qquad (2.85)$$

The equations of motion for the annihilation operators $a_j(t)$ are

$$\dot{a}_1(t) = -i\omega_1 a_1(t) - i\lambda a_2(t), \qquad \dot{a}_2(t) = -i\omega_2 a_2(t) - i\lambda a_1(t), \qquad (2.86)$$

which are formally identical to those in Equation (2.47). They can be solved imposing the initial conditions $a_1(0) = a_1$ and $a_2(0) = a_2$, and the solution is

$$a_1(t) = \frac{1}{2\delta} \left(a_1 \left((\omega_1 - \omega_2)\Phi_-(t) + \delta\Phi_+(t) \right) + 2\lambda a_2 \Phi_-(t) \right),$$
$$a_2(t) = \frac{1}{2\delta} \left(a_2 \left(-(\omega_1 - \omega_2)\Phi_-(t) + \delta\Phi_+(t) \right) + 2\lambda a_1 \Phi_-(t) \right), \qquad (2.87)$$

where

$$\delta = \sqrt{(\omega_1 - \omega_2)^2 + 4\lambda^2},$$
$$\Phi_+(t) = 2 \exp\left(-\frac{it(\omega_1 + \omega_2)}{2} \right) \cos\left(\frac{\delta t}{2} \right), \qquad (2.88)$$
$$\Phi_-(t) = -2i \exp\left(-\frac{it(\omega_1 + \omega_2)}{2} \right) \sin\left(\frac{\delta t}{2} \right).$$

Then, the functions $n_j(t) := \langle \varphi_{n_1,n_2}, \hat{n}_j(t) \varphi_{n_1,n_2} \rangle$ are

$$
\begin{aligned}
n_1(t) &= \frac{n_1(\omega_1 - \omega_2)^2}{\delta^2} + \frac{4\lambda^2}{\delta^2} \left(n_1 \cos^2 \left(\frac{\delta t}{2} \right) + n_2 \sin^2 \left(\frac{\delta t}{2} \right) \right), \\
n_2(t) &= \frac{n_2(\omega_1 - \omega_2)^2}{\delta^2} + \frac{4\lambda^2}{\delta^2} \left(n_2 \cos^2 \left(\frac{\delta t}{2} \right) + n_1 \sin^2 \left(\frac{\delta t}{2} \right) \right),
\end{aligned}
\tag{2.89}
$$

which oscillate in time as deduced for their bosonic counterparts in (2.47).

These functions could be interpreted, in agreement with other similar applications, as the densities of two species, S_1 and S_2, interacting as in Equation (2.82) in a given (small) region.[17] The interaction Hamiltonian H_I in Equation (2.82) describes a sort of predator-prey mechanism, and this is reflected by the solutions in Equation (2.89), which show how the two densities, because of the interaction between S_1 and S_2, oscillate in the interval $[0, 1]$. Otherwise, if $\lambda = 0$, $n_j(t) = n_j$: the densities stay constant, and nothing interesting happens in S. We observe that the formulas in (2.89) automatically imply that $n_1(t) + n_2(t) = n_1 + n_2$, independently of t and λ: the oscillations are such that they sum up to zero. We refer to [7] for the role of this Hamiltonian in modeling migration, which is achieved considering a 2D version of the H in Equation (2.82), with an additional term responsible for the diffusion of the two species along a rectangular lattice. Here, we exploit the possibility of getting some limiting values for $n_1(t)$ and $n_2(t)$, or for some functions naturally related to these, for large values of t, when $\lambda \neq 0$.

The first trivial remark is that the functions $n_1(t)$ and $n_2(t)$ in Equation (2.89) do not admit any asymptotic limit, except when $n_1 = n_2$ (or when $\lambda = 0$, which is excluded here). In this case, clearly, $n_1(t) = n_2(t) = n_1 = n_2$. On the other hand, if $n_1 \neq n_2$, then both $n_1(t)$ and $n_2(t)$ always oscillate in time. This is not surprising since we have already proved that, if S is a system living in a finite-dimensional Hilbert space, and if its dynamics is driven by a time-independent, self-adjoint, Hamiltonian $\tilde{H}$, then its evolution is necessarily periodic or quasi-periodic. Then the conclusion is that, if we are interested in getting (nontrivial) asymptotic values, we need to modify the way in which the time evolution is taken, at least for large t. This is exactly what we will do next, by introducing a rule ρ in the analysis: this is much simpler, in principle, than considering the interaction of the system with a reservoir, as proposed in Section 2.7, and this simplicity makes the concept of the (H, ρ)-induced dynamics interesting, for practical purposes. Also, we will discuss later that ρ has, quite often, a specific physical interpretation, which makes the model more realistic.

---

[17] Other interpretations are also possible. For instance, they can play the role of the decision functions of two interacting agents, trying to decide on some binary question.

2.9.1 The Rule and the Existence of an Asymptotic Value

In order to see how a rule can be useful for us, we first rewrite Equation (2.89) as

$$N(t) = T_t N(0), \tag{2.90}$$

where

$$N(t) = \begin{pmatrix} n_1(t) \\ n_2(t) \end{pmatrix}, \qquad T_t = \frac{1}{\delta^2} \begin{pmatrix} \delta^2 - 4\lambda^2 \sin^2\left(\frac{\delta t}{2}\right) & 4\lambda^2 \sin^2\left(\frac{\delta t}{2}\right) \\ 4\lambda^2 \sin^2\left(\frac{\delta t}{2}\right) & \delta^2 - 4\lambda^2 \sin^2\left(\frac{\delta t}{2}\right) \end{pmatrix}. \tag{2.91}$$

Of course, the components of $N(t)$ return the expressions of $n_1(t)$ and $n_2(t)$ for all times. Let us now see what happens if we insert a certain rule ρ in the time evolution of the system.

Here, we can think of ρ as a measure of $n_1(t)$ and $n_2(t)$ repeated at time τ, 2τ, 3τ, ... We know that performing a measure on a quantum system is a delicate operation, which usually modifies the system itself [49]. Therefore, there is no reason a priori to imagine that the result of a measure at time $k\tau$ (after having measured the system at time τ, 2τ, ..., $(k-1)\tau$) would be exactly the same as the result of a single measure performed on the system at time $k\tau$. This is, in fact, what we are going to show next.

In the first case, the first measure at $t = \tau$ gives $N_1(\tau) := N(\tau) = T_\tau N(0)$, which is just the effect of a single application of our rule. Then, according to what discussed in Section 2.8.1, we let the system evolve out of this new initial condition $N_1(\tau)$ for another time interval: $N_2(\tau) := T_\tau N_1(\tau) = T_\tau^2 N(0)$, and so on. It is quite natural to call the rule ρ considered here a *stop and go* rule: apparently, in fact, ρ just stops the time evolution at τ, 2τ, 3τ and so on, and then lets the time evolution start again. Repeating this procedure again and again we produce a sequence of results

$$N_k(\tau) = T_\tau^k N(0), \tag{2.92}$$

for $k \geq 1$. On the other hand, if we perform a single measure at time $t = k\tau$, the result is $N(k\tau) = T_{k\tau} N(0)$, see Equations (2.90) and (2.91).

We want to show that $N_k(\tau)$, when k diverges, can converge to some nontrivial limit, at least under suitable conditions, while this is not possible for $N(k\tau)$: this vector never converges when k diverges, except for some trivial cases which are not interesting for us. In order to compute $N_k(\tau)$, and its limit for k diverging, we observe that T_t in (2.91) is a self-adjoint matrix, so it can be easily diagonalized. In particular, we get

$$U^{-1} T_t U = \begin{pmatrix} \lambda_1(t) & 0 \\ 0 & \lambda_2(t) \end{pmatrix} =: \Lambda_t, \tag{2.93}$$

where

$$U = \frac{1}{\sqrt{2}} \begin{pmatrix} 1 & -1 \\ 1 & 1 \end{pmatrix}, \qquad \lambda_1(t) = 1, \qquad \lambda_2(t) = \frac{1}{\delta^2} \left(\delta^2 - 8\lambda^2 \sin^2 \left(\frac{\delta t}{2} \right) \right).$$
$$\text{(2.94)}$$

Then

$$T_\tau^k = U \Lambda_\tau^k U^{-1} = U \begin{pmatrix} 1 & 0 \\ 0 & \lambda_2^k(\tau) \end{pmatrix} U^{-1}, \qquad (2.95)$$

so that $N_k(\tau) = T_\tau^k N(0)$ can converge if $\lambda_2^k(\tau)$ does converge when k diverges. This is what happens whenever the parameters δ, τ and λ satisfy the following inequalities:

$$0 < 8\lambda^2 \sin^2 \left(\frac{\delta \tau}{2} \right) < \delta^2. \qquad (2.96)$$

In fact, when this is true, $\lambda_2(\tau) \in]0, 1[$, and, therefore, $\lim_{k\to\infty} \lambda_2^k(\tau) = 0$. Hence,

$$\lim_{k\to\infty} N_k(\tau) = \begin{pmatrix} n_1(0) \\ 0 \end{pmatrix}, \qquad (2.97)$$

which clearly shows that a nontrivial equilibrium can be reached in this case. However, if the parameters do not satisfy (2.96), the asymptotic behavior of $N_k(\tau)$ can be something completely different. In fact, taking, for instance, $\tau = \frac{\pi}{\delta}$ and $\lambda = \sqrt{\frac{3}{8}} \delta$, and then fixing $\delta = 1$ for simplicity, we deduce that $\lambda_2(\tau) = -2$, so that $\lim_{\ell\to\infty} |\lambda_2(\tau)|^k = \infty$: thus, it is evident that the role of the parameters of H is in fact essential for the existence of some finite asymptotic value.

The conclusion of the analysis of this simple example is the following: even in presence of a self-adjoint Hamiltonian, a simple two-modes fermionic system admits a nontrivial asymptotic limit for a large range of values of the parameters of the model, at least if a *stop and go* rule is assumed. However, the same rule can also produce a non-converging dynamics for other choices of the parameters. Of course, if we adopt different rules we may likely obtain different results. Other examples of rules, in different contexts, will be discussed in the second part of the book.

Remarks: (1) As already observed, the rule could be seen as an alternative, and somehow simplified, way to deal with Hamiltonians with some explicit time dependence. The fact that the (H, ρ)-induced dynamics is technically simpler than the use of a Hamiltonian $H(t)$ follows from the fact that, in each interval $[k\tau, (k+1)\tau[$, the standard formulas for time-independent Hamiltonians can be used, and the only difficulty is to match the solutions for $t = (k+1)\tau$, $k \geq 0$.

(2) We stress once more that what it is discussed here is not very different from what is observed in the quantum Zeno effect, in which repeated measures are performed on a quantum system. We refer to [49] for a detailed review on this effect. However, it is not even so close! Differences are many, especially when we look at the rule as a map in the space of parameters of the Hamiltonian. In this case, in fact, we are not really modifying the state of the system, as it happens when we perform a measure (or a repeated measure) on a quantum system, but we are modifying the dynamics itself, changing the values of the parameters needed to fix the Hamiltonian. In principle, there is no reason not to modify also the analytic expression of the Hamiltonian according to the rule. The main reason why we do not consider this possibility is because, in this way, we get in general different differential equations in each time interval, and this will make the solution of the dynamics quite complicated.

Part II

Applications

3

Politics

3.1 Introduction

In the second part of the book we will show how quantum tools, and ladder operators in particular, can be used in the description of many different systems. We will begin with an application to politics. At first the analysis of political life may appear unrelated to mathematics, except perhaps for a simple analysis of the results of a political election. Of course, for this particular case, one may expect that statistical methods can be of some utility, at least to compare the new with the old results. Needless to say, this has not much to do with any mathematical modeling of our system but could be of some help to propose, out of the several agents of the related game (the parties and the electors, in particular), those elements and those interactions that are truly essential in a (sufficiently complete) analysis of the political system. Following this idea, many scientists worked in the past and in recent years to propose different models for politics, paying attention to several aspects and adopting different ideas. Classical and statistical techniques are adopted in the monograph [50] and in the papers [51–58], while quantum ideas have been used extensively in other papers, as in [12, 13, 59–65]. This suggests, in fact, that mathematics and politics have a nontrivial overlap, and this will also be evident from our analysis.

All along this chapter, the object of our analysis, what we call *the system*, is the set of parties and of their electors, including those who are not really interested in voting and those who don't have a clear idea of for whom they are going to vote (the undecided voters). The problem we will consider here can be seen as a concrete example of a decision-making process in which the interactions among the various parts of the system and the Hamiltonian constructed out of these interactions are used to describe the time evolution of some operators whose mean values are what we call *decision functions*. They describe the tendency of a party to form, or not, an alliance with any second party. In other words, alliances between different

parties are the core of this chapter. This problem originates from a political election
in Italy in 2013: The results of the elections produced no real winner. Three parties took approximately the same number of votes: the *Partito Democratico* (PD),
which was really the first party in that election but which was not strong enough to
govern alone, the *Popolo della Libertà* (PdL) and the *Movimento 5 Stelle* (M5S),
both sufficiently strong to have, if allied with PD, much more than 50% of the
votes. The weight of the other parties taking part to the same election was irrelevant. The natural choice at that stage was to try to form a (somehow) left-oriented
government PD-M5S. However, M5S decided not to ally with PD, despite the fact
that many of its electors apparently disapproved that choice. After a long impasse,
PD formed a grand coalition with PdL, even though a large part of the PD electors
was against this choice. Summarizing, in a single election we had, in just a few
weeks, two similar coalition questions: Should the M5S form a coalition with PD?
And should the PD form a coalition with PdL? This pushed us to propose a model
describing three political parties interacting with the electors and using these interactions to make their own decision on whether to form some political alliance.
As a historical comment, it should be noted that not much changed with the next
elections in 2018 in Italy, which generated a similar situation with different agents
(Lega rather than PdL, with M5S as the first party). Also, similar problems were
also experienced by other countries such as Germany.

In this chapter we will review some of our original results, all deduced by making use of ladder and number operators. We will discuss how the model can be improved by making use, among other techniques, of the (H, ρ)-induced dynamics.
We will also discuss the meaning and role of ρ in this context.

The chapter is organized as follows: In the next section we review the first
model proposed some years ago [12], in which each party communicates only
with its own electors and with the set of the undecided voters. Then we discuss
how the model can be extended, considering different strategies. For instance, in
Section 3.3 we consider the possibility that the parties also interact with the electors of other parties and that the interactions between the parties themselves are not
simply quadratic, as assumed at a first stage of our analysis (see Equation (3.3)). In
Section 3.4 we introduce a rule ρ, connected with some pools, and we discuss how
and why it affects the dynamics of the system.

3.2 A First Model

In our model we have three parties, $\mathcal{P}_1$, $\mathcal{P}_2$ and $\mathcal{P}_3$, that together form what we
call $\mathcal{S}_{\mathcal{P}}$. According to Section 3.1, they can be identified with PD, PdL and M5S
respectively. Each party has to decide whether to form an alliance with some other
party. It is not important, at this stage, who is going to be the partner. For now the

parties need to decide only whether a coalition should be formed. This is, in fact, the only aspect of the parties we are interested in, here. Each $\mathcal{P}_j$ is associated with a dichotomous choice, zero (no alliance) or one (alliance), that we associate with a time-dependent decision function ($P_j(t)$ in (3.13)), that takes values between zero and one for each $t \geq 0$: The closer the value of $P_j(t)$ to zero, the less the party $\mathcal{P}_j$ is attracted by the possibility of forming an alliance with an other party. However, if $P_j(t) \simeq 1$ for some t, then $\mathcal{P}_j$ is very interested in forming such an alliance. Of course its decision is sharp if $P_j(t) = 0$ or when $P_j(t) = 1$. Otherwise the value of $P_j(t)$ is taken as an indication of the *mood* of the party concerning this aspect.

Considering the system $\mathcal{S}_\mathcal{P}$ as a whole, and assuming that at $t = 0$ each party can chose only zero or one, we have eight different possibilities, which we associate with eight different and mutually orthogonal vectors in an eight-dimensional Hilbert space $\mathcal{H}_\mathcal{P}$. These vectors are called $\varphi_{i,k,l}$, with $i, k, l = 0, 1$. The three subscripts refer to whether the three parties of the model want to form a coalition at time $t = 0$. Hence, for example, the vector $\varphi_{0,0,0}$ describes the fact that, at $t = 0$, no party wants to ally with any other party. Of course, this attitude can change during the time evolution, and deducing these changes is, in fact, what is interesting for us. Analogously, for instance, $\varphi_{0,1,0}$ describes the fact that at $t = 0$, $\mathcal{P}_1$ and $\mathcal{P}_3$ don't want to form any coalition although $\mathcal{P}_2$ does. The set $\mathcal{F}_\varphi = \{\varphi_{i,k,l}, i, k, l = 0, 1\}$ is an o.n. basis for $\mathcal{H}_\mathcal{P}$. A generic vector of $\mathcal{S}_\mathcal{P}$, for $t = 0$, is a linear combination of the vectors $\varphi_{i,k,l}$'s:

$$\Psi = \sum_{i,k,l=0}^{1} \alpha_{i,k,l}\, \varphi_{i,k,l}, \tag{3.1}$$

where, to keep a sort of probabilistic interpretation of the model, we could assume that $\sum_{i,k,l=0}^{1} |\alpha_{i,k,l}|^2 = 1$ in order to normalize the total probability. In particular, for instance, $|\alpha_{1,1,1}|^2$ can be interpreted as the probability that $\mathcal{S}_\mathcal{P}$ is, at $t = 0$, in a state $\varphi_{1,1,1}$; i.e., that $\mathcal{P}_1$, $\mathcal{P}_2$ and $\mathcal{P}_3$ have all chosen one (let's form some coalition). In what follows, quite often we will assume that only one coefficient $\alpha_{i,k,l}$ in (3.1) is one, while all the others are zero. This implies that we are sure about the original intentions of the three parties.

Due to the fact that each index in $\varphi_{i,k,l}$ can only be zero or one, it is natural to use three fermionic operators, p_1, p_2 and p_3, and the functional structure related to these to build up the vectors and the Hilbert space consequently, following our construction in Section 2.2. For that we assume that p_1, p_2 and p_3, together with their adjoint, satisfy the CAR

$$\left\{p_k, p_l^\dagger\right\} = \delta_{k,l}\mathbb{1}, \qquad \text{and} \qquad \{p_k, p_l\} = 0, \tag{3.2}$$

$k, l = 1, 2, 3$. Here, as everywhere in the book, $\mathbb{1}$ is the identity operator. Now, let $\varphi_{0,0,0}$ be the vacuum of p_j, that is, a vector satisfying $p_j\varphi_{0,0,0} = 0$, $j = 1, 2, 3$.

Of course, such a nonzero vector always exists and can be explicitly constructed as the tensor product of the vacua of p_1, p_2 and p_3. The other vectors of $\mathcal{F}_\varphi$ can be constructed acting on $\varphi_{0,0,0}$ with the raising operators $p_j^\dagger$:

$$\varphi_{1,0,0} = p_1^\dagger \, \varphi_{0,0,0}, \quad \varphi_{0,1,0} = p_2^\dagger \, \varphi_{0,0,0}, \quad \varphi_{1,1,0} = p_1^\dagger \, p_2^\dagger \, \varphi_{0,0,0}, \quad \varphi_{1,1,1} = p_1^\dagger \, p_2^\dagger \, p_3^\dagger \, \varphi_{0,0,0},$$

and so on. The order is relevant here, since we have, for instance, $p_1^\dagger \, p_2^\dagger \, \varphi_{0,0,0} = -p_2^\dagger \, p_1^\dagger \, \varphi_{0,0,0}$. However, for most of our applications, these extra signs have not much of a role. Now let $\hat{P}_j = p_j^\dagger p_j$ be the *number operator* of the j-th party. As we will show later, $\hat{P}_j$ is one of the key ingredients to construct the decision functions of the various parties. This is related to the following remark: Since $\hat{P}_j \varphi_{n_1,n_2,n_3} = n_j \varphi_{n_1,n_2,n_3}$, $j = 1, 2, 3$, the eigenvalues of these operators, zero and one, correspond to the only possible choices of the three parties at $t = 0$. This is, in fact, the main reason why we are using here the fermionic operators p_j: They automatically produce only these eigenvalues. In other words, the possible decisions are labeled by the subscripts of the vectors of $\mathcal{F}_\varphi$, which are, in turn, directly connected to the eigenvalues of the operators $\hat{P}_j$ introduced previously. Of course, this is not the only possible way to construct an o.n. basis in $\mathcal{H}_\mathcal{P}$, but it surely is convenient for us, since the operators p_j and $p_j^\dagger$ will also be useful in describing the interactions occurring in our political system $\mathcal{S}$, which is defined by the union of the parties, $\mathcal{S}_\mathcal{P}$, and the electors, $\mathcal{E}$: $\mathcal{S} = \mathcal{S}_\mathcal{P} \cup \mathcal{E}$. The role of the interactions in $\mathcal{S}$ is important, since our main effort now consists in giving a dynamics to the operators $\hat{P}_j$, following the general scheme described and used several times in [1]. For that we need to construct a Hamiltonian for $\mathcal{S}$, H, deduce the equations of motion for $p_j(t)$, the time evolution of p_j, and then solve these equations. In this way $\hat{P}_j(t) = p_j^\dagger(t) p_j(t)$, $j = 1, 2, 3$, are found, and the decision functions can be deduced out of them, taking a suitable mean value of these operators (see (3.13)). In this way we can follow how the parties modify their decisions, with respect to time, regarding their intention to form some alliance. Alternatively, if H does not depend explicitly on time, we could directly compute the time evolution of the number operators as $\hat{P}_j(t) := e^{iHt} \hat{P}_j e^{-iHt}$ and then compute their mean values on some suitable state, describing the whole system $\mathcal{S}$ at $t = 0$. Choosing one approach or the other mainly depends on the form of H. In fact, for H *easy enough*, the natural choice would be the second, since the explicit computation of $e^{\pm iHt}$ is an easy task. But this task becomes harder and harder for H, *not so easy* and we are forced to use the first approach, solving the Heisenberg equations of motion. The way in which H should be constructed is discussed in [1], where a set of minimal rules is proposed and then applied in several concrete systems. The key fact here, in the analysis of $\mathcal{S}$, is that the three parties are just part of a larger system; in order to make their decisions, they need first to interact with the electors. It is mainly this interaction that creates their final decisions. For this reason, $\mathcal{S}_{\mathcal{P}}$ must be open: There must be some environment $\mathcal{E}$, or reservoir, modeling the set of electors, interacting with $\mathcal{P}_1$, $\mathcal{P}_2$ and $\mathcal{P}_3$, so to produce some sort of feedback used by $\mathcal{P}_j$ to decide what to do. The reservoir, compared with $\mathcal{S}_{\mathcal{P}}$, is expected to be very large, since the sets of the electors for $\mathcal{P}_1$, $\mathcal{P}_2$ and $\mathcal{P}_3$ are supposed to be rather big. For this reason the operators of the reservoirs will also be labeled by a continuous variable.[1] In other words, while the operators describing $\mathcal{P}_j$ are just three independent fermionic operators, those of the reservoirs, according to the literature on quantum open systems [45], are infinitely many (fermionic) operators (see (3.4) and (3.5)). Of course, this has an immediate consequence: Independent of the choice of the ladder operators we use for $\mathcal{E}$, bosonic or fermionic, its related Hilbert space, $\mathcal{H}_{\mathcal{E}}$, is infinite-dimensional. But this is not really a problem, as we will see. Incidentally we notice that, even if in physics a reservoir is usually described in terms of bosonic operators, here it makes sense to use fermionic operators, as the electors in $\mathcal{E}$ are not particularly different (from a mathematical point of view) from the parties $\mathcal{P}_j$.

The various elements of our model are described in Figure 3.1, where the various arrows show all the interactions that we assume can occur between the parties and the electors, and among the parties themselves.

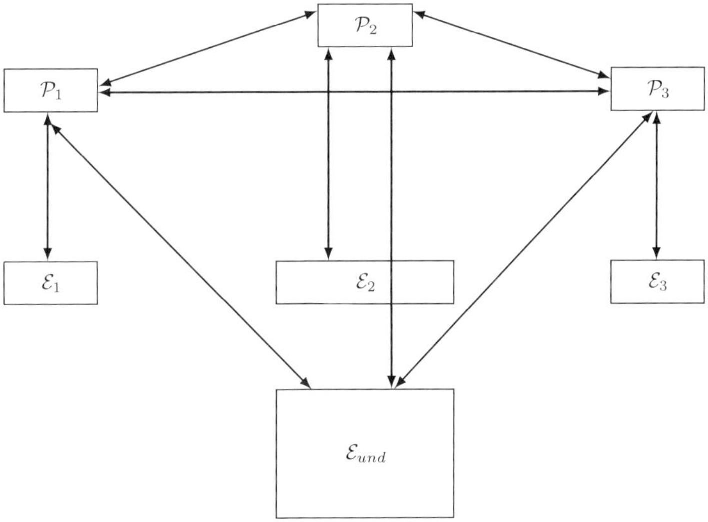

Figure 3.1 The system and its multicomponent environment (Source: Bagarello [12]).

---

[1] In fact, in view of our interpretation, it would be more natural to use a discrete index to label the operators of the reservoir. However, due to the Poisson summation formula, this is not really relevant since both discrete and continuous indexes give rise, after an infinite sum or a suitable integral, to a Dirac delta function, which is essential in our approach to produce the analytical time dependence of the decision functions.

In Figure 3.1, $\mathcal{E}_j$ represents the set of the supporters of $\mathcal{P}_j$, while $\mathcal{E}_{und}$ is the set of all the undecided electors, who do not know whom they are going to support or even if they will support any party at all. This figure also shows that, for instance, $\mathcal{P}_1$ can interact with $\mathcal{E}_1$ and $\mathcal{E}_{und}$ but not with $\mathcal{E}_2$ or with $\mathcal{E}_3$. This is a clear limit of the present model, since it is obvious that, in order to improve its position in the political scenario, each party must also try to attract the electors of the other parties. We will discuss in Section 3.3 how to remove this particular limit of the model. From Figure 3.1 we deduce that $\mathcal{P}_1$ also interacts with both $\mathcal{P}_2$ and $\mathcal{P}_3$. The Hamiltonian that describes, in our framework, the scheme in Figure 3.1 can be deduced by implementing the rules proposed in [1]:

$$
\begin{cases}
H = H_0 + H_{PBs} + H_{PB} + H_{int}, \\[4pt]
H_0 = \sum_{j=1}^3 \omega_j p_j^\dagger p_j + \sum_{j=1}^3 \int_{\mathbb{R}} \Omega_j(k) B_j^\dagger(k) B_j(k) \, dk + \int_{\mathbb{R}} \Omega(k) B^\dagger(k) B(k) \, dk, \\[4pt]
H_{PBs} = \sum_{j=1}^3 \lambda_j \int_{\mathbb{R}} \left(p_j B_j^\dagger(k) + B_j(k) p_j^\dagger \right) \, dk, \\[4pt]
H_{PB} = \sum_{j=1}^3 \tilde{\lambda}_j \int_{\mathbb{R}} \left(p_j B^\dagger(k) + B(k) p_j^\dagger \right) \, dk, \\[4pt]
H_{int} = \mu_{12}^{ex} \left(p_1^\dagger p_2 + p_2^\dagger p_1 \right) + \mu_{12}^{coop} \left(p_1^\dagger p_2^\dagger + p_2 p_1 \right) + \mu_{13}^{ex} \left(p_1^\dagger p_3 + p_3^\dagger p_1 \right) \\[4pt]
\qquad\quad + \mu_{13}^{coop} \left(p_1^\dagger p_3^\dagger + p_3 p_1 \right) + \mu_{23}^{ex} \left(p_2^\dagger p_3 + p_3^\dagger p_2 \right) + \mu_{23}^{coop} \left(p_2^\dagger p_3^\dagger + p_3 p_2 \right).
\end{cases}
$$

$$(3.3)$$

Here ω_j, λ_j, $\tilde{\lambda}_j$, μ_{ij}^{ex} and μ_{ij}^{coop} are real quantities, while $\Omega_j(k)$ and $\Omega(k)$ are real-valued functions. We will come back on their meaning in a moment. First of all we observe that, analogous to our choice for the operators of the parties, which have been chosen to be fermionic since they have to describe a dichotomous situation, the following CARs for the operators of the reservoir are (somewhat naturally) assumed:

$$
\left\{ B_i(k), B_l^\dagger(q) \right\} = \delta_{i,l}\, \delta(k-q) \, \mathbb{1}, \qquad \{ B_i(k), B_l(k) \} = 0, \tag{3.4}
$$

for the operators of $\mathcal{E}_j$, as well as

$$
\left\{ B(k), B^\dagger(q) \right\} = \delta(k-q)\, \mathbb{1}, \qquad \{ B(k), B(k) \} = 0, \tag{3.5}
$$

for those of $\mathcal{E}$. Here $i, l = 1, 2, 3$, and $k, q \in \mathbb{R}$. Moreover, each $p_j^\sharp$ anticommutes with each $B_l^\sharp(k)$ and with $B^\sharp(k)$: $\left\{ b_j^\sharp, B_l^\sharp(k) \right\} = \left\{ b_j^\sharp, B^\sharp(k) \right\} = 0$ for all j, l and for all k, and we further assume that $\left\{ B^\sharp(q), B_l^\sharp(k) \right\} = 0$. Here $X^\sharp$ stands for X or $X^\dagger$.

The quantities λ_j, $\tilde{\lambda}_j$, μ_{jk}^{ex} and μ_{jk}^{coop} are all interaction parameters, measuring respectively the strength of the interaction of $\mathcal{P}_j$ with $\mathcal{E}_j$, with $\mathcal{E}_{und}$ and with the other parties $\mathcal{P}_k$. In particular, these last are assumed to be of two opposite natures (see Section 3.2.2), cooperative and not-cooperative. The parameters and the functions appearing in the free Hamiltonian H_0, ω_j, $\Omega_j(k)$ and $\Omega(k)$, are related to a sort of inertia of $\mathcal{P}_j$, $\mathcal{E}_j$ and $\mathcal{E}_{und}$ respectively, as widely discussed in [1]: In a very schematic way, we could say that the larger their values, the smaller the amplitude of the variations in time for some related dynamical variable of the model.[2] This is quite a general result, suggested by many numerical and analytical computations performed along the years on similar systems (see [1, 8, 9, 66], for example). The operator H_0 in Equation (3.3) describes the free evolution of the operators of $\mathcal{S} = \mathcal{S}_{\mathcal{P}} \otimes \mathcal{E}$, where $\mathcal{E}$, which was already introduced, can now be identified with the following tensor product: $\mathcal{E} = (\mathcal{E}_1 \otimes \mathcal{E}_2 \otimes \mathcal{E}_3) \otimes \mathcal{E}_{und}$. When all the interaction parameters λ_j, $\tilde{\lambda}_j$, μ_{ij}^{ex} and μ_{ij}^{coop} are zero, then H simply coincides with H_0. Hence, since in this case $\left[H, \hat{P}_j\right] = 0$, the number operators describing the choices of the three parties and their related decision functions stay constant in time. In other words, in this case the original idea of each $\mathcal{P}_j$ is not modified during its time evolution. Each $\hat{P}_j$ is an integral of motion, in this simple situation. This is in agreement with our idea that all the decisions are driven by the interactions. In fact, the situation becomes more interesting when some of the interaction parameters, if not all, are taken to be not zero. This is connected to the meaning of the other terms in H, which describe interactions of different kind. For instance, H_{PBs} describes the interaction between the three parties and their related groups of electors, and in fact *PBs* stands for *Parties $\leftrightarrow$ Backgrounds* (the $\mathcal{E}'_j$s). The contribution $p_j B_j^{\dagger}(k)$ in H_{PBs} describes the fact that, when the electors in $\mathcal{E}_j$ are against any alliance, what we call a *global reaction against alliance* (GRAA) increases (because of $B_j^{\dagger}(k)$), and then $\mathcal{P}_j$ tends to choose 0 (no coalition) because of p_j. However $B_j(k)p_j^{\dagger}$ describes the fact that $\mathcal{P}_j$ is more interested in forming a coalition when the GRAA decreases. The presence of the integrals over k in H_{PBs} justifies why we call the effect of each $\mathcal{E}_j$ *global*: This is not linked just to a single elector but to the set of electors. A similar phenomenon is described by H_{PB}, where now *PB* stands for *Parties $\leftrightarrow$ Background* (i.e., $\mathcal{E}_{und}$), with the difference that the interaction is now between the parties and the set of undecided voters, who communicate with all the parties. The last contribution in H, H_{int}, is introduced to describe the fact that the parties try also to talk to one an other, possibly to reach some agreement. Two possibilities are allowed, one in which the parties act cooperatively (they tend to make the same choice, and, in fact, we have terms like $p_j^{\dagger}p_k^{\dagger}$), and one in which they tend to make opposite choices; for instance, $\mathcal{P}_1$ tries to form some alliance, while $\mathcal{P}_2$ excludes this possibility (and we have terms like $p_1^{\dagger}p_2$).

---

[2] This is why, analogous to classical mechanics, we have introduced the word *inertia* in connection with these quantities: The larger the inertia of a material point, the greater the tendency to remain still or to change its position only slightly.

Not all these contributions are assumed to be of the same strength, and in fact the relative magnitude of the coefficients μ_{jk}^{ex} and μ_{jk}^{coop} decides which are the leading contributions in H_{int}.

> *Remark:* It is clear that the expression for H does not include all the possible interactions one can imagine for S. However, in our opinion, it is already sufficiently rich to be considered an interesting starting point, and in fact it produces a nontrivial dynamics for the decision functions. This is true despite the fact that H contains only quadratic contributions, which give rise (see (3.7)) to a linear system of differential equations for which an analytical solution can be obtained. Both these limitations will be removed later in this chapter.

Before deducing the equations of motion for the relevant observables of the system, it is interesting to discuss the presence, or the absence, of some integrals of motion for the model. In our context (see [1] and Section 2.4), these are (self-adjoint) operators that commute with the Hamiltonian. Then, because of rather general reasons, these operators do not change in time. In many concrete situations, the existence of these kinds of operators can be used to suggest what the Hamiltonian should look like. Moreover, integrals of motion can also be used to check how realistic our model is or to simplify the analysis of the system while keeping the full generality of the results, as in [9], where one such integral guarantees the existence of a finite-dimensional effective Hilbert space where, de facto, the system lives during its time evolution. Let us introduce here the operator

$$\hat{N} = \sum_{j=1}^{3} \hat{N}_j = \sum_{j=1}^{3} \left(p_j^\dagger p_j + \int_{\mathbb{R}} B_j^\dagger(k) B_j(k)\, dk \right), \tag{3.6}$$

with obvious notation. It is clear that $\hat{N}$ is the sum of the decision operators $p_j^\dagger p_j$ plus the total GRAA. It is easy to check that $[\hat{N}, H] = 0$ if the μ_{jk}^{coop} and the $\tilde{\lambda}_j$ are all zero; that is, if the only allowed interaction between parties is noncooperative. However, if at least one of the μ_{jk}^{coop}'s is not zero, then $\hat{N}$ is no longer an integral of motion, since in this case $[\hat{N}, H] \neq 0$: Cooperation gives a nontrivial dynamics to $\hat{N}$. Stated differently, if we have some reason to believe that some contribution of the form $p_k^\dagger p_l^\dagger + p_l p_k$ must appear in H, then no integral of motion exists for S. However $\hat{N}$ stays constant in time if no such term contributes to H, which is the case when $\mu_{jk}^{coop} = 0$ for all j, k, if there is further no interaction between $\mathcal{P}_j$ and the undecided electors.

We can now go back to the analysis of the dynamics of the system. The Heisenberg equations of motion $\dot{X}(t) = i[H, X(t)]$ can be deduced by using the CARs in Equations (3.2), (3.4) and (3.5):

$$
\left\{
\begin{aligned}
\dot{p}_1(t) &= -i\omega_1 p_1(t) + i\lambda_1 \int_{\mathbb{R}} B_1(q,t)\,dq + i\tilde{\lambda}_1 \int_{\mathbb{R}} B(q,t)\,dq - i\mu_{12}^{ex} p_2(t) \\
&\quad - i\mu_{12}^{coop} p_2^{\dagger}(t) - i\mu_{13}^{ex} p_3(t) - i\mu_{13}^{coop} p_3^{\dagger}(t), \\
\dot{p}_2(t) &= -i\omega_2 p_2(t) + i\lambda_2 \int_{\mathbb{R}} B_2(q,t)\,dq + i\tilde{\lambda}_2 \int_{\mathbb{R}} B(q,t)\,dq - i\mu_{12}^{ex} p_1(t) \\
&\quad + i\mu_{12}^{coop} p_1^{\dagger}(t) - i\mu_{23}^{ex} p_3(t) - i\mu_{23}^{coop} p_3^{\dagger}(t), \\
\dot{p}_3(t) &= -i\omega_3 p_3(t) + i\lambda_3 \int_{\mathbb{R}} B_3(q,t)\,dq + i\tilde{\lambda}_3 \int_{\mathbb{R}} B(q,t)\,dq - i\mu_{13}^{ex} p_1(t) \\
&\quad + i\mu_{13}^{coop} p_1^{\dagger}(t) - i\mu_{23}^{ex} p_2(t) + i\mu_{23}^{coop} p_2^{\dagger}(t), \\
\dot{B}_j(q,t) &= -i\Omega_j(q) B_j(q,t) + i\lambda_j p_j(t), \qquad j = 1,2,3, \\
\dot{B}(q,t) &= -i\Omega(q) B(q,t) + i \sum_{j=1}^{3} \tilde{\lambda}_j p_j(t).
\end{aligned}
\right.
\tag{3.7}
$$

In particular, these last two equations can be rewritten as

$$
B_j(q,t) = B_j(q) e^{-i\Omega_j(q)t} + i\lambda_j \int_0^t p_j(t_1) e^{-i\Omega_j(q)(t-t_1)}\,dt_1
$$

and

$$
B(q,t) = B(q) e^{-i\Omega(q)t} + i \int_0^t \sum_{j=1}^{3} \tilde{\lambda}_j p_j(t_1) e^{-i\Omega(q)(t-t_1)}\,dt_1,
$$

which, assuming that $\Omega_j(k) = \Omega_j\, k$ and $\Omega(k) = \Omega\, k$, $\Omega, \Omega_j > 0$, produce the following equalities:

$$
\int_{\mathbb{R}} B_j(q,t)\,dq = \int_{\mathbb{R}} B_j(q) e^{-i\Omega_j qt}\,dq + i\pi \frac{\lambda_j}{\Omega_j} p_j(t),
\tag{3.8}
$$

and

$$
\int_{\mathbb{R}} B(q,t)\,dq = \int_{\mathbb{R}} B(q) e^{-i\Omega qt}\,dq + i\pi \frac{\sum_{j=1}^{3} \tilde{\lambda}_j\, p_j(t)}{\Omega}.
\tag{3.9}
$$

We refer to [1] and [45] for more details on this and similar computations. Here we just want to mention that the choice $\Omega_j(k) = \Omega_j k$, for example, is rather common when dealing with quantum open systems. If we now replace Equations (3.8) and (3.9) in the three equations for $\dot{p}_j(t)$, Equation (3.7), after some computations we can write them in a simple matrical form

$$
\dot{q}(t) = -U\, q(t) + \rho(t),
\tag{3.10}
$$

where we have defined the vectors

$$q(t) = \begin{pmatrix} p_1(t) \\ p_2(t) \\ p_3(t) \\ p_1^\dagger(t) \\ p_2^\dagger(t) \\ p_3^\dagger(t) \end{pmatrix}, \quad \rho(t) = \begin{pmatrix} \eta_1(t) \\ \eta_2(t) \\ \eta_3(t) \\ \eta_1^\dagger(t) \\ \eta_2^\dagger(t) \\ \eta_3^\dagger(t) \end{pmatrix},$$

and the symmetric matrix

$$U = \begin{pmatrix} \hat{\omega}_1 & \gamma_{1,2} & \gamma_{1,3} & 0 & i\mu_{1,2}^{coop} & i\mu_{1,3}^{coop} \\ \gamma_{1,2} & \hat{\omega}_2 & \gamma_{2,3} & -i\mu_{1,2}^{coop} & 0 & i\mu_{2,3}^{coop} \\ \gamma_{1,3} & \gamma_{2,3} & \hat{\omega}_3 & -i\mu_{1,3}^{coop} & -i\mu_{2,3}^{coop} & 0 \\ 0 & -i\mu_{1,2}^{coop} & -i\mu_{1,3}^{coop} & \overline{\hat{\omega}_1} & \overline{\gamma_{1,2}} & \overline{\gamma_{1,3}} \\ i\mu_{1,2}^{coop} & 0 & -i\mu_{2,3}^{coop} & \overline{\gamma_{1,2}} & \overline{\hat{\omega}_2} & \overline{\gamma_{2,3}} \\ i\mu_{1,3}^{coop} & i\mu_{2,3}^{coop} & 0 & \overline{\gamma_{1,3}} & \overline{\gamma_{2,3}} & \overline{\hat{\omega}_3} \end{pmatrix}.$$

Here we have introduced the following simplifying notation:

$$\mu_l := \frac{\lambda_l^2}{\Omega_l} + \frac{\tilde{\lambda}_l^2}{\Omega}, \quad \hat{\omega}_l := i\omega_l + \pi\mu_l, \quad \gamma_{k,l} := i\mu_{k,l}^{ex} + \frac{\pi}{\Omega}\tilde{\lambda}_k\tilde{\lambda}_l, \qquad (3.11)$$

for $k, l = 1, 2, 3$, as well as the operator-valued functions:

$$\eta_j(t) = i\left(\lambda_j\beta_j(t) + \tilde{\lambda}_j\beta(t)\right), \quad \beta_j(t) = \int_{\mathbb{R}} B_j(q)e^{-i\Omega_j qt}dq,$$

$$\beta(t) = \int_{\mathbb{R}} B(q)e^{-i\Omega qt}dq.$$

We see that U is fully defined in terms of the parameters of H and that $\rho(t)$ is the only part of the equation where the operators of the environment appear. The solution of Equation (3.10) is easily found in a matrix form:

$$q(t) = e^{-U\,t}q(0) + \int_0^t e^{-U\,(t-t_1)}\,\rho(t_1)\,dt_1, \qquad (3.12)$$

which is the starting point for our analysis below.

Remark: A particularly simple situation occurs when there is no interaction at all, i.e., when $\lambda_j = \tilde{\lambda}_j = \mu_{k,l}^{coop} = \mu_{k,l}^{ex} = 0$ for all $j, k, l = 1, 2, 3$. In this case, we trivially have $H = H_0$, and no interesting dynamics is expected, as we have already noticed. Indeed, this is reflected by the fact that U becomes a diagonal matrix with purely imaginary elements, while $\rho(t)$ is just identically zero. Then, the equations for each fermionic mode produce oscillations for the creation and annihilation operators $p_j^\dagger(t)$ and $p_j(t)$ and constant values for their products, the number operators $\hat{P}_j(t) = p_j^\dagger(t)p_j(t)$, $j = 1, 2, 3$. Because of (3.13) below, in this case all the decision functions are constant in time, exactly as expected. Interactions are the main (and only, in our present case) sources of variations in time of the decision of the parties.

Once we have obtained $q(t)$, we need to compute the decision functions $P_j(t)$, which are defined as follows:

$$P_j(t) := \left\langle \hat{P}_j(t) \right\rangle = \left\langle p_j^\dagger(t)p_j(t) \right\rangle, \tag{3.13}$$

$j = 1, 2, 3$. Here $\langle . \rangle$ is a state over the full system $\mathcal{S}$. These states [1] are taken to be suitable tensor products of vector states on $\mathcal{S}_{\mathcal{P}}$ and states on the reservoir that obey some standard rules (see Equations (3.15) and (3.16)). More specifically, for each operator of the form $X_{\mathcal{S}} \otimes Y_{\mathcal{E}}$, $X_{\mathcal{S}}$ being an operator of $\mathcal{S}_{\mathcal{P}}$ and $Y_{\mathcal{E}}$ an operator of the reservoir, we put

$$\langle X_{\mathcal{S}} \otimes Y_{\mathcal{E}} \rangle := \left\langle \varphi_{n_1,n_2,n_3}, X_{\mathcal{S}} \varphi_{n_1,n_2,n_3} \right\rangle \omega_{\mathcal{E}}(Y_{\mathcal{E}}). \tag{3.14}$$

Here φ_{n_1,n_2,n_3} is one of the vectors introduced at the beginning of this section, and each n_j measures, as discussed previously, the tendency of $\mathcal{P}_j$ to form or not some coalition at $t = 0$. As for $\omega_{\mathcal{E}}(.)$, this is a state on $\mathcal{E}$ satisfying the following standard properties, [1]:

$$\omega_{\mathcal{E}}(\mathbb{1}_{\mathcal{E}}) = 1, \quad \omega_{\mathcal{E}}\left(B_j(k)\right) = \omega_{\mathcal{E}}\left(B_j^\dagger(k)\right) = 0,$$

$$\omega_{\mathcal{E}}\left(B_j^\dagger(k)B_l(q)\right) = N_j(k)\ \delta_{j,l}\delta(k-q), \tag{3.15}$$

as well as

$$\omega_{\mathcal{E}}(B(k)) = \omega_{\mathcal{E}}\left(B^\dagger(k)\right) = 0, \quad \omega_{\mathcal{E}}\left(B^\dagger(k)B(q)\right) = N(k)\ \delta(k-q), \tag{3.16}$$

for some suitable functions $N_j(k)$, $N(k)$ that are related to the nature of the electors. Quite often, it is convenient, and sufficient for modeling purposes, to consider them to be constant in k: $N_j(k) = N_j$ and $N(k) = N$. This is what we will also do here. The meaning of this choice is that, for instance, the electors of each party are not really different. However, of course, the electors of different parties can be rather different, and this is reflected by the possibility that they correspond to quite different values of N_j. We assume also that $\omega_{\mathcal{E}}(B_j(k)B_l(q)) = \omega_{\mathcal{E}}(B(k)B(q)) = 0$, for all j and l. In our framework, the state in (3.14) describes the fact that, at $t = 0$, $\mathcal{P}_j$'s decision (concerning alliances) is n_j, while the overall feeling of the voters $\mathcal{E}_j$ is N_j and that of the undecided ones is N. Of course, these might appear as over-simplifying assumptions, and in fact they are, but still they produce an interesting dynamics for the model. We will return to the explicit meaning of N_j and N later on, in connection with an exactly solvable situation (see Section 3.2.1.1 and (3.20) in particular).

> *Remark:*   (1) We considered in Section 2.7 what happens when adopting different choices of the (analogous of the) functions $N_j(k)$, and other functions related to the environments of the agents in a rather general setting. In this chapter we will not consider further this possibility, which, however, could be interesting if we want to model electors of the same party that are different.
>
>   (2) Equation (3.14) shows that, rather than working with a generic normalized vector as in Equation (3.1), we are mainly interested in using eigenstates of the number operators: All the coefficients $\alpha_{i,k,l}$ are zero, except one. Hence, at least in this section, we work with initial conditions without incertitude.

Let us now call $V_t := e^{-Ut}$, and $(V_t)_{j,k}$ its (j,k)-th matrix element. Then some long but straightforward computations produce the following result:

$$P_j(t) = P_j^{(a)}(t) + P_j^{(b)}(t), \tag{3.17}$$

with

$$P_j^{(a)}(t) = \sum_{k=1}^{3} \left(\left|(V_t)_{j,k}\right|^2 n_k + \left|(V_t)_{j,k+3}\right|^2 (1 - n_k) \right)$$

and

$$P_j^{(b)}(t) = 2\pi \int_0^t dt_1 \sum_{k=1}^{3} \left(p_k^{(j)}(t - t_1) M_k + p_{k+3}^{(j)}(t - t_1) M_k^c \right)$$

$$+ 2\pi \int_0^t dt_1 \sum_{k,l=1,\, k<l}^{3} \left(p_{k,l}^{(j)}(t - t_1) \theta_{k,l} + p_{3+k,3+l}^{(j)}(t - t_1) \theta_{k,l}^c \right),$$

$j = 1, 2, 3$, where we have also introduced the shorthand notation

$$M_j := \frac{\lambda_j^2 N_j}{\Omega_j} + \frac{\tilde{\lambda}_j^2 N}{\Omega}, \qquad M_j^c := \frac{\lambda_j^2 (1 - N_j)}{\Omega_j} + \frac{\tilde{\lambda}_j^2 (1 - N)}{\Omega}, \tag{3.18}$$

as well as

$$\theta_{k,l} = \tilde{\lambda}_k \tilde{\lambda}_l \frac{N}{\Omega}, \qquad \theta_{k,l}^c = \tilde{\lambda}_k \tilde{\lambda}_l \frac{1 - N}{\Omega},$$

for $j = 1, 2, 3$ and $k, l = 1, 2, 3$ with $k < l$. We have also defined the following functions:

$$p_k^{(j)}(t) = \left|(V_t)_{j,k}\right|^2, \qquad p_{k,l}^{(j)}(t) = 2\Re\left[\overline{(V_t)_{j,k}} \, (V_t)_{j,l} \right],$$

where $\Re(z)$ stands for the real part of the complex quantity z.

Remark: In Equation (3.17) we have divided the dependence of the decision functions $P_j(t)$ in to two parts: $P_j^{(a)}(t)$ contains all the contributions coming essentially from $S_{\mathcal{P}}$, while $P_j^{(b)}(t)$ contains the contributions coming from the electors. We see that these two kinds of contributions essentially differ for the presence of some integrations in $P_j^{(b)}(t)$, while no time integral appears in $P_j^{(a)}(t)$. This has interesting consequences when computing the asymptotic values of the decision functions, as, for instance, Equations (3.20) and (3.21) clearly show. We will see that in both those formulas $P_j^{(a)}(t) \to 0$ when $t \to \infty$, while $P_j^{(b)}(t)$ does not necessarily tend to zero. In fact, this is rather general. The only nontrivial contributions in $P_j(t)$, for large t, always come from the integrals in $P_j^{(b)}(t)$ and not from $P_j^{(a)}(t)$. In other words, the final decision is always a consequence of the whole story. Memory is important in the procedure of the decision-making. However, if we want to analyze how $P_j(t)$ changes for t that is not necessarily large, $P_j^{(a)}(t)$ plays a relevant role. This will appear clear in Chapter 11, in connection with a different (but related) application to decision-making.

3.2.1 The Parties Do Not Talk to One Another

We start by considering the case in which the parties do not talk to one another, and they interact only with their own supporters. In other words, each $\mathcal{P}_j$ interacts only with $\mathcal{E}_j$, but not with $\mathcal{E}_{und}$ or with the other parties. Later in this section, we will see what happens when one, two and all the parties interact with $\mathcal{E}_{und}$ while they still do not talk to one another. We will show that a reasonably simple expression for the functions $P_j(t)$ can be deduced analytically in most of these cases. However, in some other interesting situations, suitable approximations are possibly needed. We will try to avoid these cases as much as possible here, even if it means restricting our models to only a moderately sophisticated level.

3.2.1.1 Case 1: Almost No Interaction

First of all we notice that, since we are assuming for the moment that the parties do not interact with one another, the interaction parameters $\mu_{k,l}^{ex}$ and $\mu_{k,l}^{coop}$ in H_{int} must be chosen as equal to zero: $\mu_{k,l}^{ex} = \mu_{k,l}^{coop} = 0$ for all k and l. Hence, $H_{int} = 0$. Furthermore, since none of the parties interacts with $\mathcal{E}_{und}$, we also fix $\tilde{\lambda}_k = 0$, $k = 1, 2, 3$. In this way, $H_{PB} = 0$ and the Hamiltonian in (3.3) simplifies to $H = H_0 + H_{PBs}$. The matrix U also simplifies significantly, since it becomes diagonal, with $\hat{\omega}_l = i\omega_l + \pi \frac{\lambda_l^2}{\Omega_l}$. Then V_t is diagonal, with obvious matrix elements. As a result, (3.17) becomes

$$P_j(t) = \left|(V_t)_{j,j}\right|^2 n_j + 2\pi M_j \int_0^t dt_1 \, \left|(V_{t-t_1})_{j,j}\right|^2 , \qquad (3.19)$$

$j = 1, 2, 3$. We see that the three parties behave in the same way but only with (possibly) different functions $(V_t)_{j,j}$ and with (again, possibly) different values of n_j and M_j. The reason for that is clear: Under our present working assumptions, the system $\mathcal{S}$ is just the union of three completely independent subsystems $\mathcal{S}_j = \mathcal{P}_j \cup \mathcal{E}_j$, $j = 1, 2, 3$: $\mathcal{S} = \cup_{j=1}^{3} \mathcal{S}_j$, with $\mathcal{S}_j \cap \mathcal{S}_k = \emptyset$ if $j \neq k$. Therefore, since in each $\mathcal{S}_j$ the party $\mathcal{P}_j$ experiences the same kind of interactions as $\mathcal{P}_k$ does in $\mathcal{S}_k$, it is absolutely natural that each decision function looks the same as the others. However, since the parameters of the Hamiltonian can be different in $\mathcal{S}_1, \mathcal{S}_2$ and $\mathcal{S}_3$, minor differences appear in some aspects of the $P_j(t)$. We will comment on these differences in a moment.

Now, computing the integral in (3.19), with simple algebraic manipulations we deduce that

$$
P_j(t) = n_j\, e^{-2\pi t\, \frac{\lambda_j^2}{\Omega_j}} + N_j \left(1 - e^{-2\pi t\, \frac{\lambda_j^2}{\Omega_j}} \right), \tag{3.20}
$$

which goes to N_j when t diverges: $P_j(\infty) := \lim_{t \to \infty} P_j(t) = N_j$, $j = 1, 2, 3$. The conclusion is simple: In this case each party follows the suggestions of its own electors. It is a sort of perfect democracy that, unfortunately, in real life is not realistic. In other words, $\mathcal{P}_1$ just does not care about other opinions, except those of $\mathcal{E}_1$. Again, this is quite reasonable in our particular situation since, with our choice of the parameters, each $\mathcal{S}_j$ is not linked to the other subsystems. Hence $\mathcal{S}_2$ has no possibility of producing any change in the analytic form of $P_1(t)$ or $P_3(t)$ or in their asymptotic values. It is also interesting to notice that the speed of convergence of each $P_j(t)$ to its asymptotic value depends on the ratio $\frac{\lambda_j^2}{\Omega_j}$: The higher this ratio, the higher this speed. This means that the strength of the interaction $\mathcal{P}_j \leftrightarrow \mathcal{R}_j$, measured by λ_j, is also relevant in determining the rapidity of the decision. This is an important aspect in decision-making. In principle, the final decision of $\mathcal{P}_j$ is connected with $P_j(\infty)$. This means that before making a decision, $\mathcal{P}_j$ should wait for an infinite amount of time. Of course, this is not reasonable. To be efficient, any agent (i.e., any party) should be able to make its decision in a finite (and possibly small) amount of time. Some experts working in decision-making simply consider a fixed $\tau > 0$ and then see what their model produces after τ. This produces a result that is considered to be the decision of the agent: $P_j(\tau)$, in our specific situation. For instance, this is the approach considered in [67]. Here we see that the decision time is related to the parameters of the Hamiltonian of the system $\mathcal{S}$. In fact, even if the final decision of $\mathcal{P}_j$ is $P_j(\infty)$, the analytic expression for $P_j(t)$ in (3.20) clearly shows that $P_j(t)$ already approaches $P_j(\infty)$ in a finite time $\hat{\tau}$: The more we want to get close to $P_j(\infty)$, the larger $\hat{\tau}$ must be taken to be. But, if we fix a reasonable tolerance $\epsilon > 0$, we can find a corresponding decision time τ_ϵ such that $P_j(\tau_\epsilon) \in]P_j(\infty) - \epsilon, P_j(\infty) + \epsilon[$. Not surprisingly, τ_ϵ is related to λ_j^2 and Ω_j.

In (3.20) we show that if $\lambda_j = 0$, $P_j(\infty) = P_j(t) = n_j$, for all $t \geq 0$. This is in complete agreement with the fact that, in the absence of interactions, no party modifies its initial decision about possible alliances. Furthermore, if $\Omega_j \gg \lambda_j^2$, even for a large amount of time (but not too large!), we find that $P_j(t) \simeq n_j$: the interaction is so small that, at least for small time intervals, $P_j(t)$ is not particularly different from n_j. However, if $t \to \infty$, we deduce again that $P_j(\infty) = N_j$.

Formula (3.20) and its consequence $P_j(\infty) = N_j$ are particularly useful since they give meaning to the quantity N_j that we introduced before. Because of our interpretation of $P_j(t)$, we see that $\mathcal{P}_j$ is inclined to form some coalition if $P_j(\infty)$ approaches one, which is possible in the present situation[3] only when $N_j = 1$. On the other hand, $\mathcal{P}_j$ does not really want to form any alliance if $P_j(\infty) \simeq 0$, which is possible only if $N_j = 0$. Now, we quite naturally assume that the interaction between parties and electors is constructive; that is, they communicate since they are interested in exchanging opinions, attracting electors, modifying strategies and so on. The result of these interactions is the equality $P_j(\infty) = N_j$, and therefore we are driven to the following interpretation of the N_j: A high value of N_j ($N_j = 1$) corresponds to a strong suggestion of $\mathcal{E}_j$ to $\mathcal{P}_j$ to ally. On the other hand, a small value of N_j ($N_j = 0$) corresponds to an equally strong suggestion of $\mathcal{E}_j$ to $\mathcal{P}_j$ not to ally. In this perspective it is clear that N_j could be considered as a sort of elector-version of the eigenvalues n_j of the vectors φ_{n_1,n_2,n_3} in $\mathcal{F}_\varphi$: Here, if $n_1 = 1$, $\mathcal{P}_1$ is willing to ally. Analogously, if $N_j = 1$, the electors in $\mathcal{E}_j$ are suggesting that $\mathcal{P}_j$ forms some alliance. In fact, $\mathcal{P}_j$ is listening to what its electors suggest, which is the reason why its final decision is $P_j(\infty) = N_j = 1$.

3.2.1.2 Case 2: $\mathcal{P}_1$ Interacts with $\mathcal{E}_{und}$

The second situation we want to consider is a bit richer than the previous one. Still, from an analytical point of view, it is again a simple case. We are assuming now that just one of the three parties interacts with the undecided electors, while the other two talk only with their own electors, as in Case 1. We are again assuming that the parties do not talk to one an other so that $\mu_{k,l}^{ex} = \mu_{k,l}^{coop} = 0$ for all k and l. However, assuming that the party interacting with $\mathcal{E}_{und}$ is the first one, $\tilde\lambda_1 \neq 0$, while we still have $\tilde\lambda_2 = \tilde\lambda_3 = 0$. This implies that because of the definition of $\gamma_{k,l}$, these are all zero and U is again a diagonal matrix,

$$U = diag\left(\hat\omega_1, \hat\omega_2, \hat\omega_3, \overline{\hat\omega_1}, \overline{\hat\omega_2}, \overline{\hat\omega_3}\right),$$

---

[3] Here we are assuming that the only possible values for N_j and N are simply zero or one, due to the fact that the reservoir operators obey the CAR.

and $V_t = e^{Ut}$ is a diagonal matrix as well. Repeating the same steps as in Case 1, we now deduce the following analytic expression for $P_j(t)$:

$$P_j(t) = n_j\, e^{-2\pi t \mu_j} + \frac{M_j}{\mu_j}\left(1 - e^{-2\pi t \mu_j}\right), \tag{3.21}$$

$j = 1, 2, 3$, where M_j and μ_j have been defined in (3.11) and (3.18). Of course, since $\tilde{\lambda}_2 = \tilde{\lambda}_3 = 0$, $M_2 = N_2$ and $M_3 = N_3$, and Equation (3.21) implies that $P_2(\infty) = N_2$ and $P_3(\infty) = N_3$. This is expected since there is no difference here regarding $\mathcal{P}_2$ and $\mathcal{P}_3$ with respect to what was assumed in Case 1. In fact, the subsystems $\mathcal{S}_2$ and $\mathcal{S}_3$ are still independent and non-overlapping. Moreover, they do not even overlap with $\tilde{\mathcal{S}}_1 = \mathcal{S}_1 \cup \mathcal{E}_{und}$. A clear indication of this independence of the subsystems is the fact that U is diagonal also in this case.

As for the decision function for $\mathcal{P}_1$, this is slightly more complicated, since the ratio $\frac{M_1}{\mu_1}$ does not now coincide with N_1. In fact, we deduce that

$$P_1(\infty) = \frac{M_1}{\mu_1} = \frac{\dfrac{\lambda_1^2 N_1}{\Omega_1} + \dfrac{\tilde{\lambda}_1^2 N}{\Omega}}{\dfrac{\lambda_1^2}{\Omega_1} + \dfrac{\tilde{\lambda}_1^2}{\Omega}},$$

which shows that in this situation, the final decision of $\mathcal{P}_1$ is also influenced also by $\mathcal{E}_{und}$. This is evident because of the appearance of $\tilde{\lambda}_1$, N and Ω in $P_1(\infty)$, parameters which are all related to $\mathcal{E}_{und}$. It might be interesting to compute what happens if $N_1 = N$, i.e., when $\mathcal{E}_1$ and $\mathcal{E}_{und}$ share the same opinion about what $\mathcal{P}_1$ should do. In this case, in fact, the previous formula returns $P_1(\infty) = N_1$, as if there were no $\mathcal{E}_{und}$ at all. This is in agreement with our interpretation of the model: If $N_1 = N$, then the electors in $\mathcal{E}_1$ and those in $\mathcal{E}_{und}$ are not really different, so there is no reason to put them in two different sets. This is true even if λ_1 does not coincide with $\tilde{\lambda}_1$, i.e., if the strength of the interactions of $\mathcal{P}_1$ with $\mathcal{E}_1$ and $\mathcal{E}_{und}$ are different. But again, this is not a surprise, since the asymptotic limits of the decision functions are independent of the (nonzero) interaction parameters even if the speed of convergence is not. For this reason, in our specific situation it makes no sense to look at $\mathcal{E}_1$ and $\mathcal{E}_{und}$ as separate sets. They can be considered, in fact, as two components of a single set. However, if $N_1 \neq N$, it is easy to see that $P_1(\infty)$ takes value in the open interval $]0, 1[$. The final decision is not sharp now, contrary to what happens when $\tilde{\lambda}_1 = 0$ (when $P_1(\infty) = N_1$), and it depends on the ratio between the parameters of the model and in particular on the ratio between $\tilde{\lambda}_1^2 \Omega_1$ and $\lambda_1^2 \Omega$. For instance, if we take $N_1 = 1$ and $N = 0$, we get

$$P_1(\infty) = \frac{1}{1 + \dfrac{\tilde{\lambda}_1^2 \Omega_1}{\lambda_1^2 \Omega}}, \tag{3.22}$$

which shows that the presence of $\mathcal{E}_{und}$ modifies the final decision of $\mathcal{P}_1$ when driven only by $\mathcal{E}_1$, since in this case we would get $P_1(\infty) = 1$. However, since $N = 0$, the electors in $\mathcal{E}_{und}$ are suggesting that $\mathcal{P}_1$ not form any coalition, contrary to what those in $\mathcal{E}_1$ are doing. So the two suggestions are in competition and, as a consequence, the resulting value of $P_1(\infty)$ is a number that is neither zero nor one. However, if for instance $\tilde{\lambda}_1^2 \Omega_1 \ll \lambda_1^2 \Omega$, then $P_1(\infty)$ approaches one. This is what we expect, since in this case, at least if Ω and Ω_1 are of the same order of magnitude, the interaction between $\mathcal{P}_1$ and $\mathcal{E}_1$ is much stronger than the one between $\mathcal{P}_1$ and $\mathcal{E}_{und}$ and the dynamics is mostly driven by the first rather than the second interaction. Similar conclusions, mutatis mutandis, can be deduced if $N_1 = 0$ and $N = 1$: In this case the electors in $\mathcal{E}_1$ are against any coalition, while those in $\mathcal{E}_{und}$ are supporting any possible alliance, and Equation (3.22) should be replaced by

$$P_1(\infty) = \cfrac{1}{1 + \cfrac{\lambda_1^2 \Omega}{\tilde{\lambda}_1^2 \Omega_1}}.$$

Remark: If we assume that the only party interacting with $\mathcal{E}_{und}$ is not $\mathcal{P}_1$ but, for instance, $\mathcal{P}_2$, while $\mathcal{P}_1$ and $\mathcal{P}_3$ interact only with their electors ($\mathcal{E}_1$ and $\mathcal{E}_3$ respectively), we deduce similar results and similar values for the $P_j(\infty)$. In particular, we get $P_1(\infty) = N_1$, $P_3(\infty) = N_3$, while $P_2(\infty) = \frac{M_2}{\mu_2}$, and comments similar to those given earlier could be repeated.

3.2.1.3 Case 3: $\mathcal{P}_1$ and $\mathcal{P}_2$ Interact Weakly with $\mathcal{E}_{und}$

We consider now a different situation in which, once again, the parties are not communicating among themselves. Once more $\mu_{k,l}^{ex} = \mu_{k,l}^{coop} = 0$ for every k and l. However we put now $\tilde{\lambda}_1 \neq 0$, $\tilde{\lambda}_2 \neq 0$, while $\tilde{\lambda}_3 = 0$ again. This means that, while $\mathcal{P}_3$ is still not interacting with $\mathcal{E}_{und}$, $\mathcal{P}_1$ and $\mathcal{P}_2$ are trying to convince the undecided electors to vote for them. Or, to be more optimistic, they are listening to what the unsatisfied electors are trying to indicate. From a mathematical point of view, the consequence of this fact is that U is no longer a diagonal matrix. Indeed, U looks like

$$U = \begin{pmatrix} \hat{\omega}_1 & \gamma_{1,2} & 0 & 0 & 0 & 0 \\ \gamma_{1,2} & \hat{\omega}_2 & 0 & 0 & 0 & 0 \\ 0 & 0 & \hat{\omega}_3 & 0 & 0 & 0 \\ 0 & 0 & 0 & \overline{\hat{\omega}_1} & \gamma_{1,2} & 0 \\ 0 & 0 & 0 & \gamma_{1,2} & \overline{\hat{\omega}_2} & 0 \\ 0 & 0 & 0 & 0 & 0 & \overline{\hat{\omega}_3} \end{pmatrix},$$

and V_t has a similar expression: Its diagonal elements are not zero and, moreover, the only non-diagonal elements of V_t that are nonzero are those with entries 12, 21, 45 and 54, as for the matrix U. Hence, both U and V_t are block diagonal matrices, with the same structure. In particular, they are both symmetric. We are interested here in deducing reasonably simple analytical results, using no numerical approach but just direct analytical computations. For that it is convenient to work under the assumption that $\max\{\tilde{\lambda}_1, \tilde{\lambda}_2\} \ll \min\{\lambda_1, \lambda_2, \lambda_3\}$. Hence, even if $\mathcal{P}_1$ and $\mathcal{P}_2$ interact with $\mathcal{E}_{und}$, these interactions are weak with respect to those that $\mathcal{P}_1$, $\mathcal{P}_2$ and $\mathcal{P}_3$ have with their own electors. So, electors come first! To simplify the notation, we also fix here $\Omega_1 = \Omega_2 = \Omega = 1$. Under these conditions it is possible to deduce an analytic expression for the matrix elements of V_t. In particular, the diagonal terms of V_t are always the same: $(V_t)_{jj} = e^{-\hat{\omega}_j t}$ if $j = 1, 2, 3$ and $(V_t)_{jj} = e^{-\overline{\hat{\omega}_{j-3} t}}$ if $j = 4, 5, 6$. The other nonzero terms depend on whether $\hat{\omega}_1$ is equal to $\hat{\omega}_2$. Here we restrict ourselves to the first case, since explicit formulas are simpler. In this case we have

$$P_1^{(a)}(t) = |(V_t)_{1,1}|^2 \, n_1 + |(V_t)_{1,2}|^2 \, n_2,$$

where

$$|(V_t)_{1,1}|^2 = \left|e^{-\hat{\omega}_1 t}\right|^2 = e^{-2\pi(\lambda_1^2 + \tilde{\lambda}_1^2)t},$$

$$|(V_t)_{1,2}|^2 = \gamma_{1,2}^2 t^2 \left|e^{-\hat{\omega}_1 t}\right|^2 = \gamma_{1,2}^2 t^2 e^{-2\pi(\lambda_1^2 + \tilde{\lambda}_1^2)t}.$$

Then

$$P_1^{(a)}(t) = e^{-2\pi(\lambda_1^2 + \tilde{\lambda}_1^2)t} \left(n_1 + n_2 \gamma_{1,2}^2 t^2\right) \to 0,$$

for $t \to \infty$. Hence, we deduce that also in this case the contribution of $P_1^{(a)}(t)$ to the decision function for $\mathcal{P}_1$ disappears when t is very large, even if in a more elaborated way than in Cases 1 and 2. Slightly more complicated is the computation of $P_1^{(b)}(\infty) = \lim_{t,\infty} P_1^{(b)}(t)$. With obvious notation, after some computations we get

$$P_1(\infty) = P_1^{(a)}(\infty) + P_1^{(b)}(\infty) \simeq N_1 + \left(\frac{\tilde{\lambda}_1}{\lambda_1}\right)^2 N + \frac{(\tilde{\lambda}_1 \tilde{\lambda}_2)^2}{\lambda_2^4} \left[\frac{\lambda_2^2 N_2 + \tilde{\lambda}_2^2 N}{2\lambda_2^2} - N\right].$$

A similar result can be deduced for $P_2(\infty)$. Indeed, we get

$$P_2(\infty) = P_2^{(a)}(\infty) + P_2^{(b)}(\infty) \simeq N_2 + \left(\frac{\tilde{\lambda}_2}{\lambda_2}\right)^2 N + \frac{(\tilde{\lambda}_1 \tilde{\lambda}_2)^2}{\lambda_1^4} \left[\frac{\lambda_1^2 N_1 + \tilde{\lambda}_1^2 N}{2\lambda_1^2} - N\right].$$

However, because we have fixed $\tilde{\lambda}_3 = 0$, nothing changes for the asymptotic value of $P_3(t)$: $P_3(\infty) = N_3$. In this case we can still divide $\mathcal{S}$ into three subsystems, $\mathcal{S}_1 = \mathcal{P}_1 \cup \mathcal{E}_1 \cup \mathcal{E}_{und}$, $\mathcal{S}_2 = \mathcal{P}_2 \cup \mathcal{E}_2 \cup \mathcal{E}_{und}$ and $\mathcal{S}_3 = \mathcal{P}_3 \cup \mathcal{E}_3$, but we see that, while $\mathcal{S}_3 \cap \mathcal{S}_1 = \mathcal{S}_3 \cap \mathcal{S}_2 = \emptyset$, $\mathcal{S}_1 \cap \mathcal{S}_2 = \mathcal{E}_{und} \neq \emptyset$. Therefore $\mathcal{S}_3$ is disconnected from $\mathcal{S}_1$ and $\mathcal{S}_2$, and then it behaves as in the previous cases, and in particular as in Case 1.

This is not the case for $\mathcal{S}_1$ and $\mathcal{S}_2$, at least in our perturbative scheme, and in fact we see that $P_1(\infty)$ is related not just to (the parameters of) $\mathcal{E}_1$, but also to (those of) $\mathcal{E}_{und}$. This is expected, and it is not particularly different from what we have seen in the analysis of Case 1. More interesting is to notice that $P_1(\infty)$ also depends on the parameters of $\mathcal{E}_2$. This is because $\mathcal{P}_1$ also interacts with $\mathcal{E}_2$, even if not directly: $\mathcal{P}_1$ interacts with $\mathcal{E}_{und}$, which interacts with $\mathcal{P}_2$, which interacts with $\mathcal{E}_2$!

We have to stress that these perturbative results cannot be taken too seriously and should be considered only as indicative. In fact, the above formulas for $P_1(\infty)$ and $P_2(\infty)$ show a strange feature, which cannot appear in any exact analytical solution of the problem. This can be made evident if we restrict to $N = 0$. In this case we find $P_1(\infty) \simeq N_1 + \frac{(\tilde{\lambda}_1\tilde{\lambda}_2)^2}{2\lambda_2^4} N_2$, and $P_2(\infty) \simeq N_2 + \frac{(\tilde{\lambda}_1\tilde{\lambda}_2)^2}{2\lambda_1^4} N_1$. Therefore, because of our approximation scheme, it is clear that it may happen that $P_j(\infty)$ is slightly larger than one: Let $N_1 = N_2 = 1$. Then

$$P_1(\infty) = 1 + \frac{(\tilde{\lambda}_1\tilde{\lambda}_2)^2}{2\lambda_2^4}, \qquad P_2(\infty) = 1 + \frac{(\tilde{\lambda}_1\tilde{\lambda}_2)^2}{2\lambda_1^4}.$$

These cannot be correct, since we expect that $P_j(t) \in [0, 1]$, for all j and for all t, because of our original choice to work with fermionic operators. Hence, the conclusion is that the perturbative expansion proposed here can give only some suggestions of what is going on, not rigorous results. Nevertheless, at least in principle, the exact solution could still be found because the model is linear (see Equation (3.12)), but we will not give its very complicated analytic expression, since the considerations discussed here are sufficient for us, at least at this stage.

Formulas become even more complicated if we consider the case $\hat{\omega}_1 \neq \hat{\omega}_2$. However, the numerical plots can again be easily drawn, even in this case. Just as an example, in Figure 3.2 we plot the decision functions $P_j(t)$, $j = 1, 2, 3$, for a particular choice of the parameters of the Hamiltonian and for a particular choice of the initial conditions, i.e., of n_j, N_j and N. These values are given in the caption of the figure. Similar plots can be obtained for different choices of the parameters and of the initial conditions.

From Figure 3.2 we see that $\mathcal{P}_1$ and $\mathcal{P}_2$, because of the interactions with $\mathcal{E}_j$ and $\mathcal{E}_{und}$, modify their original attitudes, going to a sort of intermediate state (i.e., they are not able to get a sharp decision). After some short transient, $\mathcal{P}_1$ appears rather interested in forming some coalition, while $\mathcal{P}_2$ loses (part of) its original interest very soon. This is essentially the effect of the presence of $\mathcal{E}_{und}$ in the time evolution of $P_1(t)$ and $P_2(t)$: $P_1(\infty)$ and $P_2(\infty)$ do not coincide with N_1 and N_2, as they should in the absence of any interaction with $\mathcal{E}_{und}$. This is similar to what we have already observed when commenting on Equation (3.22): The (extra) interaction with $\mathcal{E}_{und}$ can produce some uncertainty in the final decision of the parties. It is

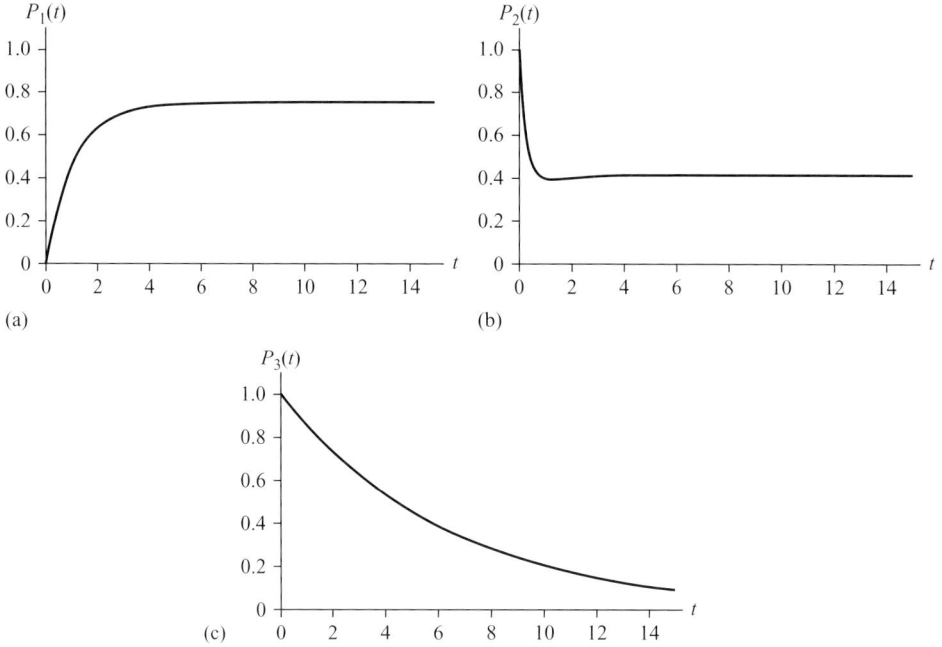

Figure 3.2 $P_1(t)$ (a), $P_2(t)$ (b) and $P_3(t)$ (c) for $\mu_{k,l}^{ex} = \mu_{k,l}^{coop} = 0$, $\omega_1 = 1$, $\omega_2 = \omega_3 = 2$, $\Omega_1 = \Omega_3 = \Omega = 0.1$, $\Omega_2 = 0.2$, $\lambda_1 = 0.1$, $\lambda_2 = 0.2$, $\lambda_3 = 0.05$, $\tilde{\lambda}_1 = 0.1$, $\tilde{\lambda}_2 = 0.2$, $\tilde{\lambda}_3 = 0$, and $n_1 = 0$, $n_2 = n_3 = 1$, $N_1 = N_2 = 1$, $N_3 = N = 0$ (Source: Bagarello [12]).

also interesting to notice that $P_2(t)$ does not tend to $P_2(\infty)$ monotonically. On the contrary, we see that this function decreases first, reaches a minimum and then starts increasing again a little bit. This behavior is not really what we expect from Equation (3.21), which is derived when only one party interacts with $\mathcal{E}_{und}$, and can be considered as the effect of the richer set of interactions we have considered in this case.

The situation is, in a sense, simpler but more extreme for $\mathcal{P}_3$, which does not interact with $\mathcal{E}_{und}$ but only with $\mathcal{E}_3$. It is simpler since the subsystem $\mathcal{S}_3$ decouples from the rest of $\mathcal{S}$ and can be treated separately, getting a result similar to Equation (3.20) but extreme, since in this case $\mathcal{P}_3$ completely modifies its original attitude, following the mood of its supporters, $\mathcal{E}_3$ (notice in fact that in Figure 3.2, $N_3 = 0$).

3.2.1.4 Case 4: All the Parties Interact (also) with $\mathcal{E}_{und}$

We conclude our preliminary analysis by considering what happens if, again, $\mu_{k,l}^{ex} = \mu_{k,l}^{coop} = 0$ for all k and l, as we have done up to this moment, but assuming now that $\tilde{\lambda}_j \neq 0$, $j = 1, 2, 3$. Then the parties still do not talk to one another, but they all communicate with their own electors and with the undecided voters in an attempt to convince the electors in $\mathcal{E}_{und}$ to vote for them or just to listen to what these electors ask for. In this case U is the following block diagonal matrix:

$$U = \begin{pmatrix} \hat{\omega}_1 & \gamma_{1,2} & \gamma_{1,3} & 0 & 0 & 0 \\ \gamma_{1,2} & \hat{\omega}_2 & \gamma_{2,3} & 0 & 0 & 0 \\ \gamma_{1,3} & \gamma_{2,3} & \hat{\omega}_3 & 0 & 0 & 0 \\ 0 & 0 & 0 & \overline{\hat{\omega}_1} & \gamma_{1,2} & \gamma_{1,3} \\ 0 & 0 & 0 & \gamma_{1,2} & \overline{\hat{\omega}_2} & \gamma_{2,3} \\ 0 & 0 & 0 & \gamma_{1,3} & \gamma_{2,3} & \overline{\hat{\omega}_3} \end{pmatrix}.$$

In Figure 3.3 we plot the decision functions $P_j(t)$ using essentially the same values of the parameters as in Figure 3.2, except for $\tilde{\lambda}_3$, which here is taken as positive and reasonably small ($\tilde{\lambda}_3 = 0.1$). As we see, the only major difference between Figures 3.2 and 3.3 is in the third function, $P_3(t)$, which does not decay (to zero) but seems to go to a positive asymptotic value. Because of the interaction with $\mathcal{E}_{und}$, there is some chance for a coalition now, at least in the time interval considered in

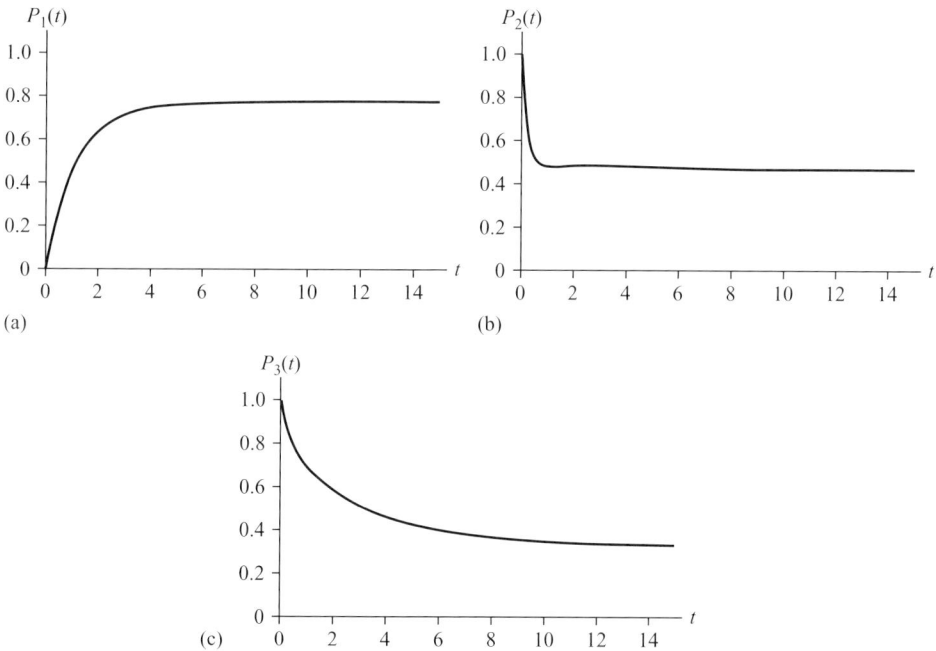

Figure 3.3 $P_1(t)$ (a), $P_2(t)$ (b) and $P_3(t)$ (c) for $\mu_{k,l}^{ex} = \mu_{k,l}^{coop} = 0$, $\omega_1 = 1$, $\omega_2 = \omega_3 = 2$, $\Omega_1 = \Omega_3 = \Omega = 0.1$, $\Omega_2 = .2$, $\lambda_1 = 0.1$, $\lambda_2 = 0.2$, $\lambda_3 = 0.05$, $\tilde{\lambda}_1 = 0.1$, $\tilde{\lambda}_2 = 0.2$, $\tilde{\lambda}_3 = 0.1$, and $n_1 = 0$, $n_2 = n_3 = 1$, $N_1 = N_2 = 1$, $N_3 = N = 0$ (Source: Bagarello [12]).

the plots. However, since both N_3 and N are zero, what we really expect is that $P_3(\infty) = 0$. However, this asymptotic value is reached after some more time: The decision time for $\mathcal{P}_3$ is larger than before! Of course, if a decision must be made under some pressure, then it might happen that $P_j(t)$ is still close to one, so that $\mathcal{P}_3$ is still in favor of allying, as it was at $t = 0$. This mood, however, decays with time, as Figure 3.3 clearly shows.

Considering other values of the initial conditions, and in particular of N_3, we arrive at the following rather general conclusion, in agreement with what we have already observed a few times: When $\tilde{\lambda}_3 = 0$, then $P_3(\infty)$ approaches, in all the cases considered in our analysis, the value N_3, so it can be only zero or one. However, if $\tilde{\lambda}_3 \neq 0$, then $P_3(\infty)$ approaches N_3 again, but not so much, and in fact the numerical values we obtain are always between zero and one. The decision of $\mathcal{P}_3$ is driven by $\mathcal{E}_3$, but not only: The interaction with $\mathcal{E}_{und}$ makes the decision less sharp.

3.2.2 What Happens When the Parties Talk to One Another?

So far we have analyzed how the decision functions behave when the parties do not interact among themselves: $H_{int} = 0$ in (3.3). Now it is natural to consider what happens to $P_j(t)$ when we do allow the parties to talk to one another. This might occur long before the election, in that temporal window in which, apparently, all the politicians seem to be interested in finding some agreement with their competitors, since they all claim to be interested to the welfare of the nation. This window does not usually last for long, at least in Italy. The closer we come to the election day, the more each party tends to attack the other parties (their enemies) for any possible reason. Moreover, quite often after the elections, the winner might be interested in collaborating with other parties only if it do not have the majority in Parliament. Otherwise, usually the party that won the elections is simply not interested in any alliance at all, since alliances mean compromises, and most of the time compromises are accepted only if they are strictly necessary. From a mathematical point of view, the main difference here with respect to what we have done in Section 3.2.1 is that we will now assume that $\mu_{k,l}^{ex}$, or $\mu_{k,l}^{coop}$, or both, are nonzero.

3.2.2.1 No Cooperative Effect

We will first consider what happens when every $\mu_{k,l}^{coop} = 0$ while $\mu_{k,l}^{ex} \neq 0$. In this way the only surviving contributions in H_{int} (see (3.3)) are those reflecting an opposite attitude of the parties regarding alliances. For instance, as we have already remarked, the term $p_1 p_2^{\dagger}$ suggests that, when the interest of $\mathcal{P}_1$ in forming some coalition decreases, that of $\mathcal{P}_2$ increases. Also in this case the matrix U turns out to be block-diagonal, and the computations are again not particularly complicated. In Figures 3.4 and 3.5 we plot, as usual, $P_1(t)$, $P_2(t)$ and $P_3(t)$ as

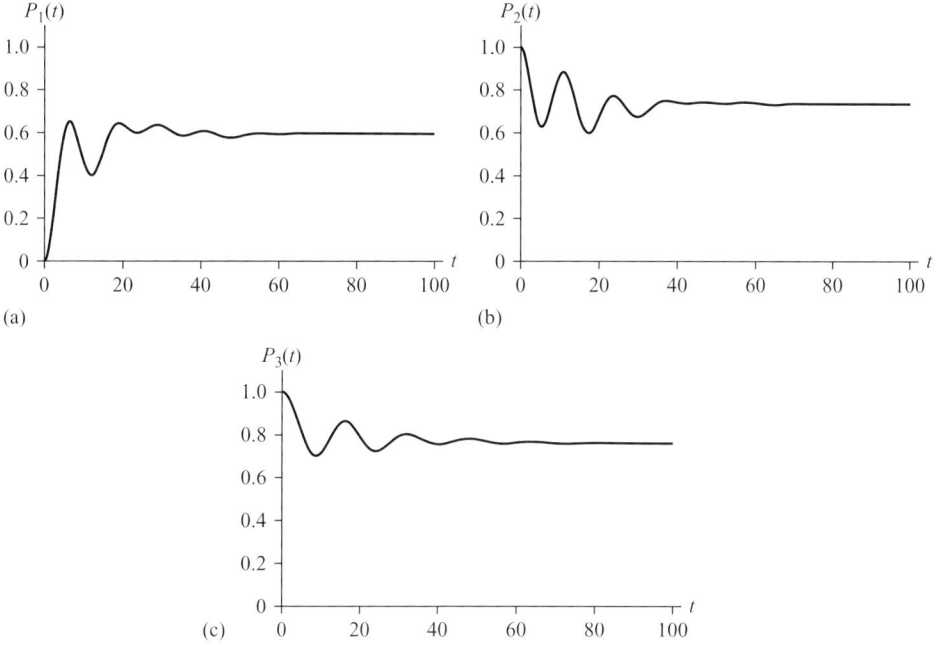

Figure 3.4 $P_1(t)$ (a), $P_2(t)$ (b) and $P_3(t)$ (c) for $\mu_{1,2}^{ex} = 0.2$, $\mu_{1,3}^{ex} = 0.1$, $\mu_{2,3}^{ex} = 0.15$, $\mu_{k,l}^{coop} = \tilde{\lambda}_j = 0$, $\omega_1 = 0.1$, $\omega_2 = \omega_3 = 0.2$, $\Omega_1 = \Omega_3 = 1$, $\Omega_2 = 2$, $\Omega = 0.1$, $\lambda_1 = 0.1$, $\lambda_2 = 0.2$, $\lambda_3 = 0.05$, and $n_1 = 0$, $n_2 = n_3 = 1$, $N_1 = 0$, $N_2 = N_3 = N = 1$ (Source: Bagarello [12]).

functions of time, for a particular choice of parameters and for two different initial conditions for $\mathcal{S}$: We fix $\mu_{1,2}^{ex} = 0.2$, $\mu_{2,3}^{ex} = 0.1$, $\mu_{2,3}^{ex} = 0.15$, $\mu_{k,l}^{coop} = \tilde{\lambda}_j = 0$, $\omega_1 = 0.1$, $\omega_2 = \omega_3 = 0.2$, $\Omega_1 = \Omega_3 = 1$, $\Omega_2 = 2$, $\Omega = 1$, $\lambda_1 = 0.1$, $\lambda_2 = 0.2$, $\lambda_3 = 0.05$. In Figure 3.4 we put $n_1 = 0$, $n_2 = n_3 = 1$, $N_1 = 0$, $N_2 = N_3 = N = 1$, while in Figure 3.5 we put $n_1 = 0$, $n_2 = n_3 = 1$, $N_1 = N_2 = 1$ and $N_3 = N = 0$.

The difference with respect to Figures 3.2 and 3.3 is evident. With this choice of parameters the parties are still able to reach a sort of final decision, but there exists a certain time interval, the transient, during which the three decision functions oscillate among different possible choices. This result, in our opinion, reflects well what we observe in real life: Many politicians say something one day and something completely different the next day. Or, to be more gentle with politicians (but less realistic!), they just need some time to reflect before they are able to make a decision. This is, in fact, what one might expect during a decision-making procedure: A decision is not always simple. The agent can change its opinion, even several times. However, going back to our parties, they need to decide sooner or later, and this is well reflected by the appearance of the asymptotic limits in our plots. Also the fact that $P_j(\infty)$'s are not really zero or one, but some intermediate

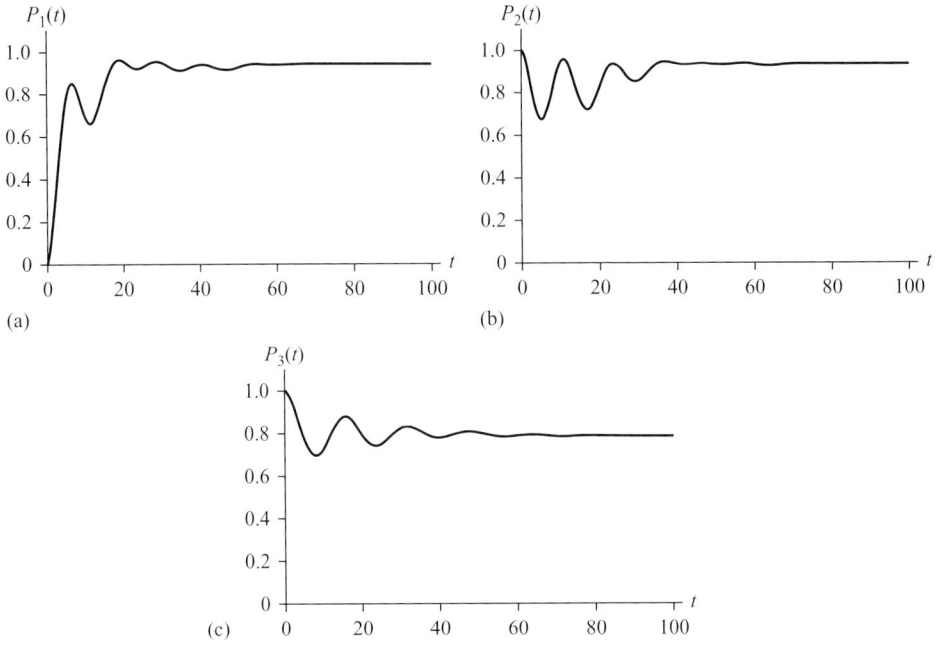

(a)

(b)

(c)

Figure 3.5 $P_1(t)$ (a), $P_2(t)$ (b) and $P_3(t)$ (c) for $\mu_{1,2}^{ex} = 0.2$, $\mu_{1,3}^{ex} = 0.1$, $\mu_{2,3}^{ex} = 0.15$, $\mu_{k,l}^{coop} = \tilde{\lambda}_j = 0$, $\omega_1 = 0.1$, $\omega_2 = \omega_3 = 0.2$, $\Omega_1 = \Omega_3 = 1$, $\Omega_2 = 2$, $\Omega = 0.1$, $\lambda_1 = 0.1$, $\lambda_2 = 0.2$, $\lambda_3 = 0.05$, and $n_1 = 0$, $n_2 = n_3 = 1$, $N_1 = N_2 = 1$, $N_3 = N = 0$ (Source: Bagarello [12]).

values, reflects the difficulty of making a decision, so there is usually no sharp position. The party reaches a sort of idea of how to behave but nothing more. The only case in which this sharp decision is made is, as we have already seen, when the parties interact only with their own electors $\mathcal{E}_j$ – and follow their suggestions.

Incidentally, it might be useful to observe that, in order to see the asymptotic behavior of the functions $P_j(t)$, in Figures 3.4 and 3.5 we have used a larger time interval than the one used in Figures 3.2 and 3.3, since asymptotic limits become evident only in larger intervals (otherwise we would have seen just oscillations). This is in agreement with our previous comment on $P_3(t)$ in Figure 3.3, which we have claimed goes to zero but only with a time scale larger than that used in that plot.

3.2.2.2 No Exchange Effect

Let us now put $\mu_{k,l}^{ex} = 0$ and $\mu_{k,l}^{coop} \neq 0$. This choice implies the fact that the only contributions in H_{int} (see (3.3)) are those representing a similar (cooperative) attitude of the parties about coalitions. For instance, the term $p_1 p_2$ in H_{int} suggests that both $\mathcal{P}_1$ and $\mathcal{P}_2$ lose interest in forming some alliance, while its adjoint $p_2^\dagger p_1^\dagger$ describes the opposite attitude.

In this case the matrix U has (almost) all nonzero entries. As a consequence, the computations are a bit harder. However, the plots we get do not differ much from those in Figures 3.4 and 3.5. In fact, Figures 3.6 and 3.7 share with those ones the same main features, i.e., an initial oscillating behavior with a subsequent convergence to a certain asymptotic value, which appears to be intermediate between zero and one. Also, in these cases the decisions of the three parties are not sharp and need some time to be reached. Further, changes of ideas are quite likely, for small values of t, even if not so evident as those found earlier. We observe that, in both figures, the values of $P_1(\infty)$ and $P_2(\infty)$ are close to one. Then, even if their decision is not strict, they have a very strong tendency to form some coalition, while $\mathcal{P}_3$ is much more against this possibility.

The parameters of Figures 3.6 and 3.7 coincide with those of Figures 3.4 and 3.5, with the only difference being that here we put $\mu_{k,l}^{ex} = 0$, whereas $\mu_{1,2}^{coop} = 0.1$, $\mu_{1,3}^{coop} = 0.08$, and $\mu_{2,3}^{coop} = 0.1$. The initial conditions are given in the captions.

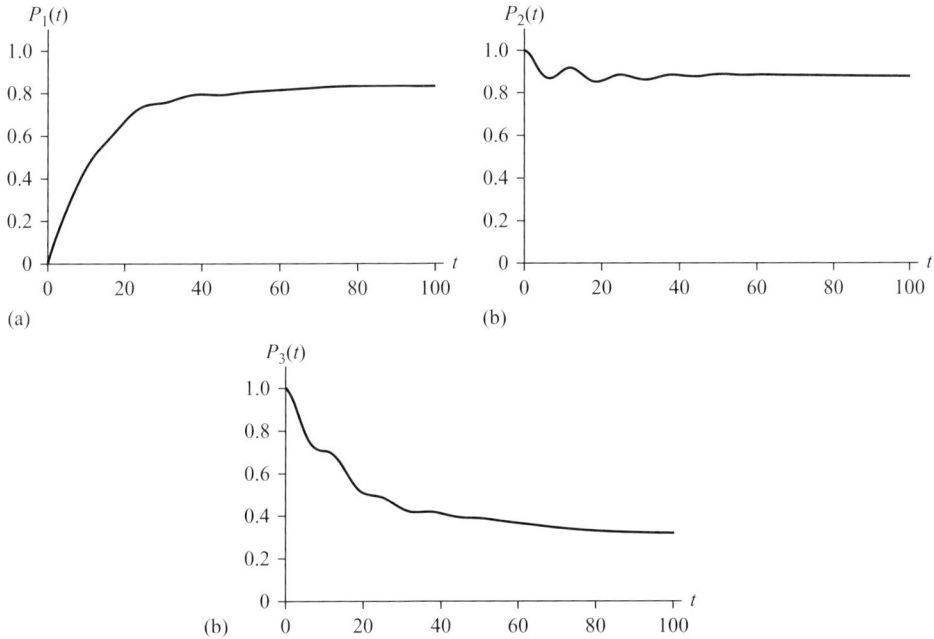

Figure 3.6 $P_1(t)$ (a), $P_2(t)$ (b) and $P_3(t)$ (c) for $\mu_{1,2}^{coop} = 0.1$, $\mu_{1,3}^{coop} = 0.08$, $\mu_{2,3}^{coop} = 0.1$, $\mu_{k,l}^{ex} = \tilde{\lambda}_j = 0$, $\omega_1 = 0.1$, $\omega_2 = \omega_3 = 0.2$, $\Omega_1 = \Omega_3 = 1$, $\Omega_2 = 2$, $\Omega = 0.1$, $\lambda_1 = 0.1$, $\lambda_2 = 0.2$, $\lambda_3 = 0.05$, and $n_1 = 0$, $n_2 = n_3 = 1$, $N_1 = N_2 = 1$, $N_3 = N = 0$ (Source: Bagarello [12]).

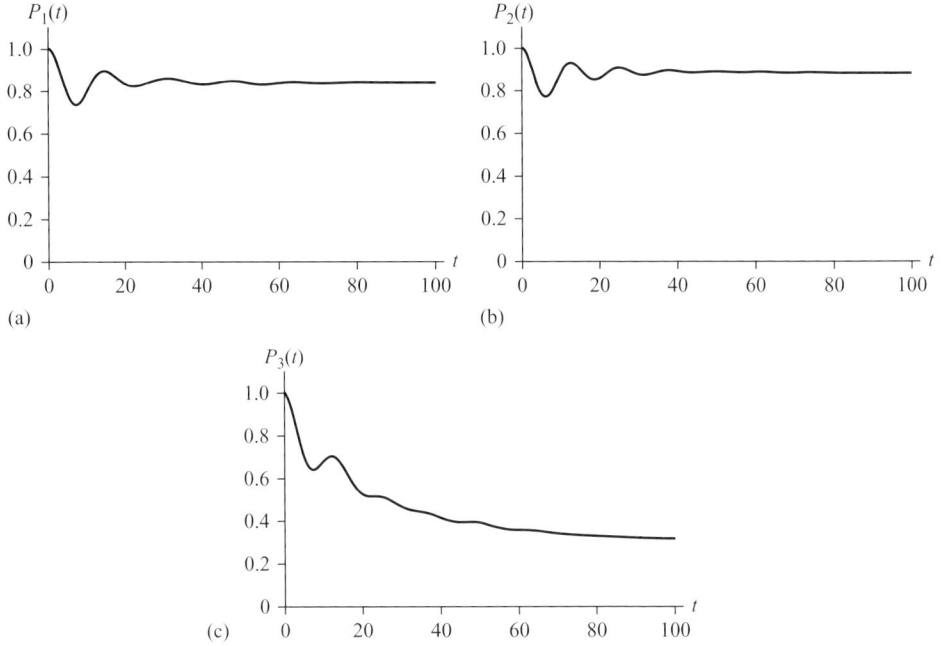

Figure 3.7 $P_1(t)$ (a), $P_2(t)$ (b) and $P_3(t)$ (c) for $\mu_{1,2}^{coop} = 0.1$, $\mu_{1,3}^{coop} = 0.08$, $\mu_{2,3}^{coop} = 0.1$, $\mu_{k,l}^{ex} = \tilde{\lambda}_j = 0$, $\omega_1 = 0.1$, $\omega_2 = \omega_3 = 0.2$, $\Omega_1 = \Omega_3 = 1$, $\Omega_2 = 2$, $\Omega = 0.1$, $\lambda_1 = 0.1$, $\lambda_2 = 0.2$, $\lambda_3 = 0.05$, and $n_1 = n_2 = n_3 = 1$, $N_1 = N_2 = 1$, $N_3 = N = 0$ (Source: Bagarello [12]).

Concluding, a comparison between the two cases considered above, and leading respectively to Figures 3.6 and 3.7 and to Figures 3.4 and 3.5, shows that the co-operative term in the Hamiltonian produces almost the same effect as the exchange term. The only difference is that, at least for our choices of parameters and initial conditions, the oscillations now look less evident.

3.2.2.3 Some Further Remarks

It might be interesting to understand how much the choice of the parameters influences our conclusions. In other words, how much do results change if we slightly modify the parameters of our simulations? The answer is easy: not much! And the reason for that is clear. Because of our choice of H, the differential equations of motions are linear. Then we cannot have any butterfly effect, and the dynamics is stable under these changes. For instance, if we plot $P_j(t)$ fixing, as in Figure 3.2, $\mu_{k,l}^{ex} = 0$, $\omega_1 = 1$, $\omega_2 = \omega_3 = 2$, $\Omega_1 = \Omega_3 = \Omega = 0.1$, $\Omega_2 = 0.2$, $\lambda_1 = 0.1$, $\lambda_2 = 0.2$, $\lambda_3 = 0.05$, $\tilde{\lambda}_1 = 0.1$, $\tilde{\lambda}_2 = 0.2$, $\tilde{\lambda}_3 = 0$, and $n_1 = 0$, $n_2 = n_3 = 1$, $N_1 = N_2 = 1$, $N_3 = N = 0$ and letting $\mu_{k,l}^{coop}$ be slightly greater than zero, the changes in the plots are extremely small, while they increase for increasing $\mu_{k,l}^{coop}$'s.

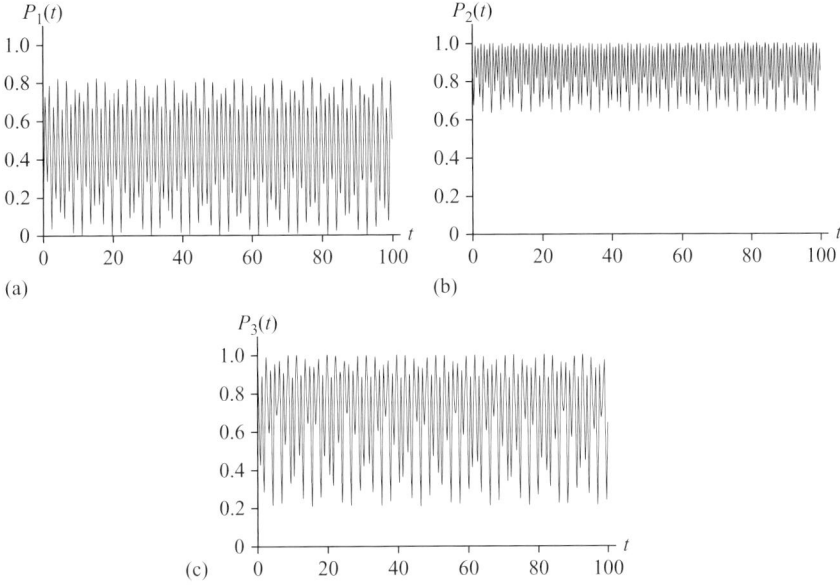

Figure 3.8 $P_1(t)$ (a), $P_2(t)$ (b) and $P_3(t)$ (c) for $\mu_{1,2}^{coop} = 0.1$, $\mu_{1,3}^{coop} = 0.08$, $\mu_{2,3}^{coop} = 0.1$, $\mu_{1,2}^{ex} = 2$, $\mu_{1,3}^{ex} = 1$, $\mu_{2,3}^{ex} = 3$, $\tilde{\lambda}_j = \lambda_j = 0$, $\omega_1 = 0.1$, $\omega_2 = \omega_3 = 0.2$, $\Omega_1 = \Omega_3 = \Omega = 0.1$, $\Omega_2 = 0.2$ and $n_1 = 0$, $n_2 = n_3 = 1$, $N_1 = N_2 = 1$, $N_3 = N = 0$ (Source: Bagarello [12]).

More interesting for us is to show what happens when, for instance, $\mu_{k,l}^{ex}$ and $\mu_{k,l}^{coop}$ are different from zero, whereas $\lambda_j = \tilde{\lambda}_j = 0, j = 1, 2, 3$. The plots are given in Figure 3.8 for a particular choice of the parameters and of the initial conditions. It is clear that there is no asymptotic value, at least for this time interval. On the contrary, the decision functions $P_j(t)$ appear to oscillate in time between some minimum and maximum values. This can be easily understood in sociological terms: Our choice on λ_j and $\tilde{\lambda}_j$, which are all chosen to be zero, is equivalent to removing the backgrounds from the system. The parties consult only each other, exchanging what we could call, with a slight abuse of language, *quanta of decision*. Also, if $\mu_{k,l}^{coop} \neq 0$ for some k and l, this operation is not compatible with the existence of any integral of motion involving the operators $\hat{P}_j(t)$. The parties have no input from their electors, and they are not able to decide what to do! Of course, as Figure 3.8 shows, there are several maxima and minima in the decision functions corresponding to opposite attitudes. This is another realistic effect observed in politics, where sometimes a decision changes with a very high frequency.[4] Of course,

---

[4] In 2013, while Mr. Letta was the prime minister in Italy, the Members of the Parliament had to vote the *Fiducia*, i.e., they had to vote to decide whether Mr. Letta should resign or not. No more than ten minutes before the vote took place, an important exponent of PdL, Mr. Brunetta, said in a TV interview that PdL was *not* going to vote in favor of Mr. Letta. However, only ten minutes later, Mr. Berlusconi, the President of the same party, talking to the other members of the House of Parliament, announced they were going to vote *in favor* of Mr. Letta. This is exactly our idea of high frequency!

in this situation we could consider a rather different point of view, following the same idea proposed in [67] and fixing a time for the decision that is completely arbitrary and independent of the oscillations we see in the figure. The idea is that, *Ok! Enough is enough: We have to decide what to do, and we have to do it right now!* But, as we discussed previously, this is not our favorite approach, and it will not be used in this book.

Despite the highly oscillating behavior shown in Figure 3.8, it is enough to add the effect of the reservoirs, by taking some λ_j or some $\tilde{\lambda}_j$ slightly larger than zero, to recover some asymptotic value (the smaller their values, the longer the time needed to reach this limit). The situation discussed here is a concrete realization, in a relevant situation, of the general effect analyzed in Section 2.7. The choice $\lambda_j = \tilde{\lambda}_j = 0, j = 1, 2, 3$ corresponds to the possibility of replacing the original system $\mathcal{S}$ with a smaller one, $\mathcal{S}_{eff} = \mathcal{P}_1 \cup \mathcal{P}_2 \cup \mathcal{P}_3$, living in a finite-dimensional Hilbert space whose dynamics is described by a Hermitian finite matrix. Therefore, only periodic or quasi-periodic motions are expected, and in fact this is what we see in Figure 3.8. The situation changes drastically when the electors are included in the treatment, which is the case when some λ_j or some $\tilde{\lambda}_j$ is nonzero. In this case the same argument cannot be repeated, and the dynamics is no longer necessarily periodic or quasi-periodic, which is exactly what we observe in Figures 3.2–3.7.

During our analysis we have also considered several other choices of parameters and initial conditions, including those in which all the parameters are nonzero; we have found few differences with respect to what has been explicitly discussed so far. The (rather general) scheme that emerges is the following:

1. If there is no reservoir at all $\left(\lambda_j = \tilde{\lambda}_j = 0\right)$, the functions $P_j(t)$ oscillate, and no asymptotic value is reached. On the other hand, it is enough that λ_j or $\tilde{\lambda}_j$ are slightly larger than zero, to recover some limiting value. This suggests that what really helps the various parties to make a decision is not the mutual interaction but the interaction with the electors, and not necessarily with those voting for them.

2. The exchange and the cooperative terms in the Hamiltonian produce a similar effect in the time evolution of the decision functions. However, while an integral of motion exists when $\mu_{k,l}^{coop}$ is zero for all k and l, this is not true if some of the $\mu_{k,l}^{coop}$ is nonzero. Another difference we have observed is that, using the same set of parameters and of initial conditions, the amplitude of the oscillations of the $P_j(t)$ may be different in these two cases.

3. The relative magnitude of the ω_j's (parameters of the parties) and of the Ω_j's (parameters of the electors) is important. In fact, we have found a strong numerical evidence of the fact that when the Ω_j's are larger than the ω_j's, the plots

oscillate much more than when the opposite happens or when they are of the same order of magnitude. In both cases, if λ_j or $\tilde{\lambda}_j$ are nonzero, $P_j(\infty)$ exists, and it is a value between zero and one, but this value is essentially reached monotonically when $\omega_j \gg \Omega_k$, $j, k = 1, 2, 3$, while it is reached after some (or many) oscillations when the opposite inequality holds.

4. Looking at the various plots and considering the analytic results (see, for example, (3.20) and (3.21)), we conclude that an essential role is played by the reservoirs. The explicit values of the other parameters, however, may change the asymptotic values of the various $P_j(\infty)$ and the speed of convergence. The parties reach some stable decision, depending on whether they interact with the electors, but the decision, in general, is related not only to N_j and N, but also to other parameters that, therefore, assume an important role in the model. This is clear, for instance, in (3.22), where we have fixed $N_1 = 1$ and $N = 0$: If $\frac{\tilde{\lambda}_1^2 \Omega_1}{\lambda_1^2 \Omega} \simeq 0$ then $P_1(\infty) \simeq 1$, and $\mathcal{P}_1$ will tend to form some coalition. However, if $\frac{\tilde{\lambda}_1^2 \Omega_1}{\lambda_1^2 \Omega} \gg 0$, then $P_1(\infty) \simeq 0$, and $\mathcal{P}_1$ will not be interested in forming any coalition. In other words, the results are *parameters-dependent*, which is quite natural in any mathematical model. This dependence can be used, and in fact it has been used in some cases, to fit the experimental data associated with some particular situation. So in principle, we could fix the numerical values of the parameters by using the known data for the 2013 or 2018 elections in Italy and see if these values also work well for future elections, to predict the behavior of the parties.

5. As we have already discussed before, even if we have considered $P_j(\infty)$ as the final decision of $\mathcal{P}_j$, this does not really mean that the parties decide how to behave only when t is extremely large! This, of course, would be rather unpleasant and not really close to the description of any realistic procedure of decision-making. However, all our plots show that an asymptotic value is reached, most of the time, sufficiently fast, at least with our choice of parameters. This means that the decisions are made reasonably soon, and this procedure can be made even faster by appropriately changing the interaction parameters, as we have shown previously in the analysis of (3.20).

3.3 Extending the Model, Part 1: More Interactions

The model discussed so far is based on a series of assumptions and approximations that can be only partially justified but that are useful in obtaining a set of differential equations that can be analytically solved. Of course, other models could be constructed with similar features, and possibly more realistically. This is exactly what we will do in the following sections, following slightly different possibilities, all based on the use of ladder operators.

One of the working assumptions in Section 3.2, which is clearly not very plausible, is the fact that the parties do not to interact with the electors of the other parties. This makes little sense, at least if, say, $\mathcal{P}_1$ wants to enlarge its electoral basin. The natural way to attract electors of a different party is by trying to communicate with them. This is also important in our model, in which the interest is not really in consensus or the results in some election but is only in the construction of the decision functions. In fact, it is not hard to imagine that each $P_j(t)$ and its asymptotic values $P_j(\infty)$ are affected in some way, not only from the electors in $\mathcal{E}_j$ and in $\mathcal{E}_{und}$, but also from those in $\mathcal{E}_k$, $k \neq j$. With this in mind, we refine the scheme of Figure 3.1 by making possible interactions not only between $\mathcal{P}_j$ and $\mathcal{E}_j$, but also between $\mathcal{P}_j$ and $\mathcal{E}_k$, with $j \neq k$. In other words, we want to see what happens if the various electors can talk with each party and how the decision functions of the parties are modified because of this possibility.

Figure 3.9 gives the schematic view of our extended system, which replaces the scheme described in Figure 3.1. The differences between the two schemes are clearly given by the arrows connecting $\mathcal{P}_j$ with $\mathcal{E}_k$, $j \neq k$, while all the other arrows coincide, meaning that all the other possible interactions already described in Section 3.2 are also a natural part of this new enlarged system.

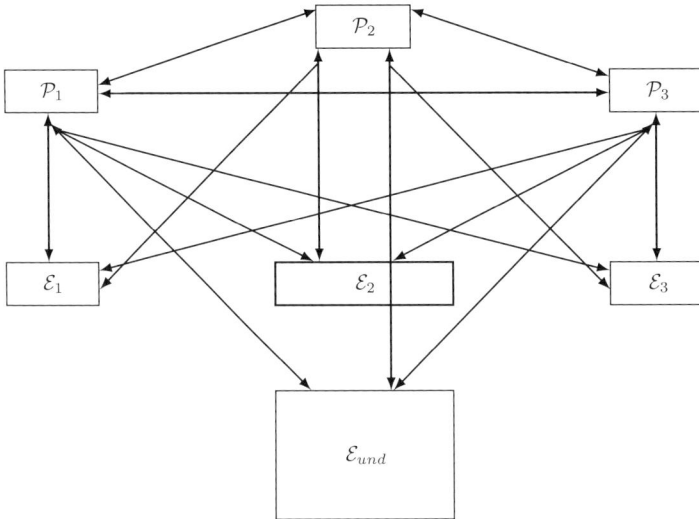

Figure 3.9 The enriched system and its multicomponent reservoir (Source: Bagarello [68]).

It is clear that the new Hamiltonian h describing the system in Figure 3.9 can be written in terms of the Hamiltonian H in (3.3), plus some new contributions describing the new arrows that we are adding to the original scheme. More explicitly, we have

$$
\begin{cases}
h = H + H_{mix}, \\
H_{mix} = H_{mix}^p + H_{mix}^{ap}, \\
H_{mix}^p = \sum_{n \neq l, 1}^{3} v_{nl}^p \int_{\mathbb{R}} dk \left(p_n B_l^\dagger(k) + B_l(k) p_n^\dagger \right), \\
H_{mix}^{ap} = \sum_{n \neq l, 1}^{3} v_{nl}^{ap} \int_{\mathbb{R}} dk \left(p_n^\dagger B_l^\dagger(k) + B_l(k) p_n \right),
\end{cases}
\tag{3.23}
$$

where H is exactly the operator introduced in (3.3). The contribution H_{mix} is the operator that describes the new interactions appearing in Figure 3.9, which were not considered in H. However, the method by which these interactions are constructed is quite similar to those used in the construction of H. In particular, all the terms in H_{mix} are quadratic, so that the set of differential equations of motion we get. see Equation (3.24), is linear. For instance, a contribution such as $p_n B_l^\dagger(k)$ means that if the party $\mathcal{P}_n$ is not willing to form any alliance, the electors of $\mathcal{P}_l$, $\mathcal{E}_l$, suggest that $\mathcal{P}_l$ does not form any alliance either; in fact, their GRAAs increase. Then $\mathcal{P}_n$ and $\mathcal{P}_l$ tend to have a similar behavior, and this explains the suffix p in H_{mix}^p: p stands for *parallel*. In a similar way, the term $p_n^\dagger B_l^\dagger(k)$ describes the fact that if the party $\mathcal{P}_n$ is now willing to form some alliance, $\mathcal{E}_l$ still suggests that $\mathcal{P}_l$ not form any such alliance: $\mathcal{P}_n$ and $\mathcal{P}_l$ tend to have opposite behaviors, and this is why we now use the suffix ap in H_{mix}^{ap}, which stands for *antiparallel*. The relative strengths of these different contributions is fixed by the explicit values of v_{nl}^{ap} and v_{nl}^p.

> *Remark:* It could be useful to notice that these behaviors are quite similar to the effects described in H_{int} in Equation (3.3), but they are obtained in a rather different way: *Cooperative* in H_{int} is like *parallel* in H_{mix}, while *exchange* in H_{int} corresponds to *antiparallel* in H_{mix}. The difference is that while in H_{int} the relative behavior of two parties is described by some direct interaction between the two, in H_{mix} this behavior arises because of the presence of the electors.

The equations in (3.7) must now be replaced by the following system, which can be deduced, as usual, by computing the Heisenberg equations of motion for the different annihilation operators appearing in the model:

$$
\begin{cases}
\dot{p}_1(t) = -i\omega_1 p_1(t) + i\lambda_1 \int_{\mathbb{R}} B_1(q,t)\,dq + i\tilde{\lambda}_1 \int_{\mathbb{R}} B(q,t)\,dq - i\mu_{12}^{ex} p_2(t) \\
\quad -i\mu_{12}^{coop} p_2^{\dagger}(t) - i\mu_{13}^{ex} p_3(t) - i\mu_{13}^{coop} p_3^{\dagger}(t) + M_1(t), \\[4pt]
\dot{p}_2(t) = -i\omega_2 p_2(t) + i\lambda_2 \int_{\mathbb{R}} B_2(q,t)\,dq + i\tilde{\lambda}_2 \int_{\mathbb{R}} B(q,t)\,dq - i\mu_{12}^{ex} p_1(t) \\
\quad +i\mu_{12}^{coop} p_1^{\dagger}(t) - i\mu_{23}^{ex} p_3(t) - i\mu_{23}^{coop} p_3^{\dagger}(t) + M_2(t), \\[4pt]
\dot{p}_3(t) = -i\omega_3 p_3(t) + i\lambda_3 \int_{\mathbb{R}} B_3(q,t)\,dq + i\tilde{\lambda}_3 \int_{\mathbb{R}} B(q,t)\,dq - i\mu_{13}^{ex} p_1(t) \\
\quad +i\mu_{13}^{coop} p_1^{\dagger}(t) - i\mu_{23}^{ex} p_2(t) + i\mu_{23}^{coop} p_2^{\dagger}(t) + M_3(t), \\[4pt]
\dot{B}_j(q,t) = -i\Omega_j(q) B_j(q,t) + i\lambda_j p_j(t) + i R_j(t), \quad j=1,2,3, \\[4pt]
\dot{B}(q,t) = -i\Omega(q) B(q,t) + i\sum_{j=1}^{3} \tilde{\lambda}_j p_j(t).
\end{cases}
\tag{3.24}
$$

where we have introduced the following quantities:

$$
\begin{cases}
M_1(t) = i\int_{\mathbb{R}} \left(v_{12}^p B_2(q,t) + v_{13}^p B_3(q,t) - v_{12}^{ap} B_2^{\dagger}(q,t) - v_{13}^{ap} B_3^{\dagger}(q,t) \right) dq \\[4pt]
M_2(t) = i\int_{\mathbb{R}} \left(v_{21}^p B_1(q,t) + v_{23}^p B_3(q,t) - v_{21}^{ap} B_1^{\dagger}(q,t) - v_{23}^{ap} B_3^{\dagger}(q,t) \right) dq \\[4pt]
M_3(t) = i\int_{\mathbb{R}} \left(v_{31}^p B_1(q,t) + v_{32}^p B_2(q,t) - v_{31}^{ap} B_1^{\dagger}(q,t) - v_{32}^{ap} B_2^{\dagger}(q,t) \right) dq \\[4pt]
R_1(t) = v_{21}^p p_2(t) + v_{21}^{ap} p_2^{\dagger}(t) + v_{31}^p p_3(t) + v_{31}^{ap} p_3^{\dagger}(t) \\[4pt]
R_2(t) = v_{12}^p p_1(t) + v_{12}^{ap} p_1^{\dagger}(t) + v_{32}^p p_3(t) + v_{32}^{ap} p_3^{\dagger}(t) \\[4pt]
R_3(t) = v_{13}^p p_1(t) + v_{13}^{ap} p_1^{\dagger}(t) + v_{23}^p p_2(t) + v_{23}^{ap} p_2^{\dagger}(t).
\end{cases}
\tag{3.25}
$$

First of all we observe that, if $v_{kl}^p = v_{kl}^{ap} = 0$ for all k and l, then $M_j(t) = R_j(t) = 0$, $j = 1, 2, 3$, and the system in Equation (3.24) gives back the one in Equation (3.7). This is obvious, since this choice of coefficients returns the scheme in Figure 3.1. Hence, Equation (3.7) can be seen as a special case of Equation (3.24). Notice also that the system in Equation (3.24) is again linear in its dynamical variables, so that an analytic solution can be found, in principle. In fact, its solution can be deduced in the usual way: We first rewrite the equations for $B_j(q,t)$ and $B(q,t)$ in their integral forms. Then we assume that $\Omega_j(q)$ and $\Omega(q)$ are linear in the variable q, $\Omega_j(q) = \Omega_j\,q$ and $\Omega(q) = \Omega\,q$ for some positive constants Ω_j and Ω, and finally we replace these integral equations in the first three equations for $\dot{p}_j(t)$ in (3.24). This approach is almost the same as in Section 3.2, and in many other models that have been considered along the years. We refer to [1] for more examples and to [45] for other applications in quantum optics.

Now, long but straightforward computations produce the following solution, which extends Equation (3.12):

$$P(t) = e^{U\,t}P(0) + \int_0^t e^{U\,(t-t_1)}\,\eta\,(t_1)\,dt_1, \tag{3.26}$$

where we have introduced the vectors

$$P(t) = \begin{pmatrix} p_1(t) \\ p_2(t) \\ p_3(t) \\ p_1^\dagger(t) \\ p_2^\dagger(t) \\ p_3^\dagger(t) \end{pmatrix}, \qquad \eta(t) = \begin{pmatrix} \eta_1(t) \\ \eta_2(t) \\ \eta_3(t) \\ \eta_1^\dagger(t) \\ \eta_2^\dagger(t) \\ \eta_3^\dagger(t) \end{pmatrix},$$

and the symmetric matrix

$$U = \begin{pmatrix} x_{1,1} & x_{1,2} & x_{1,3} & y_{1,1} & y_{1,2} & y_{1,3} \\ x_{1,2} & x_{2,2} & x_{2,3} & y_{1,2} & y_{2,2} & y_{2,3} \\ x_{1,3} & x_{2,3} & x_{3,3} & y_{1,3} & y_{2,3} & y_{3,3} \\ \hline \overline{y_{1,1}} & \overline{y_{1,2}} & \overline{y_{1,3}} & \overline{x_{1,1}} & \overline{x_{1,2}} & \overline{x_{1,3}} \\ \overline{y_{1,2}} & \overline{y_{2,2}} & \overline{y_{2,3}} & \overline{x_{1,2}} & \overline{x_{2,2}} & \overline{x_{2,3}} \\ \overline{y_{1,3}} & \overline{y_{2,3}} & \overline{y_{3,3}} & \overline{x_{1,3}} & \overline{x_{2,3}} & \overline{x_{3,3}} \end{pmatrix}.$$

This is because the set in Equation (3.24) can be rewritten, with these definitions, as

$$\dot{P}(t) = UP(t) + \eta(t).$$

The matrix U is clearly time-independent, and its entries are defined as follows:

$$
\begin{cases}
x_{1,1} = -i\omega_1 - \frac{\pi}{\Omega}\,\tilde{\lambda}_1^2 - \frac{\pi}{\Omega_1}\,\lambda_1^2 - \frac{\pi}{\Omega_2}\left(\left(v_{12}^p\right)^2 + \left(v_{12}^{ap}\right)^2\right) - \frac{\pi}{\Omega_3}\left(\left(v_{13}^p\right)^2 + \left(v_{13}^{ap}\right)^2\right) \\[2mm]
x_{1,2} = -i\mu_{12}^{ex} - \frac{\pi}{\Omega}\,\tilde{\lambda}_1\tilde{\lambda}_2 - \frac{\pi}{\Omega_1}\,\lambda_1 v_{21}^p - \frac{\pi}{\Omega_2}\,\lambda_2 v_{12}^p - \frac{\pi}{\Omega_3}\left(v_{13}^p v_{23}^p + v_{13}^{ap} v_{23}^{ap}\right) \\[2mm]
x_{1,3} = -i\mu_{13}^{ex} - \frac{\pi}{\Omega}\,\tilde{\lambda}_1\tilde{\lambda}_3 - \frac{\pi}{\Omega_1}\,\lambda_1 v_{31}^p - \frac{\pi}{\Omega_2}\left(v_{12}^p v_{32}^p + v_{12}^{ap} v_{32}^{ap}\right) - \frac{\pi}{\Omega_3}\,\lambda_3 v_{13}^p \\[2mm]
x_{2,2} = -i\omega_2 - \frac{\pi}{\Omega}\,\tilde{\lambda}_2^2 - \frac{\pi}{\Omega_1}\left(\left(v_{21}^p\right)^2 + \left(v_{21}^{ap}\right)^2\right) - \frac{\pi}{\Omega_2}\,\lambda_2^2 - \frac{\pi}{\Omega_3}\left(\left(v_{23}^p\right)^2 + \left(v_{23}^{ap}\right)^2\right) \\[2mm]
x_{2,3} = -i\mu_{23}^{ex} - \frac{\pi}{\Omega}\,\tilde{\lambda}_2\tilde{\lambda}_3 - \frac{\pi}{\Omega_1}\left(v_{21}^p v_{31}^p + v_{21}^{ap} v_{31}^{ap}\right) - \frac{\pi}{\Omega_2}\,\lambda_2 v_{32}^p - \frac{\pi}{\Omega_3}\,\lambda_3 v_{23}^p \\[2mm]
x_{3,3} = -i\omega_3 - \frac{\pi}{\Omega}\,\tilde{\lambda}_3^2 - \frac{\pi}{\Omega_1}\left(\left(v_{31}^p\right)^2 + \left(v_{31}^{ap}\right)^2\right) - \frac{\pi}{\Omega_2}\left(\left(v_{32}^p\right)^2 + \left(v_{32}^{ap}\right)^2\right) - \frac{\pi}{\Omega_3}\,\lambda_3^2,
\end{cases}
$$

and

$$
\begin{cases}
y_{1,1} = -\frac{2\pi}{\Omega_2}\,v^p_{12}v^{ap}_{12} - \frac{2\pi}{\Omega_3}\,v^p_{13}v^{ap}_{13} \\[4pt]
y_{1,2} = -i\mu^{coop}_{12} - \frac{\pi}{\Omega_1}\,\lambda_1 v^{ap}_{21} - \frac{\pi}{\Omega_2}\lambda_2 v^{ap}_{12} - \frac{\pi}{\Omega_3}\left(v^p_{13}v^{ap}_{23} + v^{ap}_{13}v^p_{23}\right) \\[4pt]
y_{1,3} = -i\mu^{coop}_{13} - \frac{\pi}{\Omega_1}\,\lambda_1 v^{ap}_{31} - \frac{\pi}{\Omega_2}\left(v^p_{12}v^{ap}_{32} + v^{ap}_{12}v^p_{32}\right) - \frac{\pi}{\Omega_3}\lambda_3 v^{ap}_{13} \\[4pt]
y_{2,2} = -\frac{2\pi}{\Omega_1}\,v^p_{21}v^{ap}_{21} - \frac{2\pi}{\Omega_3}\,v^p_{23}v^{ap}_{23} \\[4pt]
y_{2,3} = -i\mu^{coop}_{23} - \frac{\pi}{\Omega_1}\left(v^p_{21}v^{ap}_{31} + v^{ap}_{21}v^p_{31}\right) - \frac{\pi}{\Omega_2}\lambda_2 v^{ap}_{32} - \frac{\pi}{\Omega_3}\lambda_3 v^{ap}_{23} \\[4pt]
y_{3,3} = -\frac{2\pi}{\Omega_1}\,v^p_{31}v^{ap}_{31} - \frac{2\pi}{\Omega_2}\,v^p_{32}v^{ap}_{32}.
\end{cases}
$$

The components of the vector $\eta(t)$ are the following time-dependent operators:

$$
\begin{cases}
\eta_1(t) = i\tilde{\lambda}_1\beta(t) + i\lambda_1\beta_1(t) + iv^p_{12}\beta_2(t) + iv^p_{13}\beta_3(t) - iv^{ap}_{12}\beta^\dagger_2(t) - iv^{ap}_{13}\beta^\dagger_3(t) \\[4pt]
\eta_2(t) = i\tilde{\lambda}_2\beta(t) + i\lambda_2\beta_2(t) + iv^p_{21}\beta_1(t) + iv^p_{23}\beta_3(t) - iv^{ap}_{21}\beta^\dagger_1(t) - iv^{ap}_{23}\beta^\dagger_3(t) \\[4pt]
\eta_3(t) = i\tilde{\lambda}_3\beta(t) + i\lambda_3\beta_3(t) + iv^p_{31}\beta_1(t) + iv^p_{32}\beta_2(t) - iv^{ap}_{31}\beta^\dagger_1(t) - iv^{ap}_{32}\beta^\dagger_2(t),
\end{cases}
$$

where we have further introduced the operators $\beta(t) = \int_{\mathbb{R}} B(q)e^{-i\Omega qt}dq$ and $\beta_j(t) = \int_{\mathbb{R}} B_j(q)e^{-i\Omega_j qt}dq$, $j = 1, 2, 3$, both related to the sets of the electors (and not to the parties). Once again we observe that, if $v^p_{kl} = v^{ap}_{kl} = 0$ for all k and l, all these formulas, and the solution in (3.26) in particular, coincide with those deduced in Section 3.2.

Working with Equation (3.26) is not particularly different from what we have done before. Let us introduce again the matrix $V_t = e^{Ut}$. This is simply the exponential of a six-by-six matrix, with matrix elements $(V_t)_{j,l}$. Then the decision functions, which are defined as in (3.13), can be written as follows:

$$
P_j(t) = P^{(a)}_j + P^{(b)}_j, \tag{3.27}
$$

where

$$
P^{(a)}_j(t) = \sum_{l=1}^{3}\left[(V_t)_{3+j,l}\,(V_t)_{j,3+l}\,(1 - n_l) + (V_t)_{3+j,3+l}(V_t)_{j,l}n_l\right], \tag{3.28}
$$

and

$$
\begin{aligned}
P^{(b)}_j = 2\pi \sum_{k,l=1}^{3} \int_{\mathbb{R}} dt_1 \Big[& (V_{t-t_1})_{3+j,k}\,(V_{t-t_1})_{j,l}\,q^{(1)}_{k,l} \\
& + (V_{t-t_1})_{3+j,k}\,(V_{t-t_1})_{j,3+l}\,q^{(2)}_{k,l} \\
& + (V_{t-t_1})_{3+j,3+k}\,(V_{t-t_1})_{j,l}\,q^{(3)}_{k,l} + (V_{t-t_1})_{3+j,3+k}\,(V_{t-t_1})_{j,3+l}\,q^{(4)}_{k,l}\Big],
\end{aligned} \tag{3.29}
$$

for $j = 1, 2, 3$. Here, to keep the notation simple, we have introduced the following quantities:

$$
\begin{cases}
q_{1,1}^{(1)} = \frac{v_{12}^p v_{12}^{ap}}{\Omega_2} + \frac{v_{13}^p v_{13}^{ap}}{\Omega_3} \\[2mm]
q_{1,2}^{(1)} = \frac{1}{\Omega_1} \lambda_1 v_{21}^{ap}(1 - N_1) + \frac{1}{\Omega_2} \lambda_2 v_{12}^{ap} N_2 + \frac{1}{\Omega_3} \left(v_{13}^p v_{23}^{ap}(1 - N_3) + v_{13}^{ap} v_{23}^p N_3 \right) \\[2mm]
q_{1,3}^{(1)} = \frac{1}{\Omega_1} \lambda_1 v_{31}^{ap}(1 - N_1) + \frac{1}{\Omega_2} \left(v_{12}^p v_{32}^{ap}(1 - N_2) + v_{12}^{ap} v_{32}^p N_2 \right) + \frac{1}{\Omega_3} \lambda_3 v_{13}^{ap} N_3 \\[2mm]
q_{2,1}^{(1)} = \frac{1}{\Omega_1} \lambda_1 v_{21}^{ap} N_1 + \frac{1}{\Omega_2} \lambda_2 v_{12}^{ap}(1 - N_2) + \frac{1}{\Omega_3} \left(v_{13}^p v_{23}^{ap} N_3 + v_{13}^{ap} v_{23}^p (1 - N_3) \right) \\[2mm]
q_{2,2}^{(1)} = \frac{v_{21}^p v_{21}^{ap}}{\Omega_1} + \frac{v_{23}^p v_{23}^{ap}}{\Omega_3} \\[2mm]
q_{2,3}^{(1)} = \frac{1}{\Omega_1} \left(v_{21}^p v_{31}^{ap}(1 - N_1) + v_{21}^{ap} v_{31}^p N_1 \right) + \frac{1}{\Omega_2} \lambda_2 v_{32}^{ap}(1 - N_2) + \frac{1}{\Omega_3} \lambda_3 v_{23}^{ap} N_3 \\[2mm]
q_{3,1}^{(1)} = \frac{1}{\Omega_1} \lambda_1 v_{31}^{ap} N_1 + \frac{1}{\Omega_2} \left(v_{12}^p v_{32}^{ap} N_2 + v_{12}^{ap} v_{32}^p (1 - N_2) \right) + \frac{1}{\Omega_3} \lambda_3 v_{13}^{ap}(1 - N_3) \\[2mm]
q_{3,2}^{(1)} = \frac{1}{\Omega_1} \left(v_{21}^p v_{31}^{ap} N_1 + v_{21}^{ap} v_{31}^p (1 - N_1) \right) + \frac{1}{\Omega_2} \lambda_2 v_{32}^{ap} N_2 + \frac{1}{\Omega_3} \lambda_3 v_{23}^{ap}(1 - N_3) \\[2mm]
q_{3,3}^{(1)} = \frac{v_{31}^p v_{31}^{ap}}{\Omega_1} + \frac{v_{32}^p v_{32}^{ap}}{\Omega_2},
\end{cases}
$$

and

$$
\begin{cases}
q_{1,1}^{(2)} = \frac{1}{\Omega} \tilde{\lambda}_1^2 (1 - N) + \frac{1}{\Omega_1} \lambda_1^2 (1 - N_1) + \frac{1}{\Omega_2} \left((v_{12}^p)^2 (1 - N_2) + (v_{12}^{ap})^2 N_2 \right) \\[2mm]
\qquad + \frac{1}{\Omega_3} \left(\left(v_{13}^p \right)^2 (1 - N_3) + \left(v_{13}^{ap} \right)^2 N_3 \right) \\[2mm]
q_{1,2}^{(2)} = q_{2,1}^{(2)} = \frac{1}{\Omega} \tilde{\lambda}_1 \tilde{\lambda}_2 (1 - N) + \frac{1}{\Omega_1} \lambda_1 v_{21}^p (1 - N_1) + \frac{1}{\Omega_2} \lambda_2 v_{12}^p (1 - N_2) \\[2mm]
\qquad + \frac{1}{\Omega_3} \left(v_{13}^p v_{23}^p (1 - N_3) + v_{13}^{ap} v_{23}^{ap} N_3 \right) \\[2mm]
q_{1,3}^{(2)} = q_{3,1}^{(2)} = \frac{1}{\Omega} \tilde{\lambda}_1 \tilde{\lambda}_3 (1 - N) + \frac{1}{\Omega_1} \lambda_1 v_{31}^p (1 - N_1) \\[2mm]
\qquad + \frac{1}{\Omega_2} \left(v_{12}^p v_{32}^p (1 - N_2) + v_{12}^{ap} v_{32}^{ap} N_2 \right) + \frac{1}{\Omega_3} \lambda_3 v_{13}^p v_{23}^p (1 - N_3) \\[2mm]
q_{2,2}^{(2)} = \frac{1}{\Omega} \tilde{\lambda}_2^2 (1 - N) + \frac{1}{\Omega_1} \left(\left(v_{21}^p \right)^2 (1 - N_1) + \left(v_{21}^{ap} \right)^2 N_1 \right) + \frac{1}{\Omega_2} \lambda_2^2 (1 - N_2) \\[2mm]
\qquad + \frac{1}{\Omega_3} \left(\left(v_{23}^p \right)^2 (1 - N_3) + \left(v_{23}^{ap} \right)^2 N_3 \right) \\[2mm]
q_{2,3}^{(2)} = q_{3,2}^{(2)} = \frac{1}{\Omega} \tilde{\lambda}_2 \tilde{\lambda}_3 (1 - N) + \frac{1}{\Omega_1} \left(v_{21}^p v_{31}^p (1 - N_1) + v_{21}^{ap} v_{31}^{ap} N_1 \right) \\[2mm]
\qquad + \frac{1}{\Omega_2} \lambda_2 v_{32}^p (1 - N_2) + \frac{1}{\Omega_3} \lambda_3 v_{13}^p v_{23}^p (1 - N_3) \\[2mm]
q_{3,3}^{(2)} = \frac{1}{\Omega} \tilde{\lambda}_3^2 (1 - N) + \frac{1}{\Omega_1} \left(\left(v_{31}^p \right)^2 (1 - N_1) + \left(v_{31}^{ap} \right)^2 N_1 \right) \\[2mm]
\qquad + \frac{1}{\Omega_2} \left(\left(v_{32}^p \right)^2 (1 - N_2) + \left(v_{32}^{ap} \right)^2 N_2 \right) + \frac{1}{\Omega_3} \lambda_3^2 (1 - N_3).
\end{cases}
$$

The $q_{k,l}^{(3)}$'s are not listed here, since they can be deduced by the $q_{k,l}^{(2)}$'s simply by replacing $1 - N$ with N, N with $1 - N$, $1 - N_j$ with N_j and N_j with $1 - N_j$. Hence, for instance, we have

$$q_{1,1}^{(3)} = \frac{1}{\Omega}\tilde{\lambda}_1^2 N + \frac{1}{\Omega_1}\lambda_1^2 N_1 + \frac{1}{\Omega_2}\left(\left(v_{12}^p\right)^2 N_2 + \left(v_{12}^{ap}\right)^2 (1 - N_2)\right)$$

$$+ \frac{1}{\Omega_3}\left(\left(v_{13}^p\right)^2 N_3 + \left(v_{13}^{ap}\right)^2 (1 - N_3)\right)$$

and so on. Moreover, simple parity reasons allow us to conclude that $q_{k,l}^{(4)} = q_{k,l}^{(1)}$, for each k and l.

As in Section 3.2, the reason why we have written $P_j(t)$ as a sum of different contributions, $P_j^{(a)}$ and $P_j^{(b)}$, is because, in Equations (3.28) and (3.29), $P_j^{(a)}(t)$ contains all that refers to the parties, while the coefficients appearing in the definition of $P_j^{(b)}(t)$ refer only to the electors. This is exactly what we have done in Equation (3.17). Notice also that, as expected, Equation (3.27) returns the solution deduced in Section 3.2 when all the v_{kl}^p and v_{kl}^{ap} are zero, since in this case the two models coincide. Of course we are now in a position to compute the various decision functions for all interesting choices of the parameters and of the initial conditions for the parties and for the electors. A complete analysis of the various scenarios is not really useful in the present context and will not be considered. Here we restrict our analysis to a single particular case, fixing particular values of the parameters and considering several initial conditions, and plotting the functions $P_j(t)$ deduced in this way. More examples can be found in [13].

> *Remark:* It is clear that in a model like this, where so many parameters are required in the definition of the Hamiltonian,[5] a detailed sensitivity analysis would be too long, and not always meaningful or necessary; in fact, in our particular case, since the equations of motion that drive the time evolution of the system are linear, we expect that small changes in the parameters (or in the initial conditions) can cause only small changes in the decision functions. Hence, there is no reason to analyze similar small variations. We will discuss more on the parameters of the Hamiltonians throughout this book.

3.3.1 An Explicit Example

The situation we will consider here is the first natural extension of the system described in Figure 3.1. In particular, we will now consider a system with all the $v_{k,l}^p$ and $v_{k,l}^{ap}$ in H_{mix} equal to zero except for a single one, v_{12}^p, which we take different from zero. Then we are considering the possibility that $\mathcal{P}_1$ interacts with $\mathcal{E}_2$ but not

---

[5] This is unavoidable, since different terms correspond to different interactions, which can have quite different strengths. Hence, they must be proportional to different coupling constants.

with $\mathcal{E}_3$. Moreover, other *mixed* interactions are also forbidden, here. To simplify the treatment further, we will also take $\tilde{\lambda}_2 = \tilde{\lambda}_3 = 0$ and $\mu_{k,l}^{coop} = 0$ for all k and l. With these choices the only nonzero $q_{k,l}^{(j)}$ in Equation (3.29) are the following:

$$
\begin{cases}
q_{1,1}^{(2)} = \frac{1}{\Omega}\tilde{\lambda}_1^2(1-N) + \frac{1}{\Omega_1}\lambda_1^2(1-N_1) + \frac{1}{\Omega_2}\left(v_{12}^p\right)^2(1-N_2) \\
q_{1,2}^{(2)} = q_{2,1}^{(2)} = \frac{1}{\Omega_2}\lambda_2 v_{12}^p(1-N_2) \\
q_{2,2}^{(2)} = \frac{1}{\Omega_2}\lambda_2^2(1-N_2) \\
q_{1,1}^{(3)} = \frac{1}{\Omega}\tilde{\lambda}_1^2 N + \frac{1}{\Omega_1}\lambda_1^2 N_1 + \frac{1}{\Omega_2}\left(v_{12}^p\right)^2 N_2 \\
q_{2,2}^{(3)} = \frac{1}{\Omega_2}\lambda_2^2 N_2 \\
q_{3,3}^{(3)} = \frac{1}{\Omega_3}\lambda_3^2 N_3 \\
q_{1,2}^{(3)} = q_{2,1}^{(3)} = \frac{1}{\Omega_2}\lambda_2 v_{12}^p N_2
\end{cases}
$$

The matrix elements of U are also easily found: $y_{k,l} = 0$ for all k and l, while

$$
\begin{cases}
x_{1,1} = -i\omega_1 - \frac{\pi}{\Omega}\tilde{\lambda}_1^2 - \frac{\pi}{\Omega_1}\lambda_1^2 - \frac{\pi}{\Omega_2}\left(v_{12}^p\right)^2 \\
x_{1,2} = -i\mu_{12}^{ex} - \frac{\pi}{\Omega_2}\lambda_2 v_{12}^p \\
x_{1,3} = -i\mu_{13}^{ex} \\
x_{2,2} = -i\omega_2 - \frac{\pi}{\Omega_2}\lambda_2^2 \\
x_{2,3} = -i\mu_{23}^{ex} \\
x_{3,3} = -i\omega_3 - \frac{\pi}{\Omega_3}\lambda_3^2.
\end{cases}
$$

The functions $P_j^{(a)}$ and $P_j^{(b)}$ in (3.28) and (3.29), and their sums $P_j(t)$, now can be easily computed, and the decision functions are plotted in Figures 3.10–3.12 for particular values of the parameters and three different choices of the initial conditions.

More explicitly, we plot $P_1(t)$, $P_2(t)$ and $P_3(t)$ for the same choice of the parameters: $\mu_{1,2}^{ex} = 0.1$, $\mu_{1,3}^{ex} = 0.2$, $\mu_{2,3}^{ex} = 0.08$ $\mu_{k,l}^{coop} = 0$, for all k,l, $\omega_1 = \Omega_1 = \Omega = 0.1$, $\omega_2 = \omega_3 = \Omega_2 = 0.2$, $\lambda_1 = 0.1$, $\lambda_2 = 0.2$, $\lambda_3 = 0.05$, $\tilde{\lambda}_1 = 0.1$, $\tilde{\lambda}_2 = \tilde{\lambda}_3 = 0$. The three figures differ for the choice of the initial conditions on the parties and the reservoirs. In Figure 3.10 we have taken $n_1 = N_1 = 0$, $n_2 = n_3 = N_2 = N_3 = N = 1$, while in Figure 3.11 we have $n_3 = N_1 = N_2 = 0$, $n_1 = n_2 = N_3 = N = 1$ and in Figure 3.12 $n_3 = N_2 = 0$, $n_1 = n_2 = N_1 = N_3 = N = 1$. The different plots inside each picture correspond to different values of v_{12}^p. In particular we have $v_{12}^p = 0$ for the dotted lines, $v_{12}^p = 0.1$ for the small dashed lines and $v_{12}^p = 0.5$ for the large dashed lines. It is

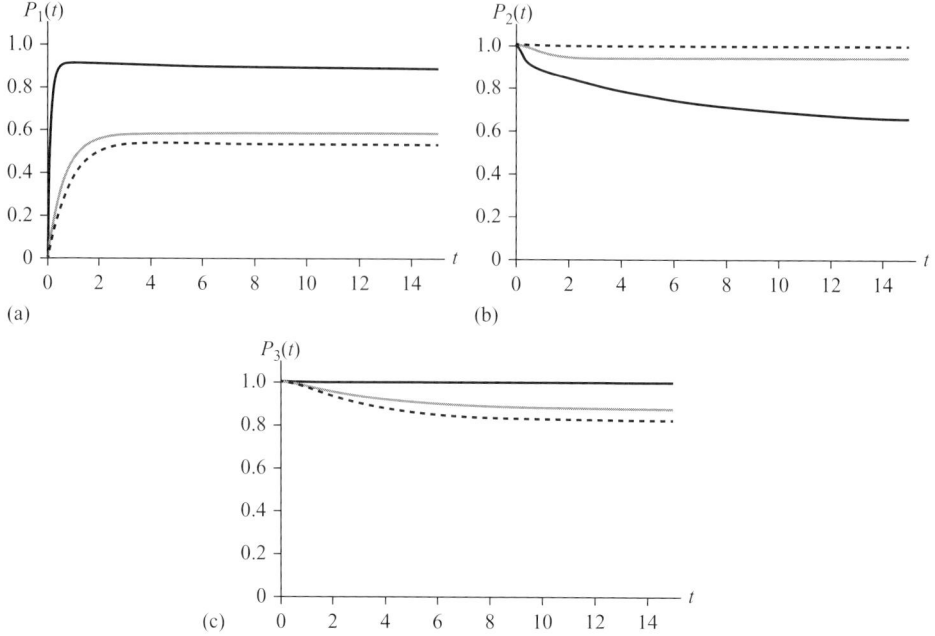

(a)   (b)

(c)

Figure 3.10 $P_1(t)$ (a), $P_2(t)$ (b) and $P_3(t)$ (c) for $\mu_{1,2}^{ex} = 0.1$, $\mu_{1,3}^{ex} = 0.2$, $\mu_{2,3}^{ex} = 0.08$ $\mu_{k,l}^{coop} = 0$, $\omega_1 = \Omega_1 = \Omega = 0.1$, $\omega_2 = \omega_3 = \Omega_2 = 0.2$, $\lambda_1 = 0.1$, $\lambda_2 = 0.2$, $\lambda_3 = 0.05$, $\tilde{\lambda}_1 = 0.1$, $\tilde{\lambda}_2 = \tilde{\lambda}_3 = 0$, and $n_1 = N_1 = 0$, $n_2 = n_3 = N_2 = N_3 = N = 1$ (Source: Bagarello [68]).

evident that $P_3(t)$ is not deeply affected from the presence of this new interaction between $\mathcal{P}_1$ and $\mathcal{E}_2$. This is not surprising, since P_3 is not directly involved in the new term in the Hamiltonian. However, since $\mathcal{P}_3$ interacts with $\mathcal{P}_1$, and since $\mathcal{P}_1$ can also interact with $\mathcal{E}_2$ when $\nu_{12}^p \neq 0$, some minor changes are expected and, in fact, this is what we observe in the plots given here for $P_3(t)$, for all choices of initial conditions.

Much more evident are the changes in $P_1(t)$ and $P_2(t)$, since $\mathcal{P}_1$ and $\mathcal{P}_2$ are directly involved by the new term in h. In particular we see in these figures that the higher the value of ν_{12}^p, the higher the tendency of $P_1(t)$ to approach the value N_2, which describes the initial status of $\mathcal{R}_2$. Of course, this is due also to the fact that the magnitude of the interaction between $\mathcal{P}_1$ and $\mathcal{E}_2$, when increasing ν_{12}^p, may become more relevant than the interaction between $\mathcal{P}_1$ and its own electors, those in $\mathcal{E}_1$. In other words, $\mathcal{P}_1$, for high values of ν_{12}^p, is more interested in the opinion of the electors in $\mathcal{E}_2$ than in that of its own electors! Hence, we recover once again the same mechanism observed before: The interactions between $\mathcal{P}_j$ and $\mathcal{E}_j$, $\mathcal{E}_k$ ($k \neq j$) and $\mathcal{E}_{und}$ produce similar effects on $P_j(t)$, but the strengths of these interactions

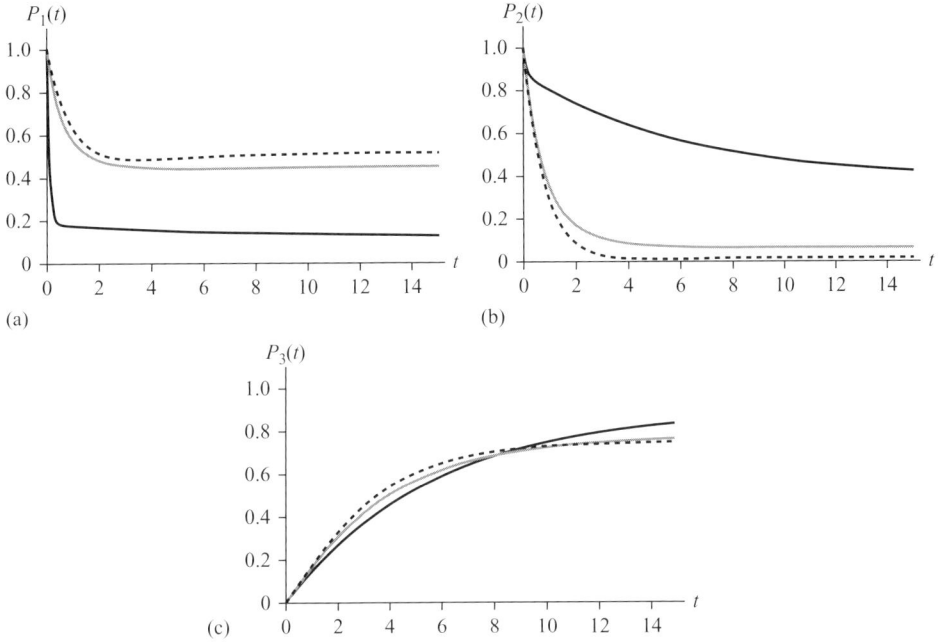

Figure 3.11 $P_1(t)$ (a), $P_2(t)$ (b) and $P_3(t)$ (c) for $\mu_{1,2}^{ex} = 0.1$, $\mu_{1,3}^{ex} = 0.2$, $\mu_{2,3}^{ex} = 0.08$ $\mu_{k,l}^{coop} = 0$, $\omega_1 = \Omega_1 = \Omega = 0.1$, $\omega_2 = \omega_3 = \Omega_2 = 0.2$, $\lambda_1 = 0.1$, $\lambda_2 = 0.2$, $\lambda_3 = 0.05$, $\tilde{\lambda}_1 = 0.1$, $\tilde{\lambda}_2 = \tilde{\lambda}_3 = 0$, and $n_3 = N_1 = N_2 = 0$, $n_1 = n_2 = N_3 = N = 1$ (Source: Bagarello [68]).

decide which one is the most relevant for the time evolution of $P_j(t)$ and for its limiting value $P_j(\infty)$ in particular.

We refer the curious reader to [13] for a detailed analysis of several different choices of the parameters. One will see that the results in [13] confirm our conclusions here.

3.4 Extending the Model, Part 2: Adding a *Rule* ρ

In this section we show how to enrich the model described in Section 3.3 by considering a suitable rule, and its related (H, ρ)-induced dynamics, in the sense of Section 2.8. In other words, what is interesting for us now is to include in the dynamics of the system some extra effect, not easily described in terms of any self-adjoint Hamiltonian, and discuss the consequences of this effect.

As a first step we fix a positive time $\tau > 0$, and we imagine that, at each instant of time $k\tau$, $k = 1, 2, 3, \dots$, some kind of information reaches the electors. They, as a consequence of this input, can modify their opinions on whether the parties should form an alliance. More explicitly, we suppose that the values N_1, N_2 and N_3 in

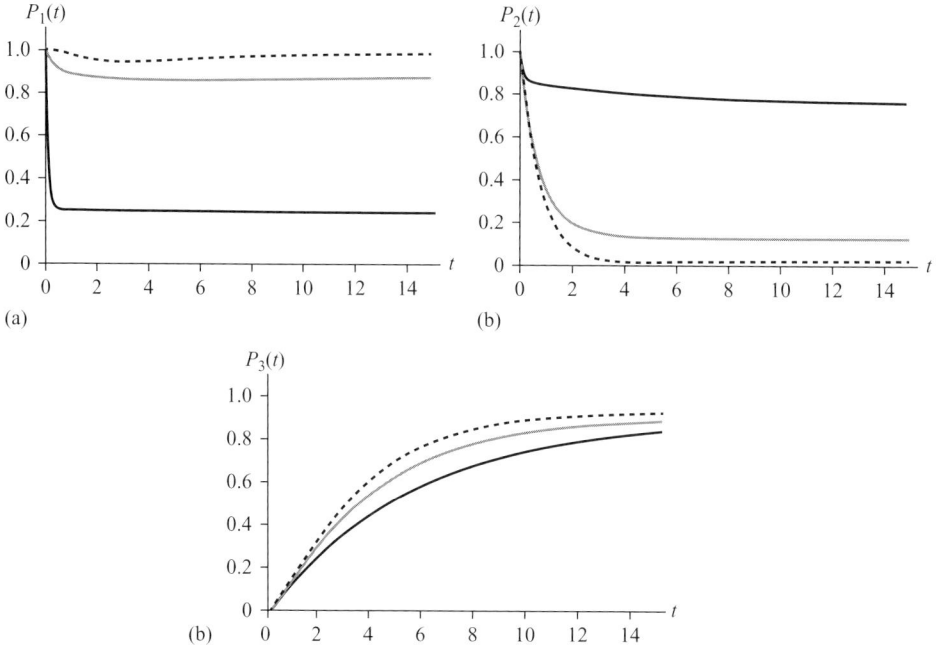

Figure 3.12 $P_1(t)$ (a), $P_2(t)$ (b) and $P_3(t)$ (c) for $\mu_{1,2}^{ex} = 0.1$, $\mu_{1,3}^{ex} = 0.2$, $\mu_{2,3}^{ex} = 0.08$ $\mu_{k,l}^{coop} = 0$, $\omega_1 = \Omega_1 = \Omega = 0.1$, $\omega_2 = \omega_3 = \Omega_2 = 0.2$, $\lambda_1 = 0.1$, $\lambda_2 = 0.2$, $\lambda_3 = 0.05$, $\tilde{\lambda}_1 = 0.1$, $\tilde{\lambda}_2 = \tilde{\lambda}_3 = 0$, and $n_3 = N_2 = 0$, $n_1 = n_2 = N_1 = N_3 = N = 1$ (Source: Bagarello [68]).

Equation (3.15), and possibly N in Equation (3.16), can be modified after each time interval of length τ, because of this (public or private) information.[6] Then, with these new values, the time evolution of the system starts again for a second time interval of length τ, when the rule is applied again – and so on. This is, in synthesis, our rule. Of course we have to clarify how N_j and N are modified, and this will be done later. We notice that this is exactly what we have discussed, at a more abstract level, in Section 2.8. However, while in that section the rule was applied to a vector Ψ of the system, here it is applied to the state of the environment, ω, keeping unchanged the parameters of the Hamiltonian. We will make the rule more detailed later in this section.

The initial conditions for the decision functions are the following: $P_{j,0}(0) = n_{j,0}$, while the initial conditions for the electors are given by $(N_{j,[0]}, N_{[0]})$. These numbers, in particular, should be understood as the ones in Equation (3.15) and in Equation (3.16) before any rule is applied, which is reflected by the presence of

---

[6] Information is an important aspect in several dynamical systems. It is used, for instance, to perform optimal choices. We will say more on the role of information in, for example, Chapters 10 and 11.

"0" in $P_{j,0}(0)$ and of [0] in $N_{j,[0]}$ and $N_{[0]}$. Hence, we are considering the possibility that both $\mathcal{E}_j$ and $\mathcal{E}_{und}$ are modified by the rule.[7] However, as we will see, the concrete situation described by our rule in Equations (3.31) and (3.32) below modifies $\mathcal{E}_j$, while $\mathcal{E}_{und}$ is left unchanged.

During the time interval $[0, \tau[$ the time evolution of each decision function $P_{j,0}(t)$ is deduced, given the Hamiltonian (3.23), through Equation (3.27). We stress once more that the two subscripts which we are using to label the decision functions refer to the three parties ($j = 1, 2, 3$) and to the number of iteration. For instance, $P_{1,0}(t)$ is the decision function of $\mathcal{P}_1$ during the zeroth iteration, i.e. when the rule has not yet acted. With $P_{2,1}(t)$ we indicate the decision function of $\mathcal{P}_2$ after a single action of the rule. At $t = \tau$ the values $P_{j,0}(\tau)$ are taken as input for a set of rule ρ (specified later), together with the original values of the parameters of the reservoir, ($N_{j,[0]}, N_{[0]}$). The rule returns as output the new values ($N_{j,[1]}, N_{[1]}$), in general different from ($N_{j,[0]}, N_{[0]}$), for the reservoirs ($\mathcal{E}_j, \mathcal{E}_{und}$). Summarizing, the rule ρ checks the values of the $P_{j,0}(\tau)$, and changes consequently the parameters of the reservoirs from ($N_{j,[0]}, N_{[0]}$) to ($N_{j,[1]}, N_{[1]}$).[8] This is because, for instance, at time τ, the electors may react in different ways depending on whether they are satisfied or not with what the parties are doing, and in particular with the (temporary) decision the parties have taken on the possibility of giving rise to some coalition. Then we start a new iteration with the initial conditions $P_{j,1}(0) = P_{j,0}(\tau)$ for the decision functions, and with $\left(N_{j,[1]}, N_{[1]}\right)$ for the parameters of the reservoirs, and the evolution of $P_{j,1}(t)$ is ruled, for a second time interval of length τ, again by the Hamiltonian (3.23). Then ρ is applied a second time, and the rule checks the values $P_{j,1}(\tau)$, and $\left(N_{j,[2]}, N_{[2]}\right)$ are deduced out of them and of the quantities $\left(N_{j,[1]}, N_{[1]}\right)$. This process is iterated M times so that, at the end, we obtain a (finite) sequence of decision functions $P_{j,k}(t), j = 1, 2, 3, k = 0 \ldots M - 1, t \in [0, \tau]$ that can be used to reconstruct the modified decision functions $P_j^{(\rho)}(t), t \in [0, M\tau[$ by merging all these functions as follows:

$$P_j^{(\rho)}(t) = \sum_{k=0}^{M-1} \chi_k(t) P_{j,k}(t - k\tau), \quad j = 1, 2, 3. \tag{3.30}$$

Here $\chi_k(t)$ is the characteristic function defined as follows: $\chi_k(t) = 1$ if $t \in [k\tau, (k + 1)\tau[$, and $\chi_k(t) = 0$ otherwise. Of course, $P_j^{(\rho)}(t)$ is, in general, different from the original function $P_j(t)$, and it is strongly dependent on the explicit form of the rule we are adopting. We will propose in the next section one possibility, and we will describe in some details the meaning of that rule and its consequences.

---

[7] In fact, what is modified by the rule are the numbers N_j and N in Equations (3.15) and (3.16). These are particularly simple choices of the functions $N_j(k)$ and $N(k)$ already adopted in Sections 3.2 and 3.3. But this concretely means that some of the quantities that in practice define the reservoirs are modified. This is why we write that "both $\mathcal{E}_j$ and $\mathcal{E}_{und}$ are modified by the rule."

[8] In fact, as we have already observed, we will see that $N_{[1]} = N_{[0]}$.

Of course, other possibilities could be considered, but the one we will introduce describes some of the feature we want to insert in the model, and for this reason it is interesting to us.

3.4.1 The Rule: An Example

Of course, the key ingredient for the whole procedure outlined above is the concrete definition of the rule ρ. In what follows, ρ checks the values of the decision functions at the end of some time integral of length τ and, using also the old values of the reservoirs $N_{j,[k]}$ and $N_{[k]}$, produces their new values, $N_{j,[k+1]}$ and $N_{[k+1]}$, which are further taken to again start the Heisenberg time evolution of the decision functions. Hence ρ is a map between $\left(P_{j,k}(\tau), P_{j,k}(0), N_{j,[k]}, N_{[k]}\right)$ and $\left(N_{j,[k+1]}, N_{[k+1]}\right)$.

As we have already discussed in Section 2.8, this procedure is certainly not unique. We describe here just one possible choice, originally proposed in [13]. Other rules applied to politics have been discussed by different authors recently [59, 60]. The results contained there are particularly relevant. The authors are able to explain, by adopting a suitable rule, some experimental data related to people in the Italian parliament who, elected with one party, modified their position and moved to a different party during the same legislature. This is a first clear indication that the approach proposed in this book works well, at least in some realistic situations. More examples of models in good agreement with real data will be discussed later, and in Chapter 9 in particular.

The rule we adopt here is based on the following formula:

$$N_{j,[k+1]} = \rho\left(N_{j,[k]}, P_{j,k}(\tau), P_{j,k}(0)\right) = N_{j,[k]} + \delta N_{j,[k]} \tag{3.31}$$

where

$$\delta N_{j,[k]} = \begin{cases} \left(1 - P_{j,k}(\tau)\right)^s \left(1 - N_{j,[k]}\right)^{\frac{1-\delta_{j,k}}{1+|\delta_{j,k}|}}, & \text{if } N_{j,[k]} \geq 0.5 \\ -\left(P_{j,k}(\tau)\right)^s N_{j,[k]}^{\frac{1+\delta_{j,k}}{1+|\delta_{j,k}|}}, & \text{if } N_{j,[k]} < 0.5. \end{cases} \tag{3.32}$$

Here $\delta_{j,k} = P_{j,k}(\tau) - P_{j,k}(0)$ measures the change of the decision function in a time interval, and s is a nonnegative parameter that magnifies (or not) the relevance of the value of $P_{j,k}(\tau)$ in our computations. The smaller the value of s, the higher the values of both $\left(P_{j,k}(\tau)\right)^s$ and $\left(1 - P_{j,k}(\tau)\right)^s$, which can, at most, be equal to one. Equation (3.31) describes how the values of the $N_{j,[k]}$ are modified at each iteration and shows, in particular, that the value of $N_{[k]}$ for $\mathcal{E}_{und}$ is not affected by ρ. This is clear since $N_{[k]}$ does not appear in the formula at all, and this is equivalent to saying that $N_{[k+1]} = N_{[k]}$, for all k.

A consequence of our rule is that if $N_{j,[k]} \in [0, 1]$, then $N_{j,[k+1]} \in [0, 1]$ as well. So the parameters of $\mathcal{E}_j$ stay always in the range $[0, 1]$, as we expect because of the fermionic nature of the operators of the reservoir. This claim follows from some easy estimates. First we notice that, in any possible situation, $\frac{1-\delta_{j,k}}{1+|\delta_{j,k}|}$, $\left(1 - P_{j,k}(\tau)\right)^s$ and $\left(P_{j,k}(\tau)\right)^s$ all belong to the interval $[0, 1]$. Then, if $N_{j,[k]} \geq 0.5$, from Equation (3.31) and (3.32), we get

$$N_{j,[k+1]} \leq N_{j,[k]} + \left(1 - N_{j,[k]}\right) = 1,$$

while $N_{j,[k+1]} \geq N_{j,[k]} \geq 0$. Analogously, if $N_{j,[k]} < 0.5$, it turns out that $N_{j,[k+1]} \leq N_{j,[k]} \leq 1$, while

$$N_{j,[k+1]} = N_{j,[k]} \left(1 - \left(P_{j,k}(\tau)\right)^s \frac{1+\delta_{j,k}}{1+|\delta_{j,k}|}\right) \geq 0,$$

since $(P_{j,k}(\tau))^s \frac{1+\delta_{j,k}}{1+|\delta_{j,k}|} \leq 1$ for all possible values of $\delta_{j,k}$.

Now, to understand further the meaning of this rule, let us imagine that, at a certain iteration, the electors in $\mathcal{E}_j$ are satisfied with what $\mathcal{P}_j$ is doing. Then, in view of what we have discussed previously, the difference between $N_{j,[k]}$ and $P_{j,k}(\tau)$ is expected to be small (or even zero) and, since the electors are satisfied by $\mathcal{P}_j$, $N_{j,[k+1]}$, is not expected to be particularly different from $N_{j,[k]}$. The iteration should not really modify the mood of the electors, in this case. This is, in fact, what Equation (3.32) expresses. To show this aspect of the rule, we consider four different situations: (*i*) $N_{j,[k]} \simeq 1$ and $P_{j,k}(\tau) \simeq 1$; (*ii*) $N_{j,[k]} \simeq 0$ and $P_{j,k}(\tau) \simeq 0$; (*iii*) $N_{j,[k]} = 1$ and (*iv*) $N_{j,[k]} = 0$. Notice that in cases (*iii*) and (*iv*) we are not saying anything about $P_{j,k}(\tau)$. We will discuss why in a moment.

If we are in case (*i*), (3.32) implies that

$$\delta N_{j,[k]} = \left(1 - P_{j,k}(\tau)\right)^s \left(1 - N_{j,[k]}\right) \frac{1-\delta_{j,k}}{1+|\delta_{j,k}|},$$

which is the product of two quantities that are very small, $\left(1 - P_{j,k}(\tau)\right)^s$ and $\left(1 - N_{j,[k]}\right)$, and a third quantity that is less than or equal to one. Hence, $\delta N_{j,[k]}$ is positive and small (or even very small). Then $N_{j,[k+1]}$ is just a little bit larger than $N_{j,[k]}$ but always less than or equal to one, as we have shown before. This shows that $\mathcal{E}_j$ does not really change its state; it only strengthen a little bit its *alliance message* to the party.

If we are in case (*ii*), then

$$\delta N_{j,[k]} = - \left(P_{j,k}(\tau)\right)^s N_{j,[k]} \frac{1+\delta_{j,k}}{1+|\delta_{j,k}|},$$

and again the right-hand side of this equation is small, in absolute value, but negative. Hence $N_{j,[k+1]}$ is just a little bit smaller than $N_{j,[k]}$ but still greater than or equal to zero. This shows that, even in this case, $\mathcal{E}_j$ does not modify particularly its state; it strengthens only a little bit its *non-alliance message*.

The case (*iii*) can be understood in a slightly different way: Since $N_{j,[k]} = 1$, $\mathcal{E}_j$ is already using all its strength to communicate to the party they have to form some coalition. So it is not possible to make this message stronger, despite what $\mathcal{P}_j$ is doing. And, in fact, this is reflected in Equation (3.32), which returns $\delta N_{j,[k]} = 0$, so that $N_{j,[k+1]} = N_{j,[k]} = 1$.

Finally, case (*iv*) works in a similar way: $\mathcal{E}_j$ is doing its best to suggest to $\mathcal{P}_j$ not to form any coalition. So, again, it is not possible to make this message stronger, despite what $\mathcal{P}_j$ is really doing. And, in fact, we get $\delta N_{j,[k]} = 0$ also in this case. Hence, $N_{j,[k+1]} = N_{j,[k]} = 0$.

We end the analysis (and the justification) of our expression of ρ by considering what happens for intermediate values of $N_{j,[k]}$ and how, in this case, $\delta N_{j,[k]}$ is also connected with the temporary decision of $\mathcal{P}_j$. Just to fix the ideas, let us suppose that $N_{j,[k]} \simeq 0.6$. Then, $\mathcal{E}_j$ is gently suggesting that $\mathcal{P}_j$ should ally. Using Equation (3.32), we have

$$\delta N_{j,[k]} = 0.4 \left(1 - P_{j,k}(\tau) \right)^s \frac{1 - \delta_{j,k}}{1 + |\delta_{j,k}|},$$

which of course depends on how far $P_{j,k}(\tau)$ is from one and on the value of s. In particular, if $P_{j,k}(\tau)$ is small (close to zero, for instance), $\mathcal{P}_j$ is not trying to ally at all! Then, especially for small values of s, $\delta N_{j,[k]}$ can be large (and positive). The result is that $N_{j,[k+1]}$ is significantly larger than $N_{j,[k]}$, and this reflects the fact that the electors in $\mathcal{E}_j$ are not satisfied with $\mathcal{P}_j$, so that they repeat their suggestion but not as gently as they did in the previous iteration.

As already stated, other possible general ideas could be assumed, maybe even more reasonable than those assumed here. For instance, we could also consider some effect of the rule on the undecided electors, changing the value of $N_{[k]}$ at each iteration. In fact, a huge set of possible choices could be considered. This is due to the freedom that is intrinsic to the (H, ρ)-induced dynamics. The only constraints to be really considered are given by the (technical) possibility of implementing these rules and by the obvious fact that they should be connected to the real phenomena occurring in our system. Just to cite a different possibility, another rule could modify, at each step τ, the parameters of the Hamiltonian h in Equation (3.23) according to some external factors such as opinion polls, international political facts, economical crises and so on. The possibility of having rules of this kind has been discussed in Section 2.8.2. Some results in this direction can be found in [59] and in [60].

3.4.2 Consequences of the Rule

Let us now consider a concrete application of the rule in Equations (3.31) and (3.32), looking for the time dependence of the various decision functions for four iterations. Hence, $M = 3$. Moreover, in this section the parameter s is fixed to be $\frac{1}{4}$, while the time interval of each iteration is taken to be $\tau = 25$.

To make the situation as simple as possible, we discuss here an application in which the various parties interact only with their electors and not with the other parties or with the other electors. This case, as we have seen in Section 3.2, corresponds to the following choice of the parameters $\mu_{k,l}^{ex} = \mu_{k,l}^{coop} = \tilde{\lambda}_k = \nu_{k,l}^{p} = \nu_{k,l}^{ap} = 0$, for all k and l. This choice is also useful, since we have an analytic expression for the decision functions in absence of rule, in Equation (3.20). We have shown that, in this case, the matrix U is diagonal and the decision functions at each iteration have the following simple expression, which is deduced by adapting Equation (3.20) to the present situation:

$$P_{j,k}(t) = P_{j,k}(0)e^{-2\pi t \lambda_j^2 / \Omega_j} + N_{j,[k]}\left(1 - e^{-2\pi t \lambda_j^2 / \Omega_j}\right), \qquad (3.33)$$

for $j = 1, 2, 3$ and for all k. Of course, this equation implies that $P_{j,k}(t)$ would tend asymptotically to $N_{j,[k]}$, if $\lambda_j \neq 0$, with a convergence speed that depends on the value of the ratio λ_j^2 / Ω_j. However, it makes little sense to talk here of asymptotic convergence, since $t \leq \tau$ and $\tau = 25$, because of our choice! So what could be true but is not guaranteed at all is that $P_{j,k}(t)$ approaches some asymptotic value already for $t < \tau$. This depends, of course, on the explicit values of λ_j^2 and Ω_j and on their ratio. However, even when this does not happen, the function $P_j^{(\rho)}(t)$ in Equation (3.30) can admit some limiting values, although usually some iterations are needed.

The decision functions in Equation (3.30), deduced after the four iterations, are shown in Figure 3.13, where we have taken the initial conditions to be $n_{1,0} = 0.2, n_{2,0} = 0.8, n_{3,0} = 0.5$, $N_{1,[0]} = 0.1, N_{2,[0]} = 0.6, N_{3,[0]} = 1, N_{[0]} = 1$, and we have considered the following choice of the other parameters: $\omega_1 = \omega_2 = 0.1, \omega_3 = 0.2$, $\Omega_1 = \Omega_2 = \Omega_3 = \Omega = 0.1, \lambda_1 = 0.08, \lambda_2 = \lambda_3 = 0.02$. This means that, at $t = 0$, $\mathcal{P}_1$ is not particularly inclined to ally ($n_{1,0} = 0.2$), and its electors, those in $\mathcal{E}_1$, share the same mood, since $N_{1,[0]} = 0.1$ is also rather small. Hence, they will be satisfied if $\mathcal{P}_1$ does not form any alliance. Actually, in absence of any rule, Equation (3.33) predicts that $P_{1,0}(t)$ tends to 0.1 when $t \to \infty$. But this is not what we have to do, since at $t = \tau = 25$ we have to apply the rule in Equations (3.31) and (3.32). In this way we find a new value $N_{1,[1]}$, which is 0.0181, quite lower than the previous value $N_{1,[0]} = 0.1$. This means that the electors $\mathcal{E}_1$ are now suggesting that $\mathcal{P}_1$ not ally in a stronger way than before. At the end of the second iteration, at $t = 50$, $P_1^{(\rho)}(50) = P_{1,1}(25) = 0.018$, and by applying the rule we obtain $N_{1,[2]} = 0.00275$, still quite lower than the previous value $N_{1,[1]}$. By continuing this procedure, at the end of the fourth iteration the value of the decision function for $\mathcal{P}_1$ is $P_1^{(\rho)}(100) = 0.0000835$, which is much smaller than 0.1, the value we should get in absence of the rule. So, we see that the party is really reacting as the electors are

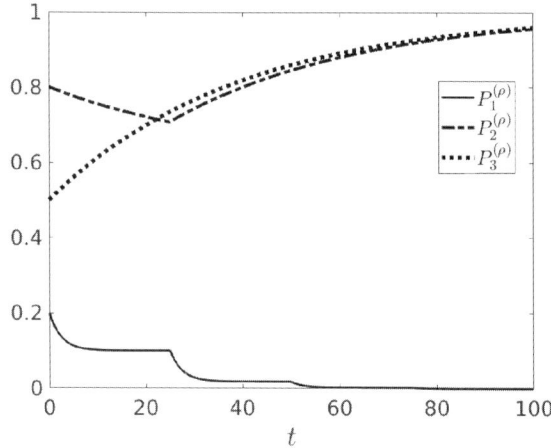

Figure 3.13 Initial conditions are $n_{1,0} = 0.2, n_{2,0} = 0.8, n_{3,0} = 0.5$, $N_{1,[0]} = 0.1, N_{2,[0]} = 0.6, N_{3,[0]} = 1, N_{[0]} = 1$. The parameters are $\omega_1 = \omega_2 = 0.1, \omega_3 = 0.2, \Omega_1 = \Omega_2 = \Omega_3 = \Omega = 0.1, \lambda_1 = 0.08, \lambda_2 = \lambda_3 = 0.02$. The other parameters are set to 0 (Source: Bagarello and Gargano [13]).

suggesting: $\mathcal{E}_1$ suggests to avoid alliances, and $\mathcal{P}_1$ behaves exactly as its electors require, as the value of $P_1^{(\rho)}(4\tau)$ reflects. We can then say that the rule in Equation (3.32) is really efficient and creates a very good feeling between $\mathcal{P}_1$ and $\mathcal{E}_1$.

Similar conclusions can be repeated for the decision functions of the other parties. Incidentally, Figure 3.13 shows that the decision functions are continuous but, in general, not differentiable in the points where the rule is applied. In fact, continuity of each $P_j(t)$ is clearly due to the method we have applied, a method that does not guarantee, without extra requirements, any higher regularity of these functions.

A second situation we want to describe here arises if we add the effect of the interactions between the parties and the other parties' electors, so that the parameters $\lambda_j, \tilde{\lambda}_j, \nu_{kl}^p, \nu_{kl}^{ap}$ can be different from 0. In particular, we put $\lambda_1 = 0.08, \lambda_2 = \lambda_3 = 0.02, \tilde{\lambda}_1 = 0.01$ and $\nu_{12}^p = \nu_{12}^{ap} = \nu_{13}^{ap} = 0.001$. The other parameters and the initial conditions are fixed as in the previous example.

These parameters show that $\mathcal{P}_1$ interacts with its own electors more strongly than it does with the others. Hence, we expect that $\mathcal{P}_1$ should decide to essentially follow what its electors in $\mathcal{E}_1$ are suggesting; that is, no alliance at all (recall that $N_{1,[0]} = 0.1$). However, the choice $\nu_{12}^p, \nu_{12}^{ap}, \nu_{13}^p \neq 0$ implies that $\mathcal{P}_1$ is also (but less!) influenced by the electors in $\mathcal{E}_2$ and $\mathcal{E}_3$. The results are shown in Figure 3.14. Although the various $N_{1,[k]}$ tend to zero ($N_{1,[1]} = 0.0053, N_{1,[2]} = 0.0008, N_{1,[3]} = 6 \cdot 10^{-6}$), the decision function of $\mathcal{P}_1$ seems to stabilize around the value 0.085. This is the effect of $\mathcal{E}_2$ and $\mathcal{E}_3$ on $P_1^{(\rho)}(t)$, which contrasts with the action of $\mathcal{E}_1$.

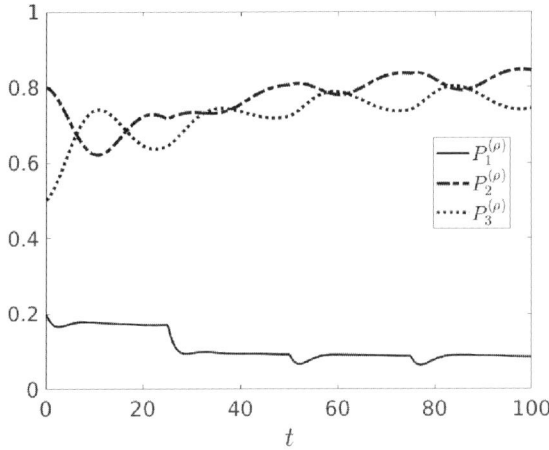

Figure 3.14 Initial conditions are $n_{1,0} = 0.2, n_{2,0} = 0.8, n_{3,0} = 0.5, \ N_{1,[0]} = 0.1,$ $N_{2,[0]} = 0.6, N_{3,[0]} = 1, N_{[0]} = 1.$ The parameters are $\lambda_1 = 0.08, \lambda_2 = \lambda_3 = 0.02,$ $\tilde{\lambda}_1 = 0.01$ and $\nu_{12}^p = \nu_{12}^{ap} = \nu_{13}^{ap} = 0.001.$ The other parameters are set to 0 (Source: Bagarello and Gargano [13]).

We see that, in this case, the decision is not as sharp as in absence of the (mixed) interactions between $\mathcal{P}_1$ and $\mathcal{E}_2$ and $\mathcal{E}_3$.

Conversely, the electors $\mathcal{E}_2$ and $\mathcal{E}_3$ suggest that $\mathcal{P}_2$ and $\mathcal{P}_3$ ally (even if with different strengths), since we have $N_{2,[0]} = 0.6$ and $N_{3,[0]} = 1$. Applying the rule, at the various iterations we obtain $N_{2,[k]} = N_{3,[k]} = 1$, $k = 1, 2, 3$, and the functions $P_2^{(\rho)}(t)$ and $P_3^{(\rho)}(t)$ increase to about 0.8, showing in this way a certain attitude of both these parties to form alliances, as their electors are suggesting. And these values are already reached very early, after the first iteration. Hence, $\mathcal{E}_1$ does not affect much $P_2^{(\rho)}$ and $P_3^{(\rho)}$.

More choices of parameters and of the initial conditions, and more general facts on the (H, ρ)-induced dynamics, can be found in [13].

3.5 Nonlinearities

The models discussed so far are all based on the existence of some interactions between the three parties and the electors. These interactions are driven only by suitable Hamiltonian, as in Sections 3.2 and 3.3, or by these Hamiltonians and by an additional set of rules, as in Section 3.4. However, in all these situations, the Hamiltonians were supposed to be quadratic in the ladder operators. The consequence of this assumption is that the differential equations of motion for the systems turn out to be linear, and this allows us to look for an explicit analytic expression for the decision functions.

The absence of any non-purely quadratic interaction can be thought as a strong limitation of the model. For this reason, in [13], a different model without this constraint has been proposed and analyzed. This section is devoted to a brief introduction to this alternative model. However, if on the one hand we want to add cubic and quartic interactions to the Hamiltonian of the system, on the other hand we would still like to keep the mathematical difficulty of the problem at a reasonable level. This result can be achieved easily if we neglect the interactions between the parties and the electors, assuming that, in the presence of these new terms, the main contribution to the time evolution of the decision functions is due to the interactions among the parties. Alternatively, we could simply imagine that the parameters λ_j and $\tilde{\lambda}_j$ in (3.3) are zero, so that H_{PBs} and H_{PB} do not contribute to H.

With this in mind, we consider now the following Hamiltonian $\hat{H}$:

$$
\begin{cases}
\hat{H} = H_0 + H_{int} + H_{nl,1} + H_{nl,2}, \\
H_0 = \sum_{j=1}^{3} \omega_j p_j^\dagger p_j, \\
H_{int} = \mu_{12}^{ex} \left(p_1^\dagger p_2 + p_2^\dagger p_1\right) + \mu_{12}^{coop} \left(p_1^\dagger p_2^\dagger + p_2 p_1\right) + \\
\qquad \mu_{13}^{ex} \left(p_1^\dagger p_3 + p_3^\dagger p_1\right) + \mu_{13}^{coop} \left(p_1^\dagger p_3^\dagger + p_3 p_1\right) + \\
\qquad \mu_{23}^{ex} \left(p_2^\dagger p_3 + p_3^\dagger p_2\right) + \mu_{23}^{coop} \left(p_2^\dagger p_3^\dagger + p_3 p_2\right), \\
H_{nl,1} = \lambda_{12}^{a} \left(p_1^\dagger \hat{P}_2 + \hat{P}_2 p_1\right) + \lambda_{13}^{a} \left(p_1^\dagger \hat{P}_3 + \hat{P}_3 p_1\right) + \lambda_{23}^{a} \left(p_2^\dagger \hat{P}_3 + \hat{P}_3 p_2\right) + \\
\qquad \lambda_{12}^{b} \left(p_2^\dagger \hat{P}_1 + \hat{P}_1 p_2\right) + \lambda_{13}^{b} \left(p_3^\dagger \hat{P}_1 + \hat{P}_1 p_3\right) + \lambda_{23}^{b} \left(p_3^\dagger \hat{P}_2 + \hat{P}_2 p_3\right), \\
H_{nl,2} = \tilde{\mu}_{12}^{ex} \hat{P}_3 \left(p_1^\dagger p_2 + p_2^\dagger p_1\right) + \tilde{\mu}_{12}^{coop} \hat{P}_3 \left(p_1^\dagger p_2^\dagger + p_2 p_1\right) + \\
\qquad \tilde{\mu}_{13}^{ex} \hat{P}_2 \left(p_1^\dagger p_3 + p_3^\dagger p_1\right) + \tilde{\mu}_{13}^{coop} \hat{P}_2 \left(p_1^\dagger p_3^\dagger + p_3 p_1\right) + \\
\qquad \tilde{\mu}_{23}^{ex} \hat{P}_1 \left(p_2^\dagger p_3 + p_3^\dagger p_2\right) + \tilde{\mu}_{23}^{coop} \hat{P}_1 \left(p_2^\dagger p_3^\dagger + p_3 p_2\right),
\end{cases} \tag{3.34}
$$

where ω_j, λ_{ij}^{ab}, μ_{ij}^{ex}, $\tilde{\mu}_{ij}^{ex}$ and $\tilde{\mu}_{ij}^{coop}$ are real quantities, so that $\hat{H} = \hat{H}^\dagger$; H_0 (in a slightly more general form) and H_{int} were already introduced in Section 3.2, where their meaning has been discussed. The presence of the terms $H_{nl,1}$ and $H_{nl,2}$ introduces some new interactions among the parties. For instance, the term $\hat{P}_3 p_1^\dagger p_2^\dagger$ in $H_{nl,2}$ describes the fact that $\mathcal{P}_1$ and $\mathcal{P}_2$ become more and more interested in forming alliances for higher values of the decision function of $\mathcal{P}_3$. Analogously, the term $p_1^\dagger \hat{P}_2$ in $H_{nl,1}$ describes the fact that $\mathcal{P}_1$ gets more interested in alliances when the attitude of $\mathcal{P}_2$ toward allying is stronger, and so on. Using our general scheme and our interpretation for the ladder operators, it is not difficult to deduce the meaning of each term in $H_{nl,1}$ and $H_{nl,2}$. Notice that some of these terms must be considered, if we want (and we do want, here!) H to be self-adjoint. Otherwise, many problems of the kind we have discussed in Section 2.7.2 occur, and for the moment we want to avoid these difficulties as much as possible. The interactions are called *nonlinear* because the Heisenberg equations of motion deduced out of $\hat{H}$ in Equation (3.34) are nonlinear. The interesting aspect is that $\hat{H}$ describes situations in which the three parties interact among them simultaneously, as in $\hat{P}_3 p_1^\dagger p_2^\dagger$, a possibility that was excluded when using the Hamiltonians in Equation (3.3) or in Equation (3.23), where, as originally proposed in [1], only *two-bodies interactions* are considered.

Our goal, as everywhere in this chapter, is to find the time evolution of the decision functions of the various parties. If Ψ_0 is the vector describing the system at $t = 0$, the decision function can be written as

$$P_j(t) = \left\langle \Psi_0, \hat{P}_j(t)\Psi_0 \right\rangle = \left\langle \Psi(t), \hat{P}_j(0)\Psi(t) \right\rangle, \tag{3.35}$$

where $\Psi(t) = e^{-i\hat{H}t}\Psi_0$ is the (Schrödinger) time evolution of Ψ_0 and $\hat{P}_j(t)$ is the (Heisenberg) time evolution of the operator $\hat{P}_j = p_j^\dagger p_j$. So far we have worked in the Heisenberg representation, looking for $\hat{P}_j(t)$ and its mean value on some suitable state of the parties and the electors. Here, there are no electors anymore, and the schemes shown in Figures 3.1 and 3.9 simplify significantly. This is why the vector Ψ_0 above describes only the state of the parties. To make the treatment more complete, we consider the possibility that Ψ_0 is not just a single eigenstate of all the operators $\hat{P}_j(0)$, $j = 1, 2, 3$, but it is rather some linear combination of these eigenstates, as in (3.1). In other words, we suppose now that $\Psi_0 = \sum_{j,k,l=0}^{1} \alpha_{j,k,l}(0)\varphi_{j,k,l}$, for some coefficients $\alpha_{j,k,l}(0)$, $j, k, l = 0, 1$ fixed by the initial conditions on the parties. In particular, if just one among all these coefficients is one and the others are zero, we go back to the situation considered in the previous sections, where the state of the parties was just a specific $\varphi_{j,k,l}$.

To compute how these coefficients $\alpha_{j,k,l}(0)$ evolve in time, and $\Psi(t)$ and $P_j(t)$ as a consequence, we use the fact that $\Psi(t) = e^{-i\hat{H}t}\Psi_0 = \sum_{j,k,l=0}^{1} \alpha_{j,k,l}(t)\varphi_{j,k,l}$ is the solution of the Schrödinger equation:

$$i\frac{\partial \Psi(t)}{\partial t} = \hat{H}\Psi(t). \tag{3.36}$$

Then we get the following system of differential equations for the time-depending coefficients $\alpha_{j,k,l}(t)$ of $\Psi(t)$:

$$i\frac{\partial \alpha_{j,k,l}(t)}{\partial_t} = i\left\langle \varphi_{j,k,l}, \frac{\partial \Psi}{\partial t} \right\rangle = \left\langle \varphi_{j,k,l}, H\Psi(t) \right\rangle. \tag{3.37}$$

The solution of this system returns the time evolution of the functions $\alpha_{j,k,l}(t)$, which are the building blocks to compute the decision functions $P_j(t)$ in Equation (3.35) as follows:

$$P_1(t) = \sum_{k,l=0}^{1} |\alpha_{1,k,l}(t)|^2, \qquad P_2(t) = \sum_{j,l=0}^{1} |\alpha_{j,1,l}(t)|^2$$

$$P_3(t) = \sum_{k,l=0}^{1} |\alpha_{k,j,1}(t)|^2. \tag{3.38}$$

These formulas can now be used to deduce the decision functions of the parties in several particular cases. For instance, let us take $H_{nl,2} = 0$, and let us assume that no competitive interactions arise among the parties (i.e., $\mu_{kj}^{coop} = 0$ for all k,l in Equation (3.34)). Moreover, to simplify further the situation, we also suppose that only $\mathcal{P}_1$ and $\mathcal{P}_2$ interact. Hence, all the interaction parameters involving the possible interactions of $\mathcal{P}_3$ with $\mathcal{P}_1$ and $\mathcal{P}_2$ are put to zero in $\hat{H}$. We also consider $H_0 = 0$ here, since its role in the dynamics of a system is nowadays well understood [1], and it is related to the inertia of the various agents of the model. In this way, we can significantly reduce the number of free parameters used in the model. In fact, only λ_{12}^a and μ_{12}^{ex} are taken to be different from zero.

The system of Equations (3.37) can be written as

$$\dot{\Upsilon}(t) = -iB\Upsilon(t), \tag{3.39}$$

where

$$\Upsilon(t) = \begin{pmatrix} \alpha_{0,0,0}(t) \\ \alpha_{1,0,0}(t) \\ \alpha_{0,1,0}(t) \\ \alpha_{1,1,0}(t) \\ \alpha_{0,0,1}(t) \\ \alpha_{1,0,1}(t) \\ \alpha_{0,1,1}(t) \\ \alpha_{1,1,1}(t) \end{pmatrix}, \quad B = \begin{pmatrix} 0 & 0 & 0 & 0 & 0 & 0 & 0 & 0 \\ 0 & 0 & \mu_{12}^{ex} & 0 & 0 & 0 & 0 & 0 \\ 0 & \mu_{12}^{ex} & 0 & \lambda_{12}^a & 0 & 0 & 0 & 0 \\ 0 & 0 & \lambda_{12}^a & 0 & 0 & 0 & 0 & 0 \\ 0 & 0 & 0 & 0 & 0 & 0 & 0 & 0 \\ 0 & 0 & 0 & 0 & 0 & 0 & \mu_{12}^{ex} & 0 \\ 0 & 0 & 0 & 0 & 0 & \mu_{12}^{ex} & 0 & \lambda_{12}^a \\ 0 & 0 & 0 & 0 & 0 & 0 & \lambda_{12}^a & 0 \end{pmatrix}.$$

We see that, despite the fact that $\hat{H}$ is no longer quadratic, the differential equations for $\alpha_{j,k,l}(t)$ are linear and can be solved easily, at least for the choice we have considered here. In fact, it is not difficult to find an analytic solution of Equation (3.39), and the decision functions, deduced as in Equation (3.38), turn out to be

$$P_1(t) = \left(|\alpha_{1,0,0}(0)|^2 + |\alpha_{1,0,1}(0)|^2\right)\left(\frac{\left(\lambda_{12}^a\right)^2 + \left(\mu_{12}^{ex}\right)^2\cos^2(rt)}{\left(\lambda_{12}^a\right)^2 + \left(\mu_{12}^{ex}\right)^2}\right)$$

$$+ \left(|\alpha_{0,1,0}(0)|^2 + |\alpha_{0,1,1}(0)|^2\right)\sin^2(rt)$$

$$+ \left(|\alpha_{1,1,0}(0)|^2 + |\alpha_{1,1,1}(0)|^2\right)\left(\frac{\left(\mu_{12}^{ex}\right)^2 + \left(\lambda_{12}^a\right)^2\cos^2(rt)}{\left(\lambda_{12}^a\right)^2 + \left(\mu_{12}^{ex}\right)^2}\right)$$

$$+ \left(\alpha_{1,0,0}(0)\alpha_{0,1,0}(0) + \alpha_{1,0,1}(0)\alpha_{0,1,1}(0)\right)\left(\frac{\mu_{12}^{ex}\lambda_{12}^a\sin^2(rt)}{\left(\lambda_{12}^a\right)^2 + \left(\mu_{12}^{ex}\right)^2}\right),$$

$$P_2(t) = \left(|\alpha_{1,0,0}(0)|^2 + |\alpha_{1,0,1}(0)|^2\right)\left(2\frac{\left(\lambda_{12}^a\mu_{12}^{ex}\right)^2 + (1+\cos(rt))\left(\mu_{12}^{ex}\right)^4}{\left(\left(\lambda_{12}^a\right)^2 + \left(\mu_{12}^{ex}\right)^2\right)^2}\sin^2(rt)\right)$$

$$+ \left(|\alpha_{0,1,0}(0)|^2 + |\alpha_{0,1,1}(0)|^2\right)\left(\frac{\left(\lambda_{12}^a\right)^2 + \left(\mu_{12}^{ex}\right)^2\cos^2(rt)}{\left(\lambda_{12}^a\right)^2 + \left(\mu_{12}^{ex}\right)^2}\right)$$

$$+ \left(|\alpha_{1,1,0}(0)|^2 + |\alpha_{1,1,1}(0)|^2\right)$$

$$\left(\frac{\left(\lambda_{12}^a\right)^4 + \left(\mu_{12}^{ex}\right)^4 + \left(\lambda_{12}^a\mu_{12}^{ex}\right)^2(1 + 4\cos(rt) - \cos(2rt)/2)}{\left(\left(\lambda_{12}^a\right)^2 + \left(\mu_{12}^{ex}\right)^2\right)^2}\right)$$

$$+ \left(\alpha_{1,0,0}(0)\alpha_{0,1,0}(0) + \alpha_{1,0,1}(0)\alpha_{0,1,1}(0)\right)$$

$$\left(\frac{-4\lambda_{12}^a\mu_{12}^{ex}\left(\left(\lambda_{12}^a\right)^2 + \left(\mu_{12}^{ex}\right)^2\cos^2(rt)\right)\sin^2(rt/2)}{\left(\left(\lambda_{12}^a\right)^2 + \left(\mu_{12}^{ex}\right)^2\right)^2}\right)$$

$$P_3(t) = P_3(0),$$

where, to simplify the notation, we have introduced the positive quantity $r = \sqrt{(\lambda_{12}^a)^2 + (\mu_{12}^{ex})^2}$.

The first obvious remark is that $P_3(t)$ remains constant. This is expected because of our choice of the parameters, which excludes any interaction between $\mathcal{P}_3$ and the rest of the system: $\mathcal{P}_3$ is an isolated subsystem, and for this reason its decision function is not expected to change with time and in fact this is what we find. It is also possible to see that both $P_1(t)$ and $P_2(t)$ oscillate during their evolution. This is not surprising, since $\hat{H}$ is a Hermitian matrix in a finite-dimensional Hilbert space, so that the time evolution of any observable of the system must be periodic or quasi-periodic, as we have discussed at length in Section 2.7.

However, there exist conditions that produce, at least, what could be interpreted as some sort of *asymptotic limit* (i.e., very small oscillations around certain values) for some $P_j(t)$. This is what happens for $P_2(t)$, for instance, if we increase more and more the ratio $\lambda_{12}^a/\mu_{12}^{ex}$. In this case, when t diverges, $P_2(t)$ tends toward the value

$$\left(|\alpha_{0,1,0}(0)|^2 + |\alpha_{0,1,1}(0)|^2\right) + \left(|\alpha_{1,1,0}(0)|^2 + |\alpha_{1,1,1}(0)|^2\right),$$

which coincides with the initial value of $P_2(t)$ itself, $P_2(0)$. It is interesting to notice that also the amplitude of the oscillations of $P_2(t)$ depends on the same ratio, $\lambda_{12}^a/\mu_{12}^{ex}$. The situation is not so clear for $P_1(t)$, for which it is not evident which choice should be taken for the parameters to guarantee the existence of any asymptotic limit, at least if $\alpha_{0,1,0}(0)$ or $\alpha_{0,1,1}(0)$ are nonzero. In fact, in this case, the analytic expression of $P_1(t)$ shows that there is an always oscillating term that prevents convergence of the decision function to any asymptotic value (even in our generalized sense, as small oscillations around some fixed value).

Just to be concrete, let us take $\alpha_{1,0,0}(0) = 1$, whereas all the other $\alpha_{k,j,l}(0)$ are zero. We observe that, with this particular choice, the oscillating term in $P_2(t)$ does not contribute to its final expression. In fact, the formulas above simplify, producing

$$P_1(t) = \frac{\left(\lambda_{12}^a\right)^2 + \left(\mu_{12}^{ex}\right)^2 \cos^2(rt)}{\left(\lambda_{12}^a\right)^2 + \left(\mu_{12}^{ex}\right)^2},$$

$$P_2(t) = 2\,\frac{\left(\lambda_{12}^a\mu_{12}^{ex}\right)^2 + (1 + \cos(rt))\left(\mu_{12}^{ex}\right)^4}{\left(\left(\lambda_{12}^a\right)^2 + \left(\mu_{12}^{ex}\right)^2\right)^2}\,\sin^2(rt).$$

Hence both $P_1(t)$ and $P_2(t)$ oscillate with amplitudes that decrease if we increase the rate $\lambda_{12}^a/\mu_{12}^{ex}$. In particular, as $\lambda_{12}^a/\mu_{12}^{ex} \to \infty$, the decision functions $P_1(t)$ and $P_2(t)$ approach, for large t, the values $P_1(0) = 1$ and $P_2(0) = 0$ respectively, so that both $P_1(t)$ and $P_2(t)$ converge to their initial values. This suggests that the addition of the term $\lambda_{kj}^a\hat{P}_j\left(p_k^\dagger + p_k\right)$ in the Hamiltonian forces the decision functions to stay close to their initial values, at least for sufficiently large values of $\lambda_{kj}^a/\mu_{kj}^{ex}$ and for large t. In a certain sense, this term has the same effect as the free Hamiltonian H_0. In fact (see [1]), both H_0 (via its parameters) and this new term $\lambda_{kj}^a\hat{P}_j\left(p_k^\dagger + p_k\right)$ are related to the amplitude of the oscillations of the mean values of the number operators representing the observables of the system. This is the main reason why we have taken $H_0 = 0$ in the Hamiltonian $\hat{H}$ considered in this section. This was useful to avoid overlapping effects, which could have hidden this result.

No particular difference arises if we now suppose that $\mathcal{P}_1$ and $\mathcal{P}_2$ have a co-operative attitude (i.e., if we assume that $\mu_{12}^{coop} \neq 0$). As before, all the parameters regarding the interactions of the party $\mathcal{P}_3$ with $\mathcal{P}_1$ and $\mathcal{P}_2$ are put equal to zero, so that $\mathcal{P}_3$ is again an isolated subsystem, which, as such, is not expected to have any significant dynamics. We further fix $\omega_1 = \omega_2 = 0$ and $\mu_{12}^{ex} = 0$. Hence, λ_{12}^a and μ_{12}^{coop} are the only parameters different from zero, and the matrix B in Equation (3.39) is

$$B = \begin{pmatrix} 0 & 0 & 0 & \mu_{12}^{coop} & 0 & 0 & 0 & 0 \\ 0 & 0 & 0 & 0 & 0 & 0 & 0 & 0 \\ 0 & 0 & 0 & \lambda_{12}^a & 0 & 0 & 0 & 0 \\ \mu_{12}^{coop} & 0 & \lambda_{12}^a & 0 & 0 & 0 & 0 & 0 \\ 0 & 0 & 0 & 0 & 0 & 0 & 0 & \mu_{12}^{coop} \\ 0 & 0 & 0 & 0 & 0 & 0 & 0 & 0 \\ 0 & 0 & 0 & 0 & 0 & 0 & 0 & \lambda_{12}^a \\ 0 & 0 & 0 & 0 & \mu_{12}^{coop} & 0 & \lambda_{12}^a & 0 \end{pmatrix}.$$

This is, again, a *sparse* matrix, with only a few nonzero entries. Therefore, as before, it is not difficult to find an exact solution of the system in Equation (3.39), for any choice of the initial coefficients $\alpha_{j,k,l}(0)$. However, we just consider here what happens if we take $\alpha_{1,1,0}(0) = 1$, with all the other $\alpha_{k,j,l}(0)$ equal to zero. This implies that Ψ_0 is an eigenstate of the $\hat{P}_j(0)$, $j = 1, 2, 3$, $\Psi_0 = \varphi_{1,1,0}$. Then, putting $\tilde{r} = \sqrt{\left(\lambda_{12}^a\right)^2 + \left(\mu_{12}^{coop}\right)^2}$, we find

$$P_1(t) = \cos^2(\tilde{r}t),$$

$$P_2(t) = \frac{\left(\lambda_{12}^a\right)^2 + \left(\mu_{12}^{coop}\right)^2 \cos^2(\tilde{r}t)}{\left(\lambda_{12}^a\right)^2 + \left(\mu_{12}^{coop}\right)^2},$$

while $P_3(t) = P_3(0)$ for all t, as expected. Hence P_1 oscillates between 0 and 1, and P_2 oscillates between 1 and $\frac{\left(\lambda_{12}^a\right)^2}{\left(\lambda_{12}^a\right)^2 + \left(\mu_{12}^{coop}\right)^2}$. If we increase $\lambda_{12}^a / \mu_{12}^{coop} \rightarrow \infty$, then $P_2(t) \rightarrow P_2(0) = 1$, so that a sort of asymptotic limit (and final decision) can be reached. On the other hand, $\mathcal{P}_1$ keeps on changing ideas, independently of the values of the parameters that are needed to fix $\hat{H}$, at least if λ_{12}^a or μ_{12}^{coop} are not zero. They affect only the frequency of these changes.

We close this section by noticing that, in principle, the model considered here is essentially simpler than the ones analyzed in the previous sections, mainly because the Hilbert space of the system here is finite-dimensional, while we need an infinite-dimensional space to model the electors, with or without the rule. How-ever, if we compare the results deduced now with those found in the previous sections, the ones here appear less useful, possibly because of the nonlinearities we have considered in $\hat{H}$. We can just say that we get oscillating decision functions, but we cannot say a priori if we expect some sort of asymptotic limit to be reached or not. This depends on the initial conditions and on the relation between some of the parameters of the model. Of course, we expect the situation becomes even more complicated (or richer, to use a different word), if we consider more interactions in $\hat{H}$, i.e., if we consider more parameters of $\hat{H}$ to be nonzero. This will be reflected by a different expression for B in Equation (3.39), which will possibly make the analysis of the solution a bit harder but still possible. We refer to [13] for more on this model and for other choices of the parameters.

3.6 Conclusions

It is clear that our original model and its extensions describe only a few possible applications of mathematical modeling to politics. This is obvious already from the literature on this subject that we have cited at the beginning of this chapter, and it also should also be evident from what we have discussed so far. For instance, a rather natural improvement of our models should be able to answer the question *Who is going to form an alliance with whom?* This question, we believe, is not hard to answer, within our scheme. In fact, it is sufficient to split the ladder operators of each party into two parts: p_1 should be replaced by $p_{1,2}$ and $p_{1,3}$ and so on. In this way, the decision function $P_1(t)$ is also doubled: $P_{1,2}(t)$ and $P_{1,3}(t)$ describe respectively the attitude of $\mathcal{P}_1$ to form an alliance with $\mathcal{P}_2$ or with $\mathcal{P}_3$. And the same doubling should be repeated for the operators of the other parties. Of course, these new ladder operators should be used to define a new Hamiltonian that extends those considered throughout this chapter, with or without rules and with or without nonlinearities. They should also be used to check what this new Hamiltonian produces for the new decision functions.

A different and much more complicated problem would be to construct a model able to reproduce, or even to predict, the results of a political election. Some steps in this direction have already been done, but there is still a lot of work to do.

4

Desertification

4.1 Introduction

In Chapter 3, we have shown how fermionic ladder operators, and Hamiltonians constructed out of them, can be used to analyze a very explicit problem in decision-making. We have discussed how they provide the time evolution of three decision functions, measuring the interest of three different parties to form, or not, some political coalition, under suitable conditions.

This chapter is devoted to an application of the same techniques to a completely different system: we have a system S made of three different *agents*, and they mutually interact. The first agent is the *soil* where seeds and plants are distributed. The only aspect of the soil that is interesting to us is its quality, which we measure using a number between zero (extremely poor quality) and one (very good quality). The soil occupies a two-dimensional region, which we consider as a discrete lattice $\mathcal{R}$, divided in cells of some fixed size that can be different, in principle, from one another. The second agent of S is the *set of seeds* of one or more species of plants, which are only described in terms of some dimensionless density, again between zero and one. If, in a given cell, the density of the seeds is zero we have no seeds at all, or very few seeds, while a density one corresponds to the maximum density of seeds which can reasonably occupy the cell: the *carrying capacity* is reached, in this case. Of course, so far, what should be understood for *very few* or *reasonably* is rather vague, and it is related to several aspects of the system: the size of the cell, the type of seeds, the experience of the farmers planting the seeds and so on. The third agent of S are the *plants* which grow out of the seeds, and which produce more seeds during their lives. Again, the only aspect of the plants that is interesting for us is their densities in the various cells of $\mathcal{R}$. In particular, what we are interested in, in our analysis, is the change in the quantities describing these agents and, in particular, in the possibility of analyzing the time evolution of the quality of soil and of the densities of the plants and seeds, in the attempt to avoid these numbers going to zero. In fact, when this happens, the seeds and the plants do not receive enough nutrients from the soil to germinate or even to survive, and the region $\mathcal{R}$ goes toward a desertification process. In particular, early signs of desertification are particularly interesting since these signs can help stopping, and reversing, the process in the most *economical* way.

We don't really need to justify why the analysis of similar problems is particularly interesting and useful in real life. We just observe that the model we are going to describe in this chapter fits very well into the increasing research on ecosystems, with particular focus on changes in vegetation due to external stresses, and in particular to desertification processes in drylands. Drylands cover a large part of the Earth land surface, and desertification processes due to climatic variations or human activities can happen in a relatively short time and have severe and long-lasting consequences [69]. These processes are usually analyzed using some statistical ideas or numerical simulations, see [70,71] and references therein, or adopting techniques related to cellular automata [72]. This is different from what we will do here: The approach we follow is based on the use of (fermionic) ladder operators to construct an Hamiltonian H that describes the essential interactions we expect to take place between soil, seeds and plants localized in a two-dimensional region, and, using H, we deduce the time evolutions of the quality of soil, the density of the seeds and the density of the plants in the region. Furthermore, we show how a stress factor can be considered within our framework, and we propose a possible strategy to react successfully against this stress factor, restoring a good quality of the soil and avoiding the densities of the seeds and the plants to go to zero. Stated in different words, we will show how desertification can be modeled, and how a little (positive) reaction can help, according to the model, to stop the process. We anticipate that, in the procedure described later in the chapter, we will make use of complex parameters to define H. It will be discussed how, according to what already discussed in Section 2.7.2, and as we will see explicitly later in Section 6.4, the sign of their imaginary parts is directly connected to different effects observed during the time evolution of the system.

4.2 The Model

The system S we are interested in was originally proposed in [66], as a possible application of our operatorial techniques to a concrete ecological problem. As already discussed, S is formed by three agents: the *soil* (S_1), the *seeds* (S_2) and the *plants* (S_3). Each agent is localized in a cell of the two-dimensional lattice $\mathcal{R}$, and interacts with the other agents. Moreover, some of the agents can also move along $\mathcal{R}$, going from one to another of the $N = L^2$ cells which form $\mathcal{R}$. In our computations we fix $L = 11$, so that the lattice is made by 121 cells. This is a good compromise between the computational time required by our numerical computations and the necessity of having a sufficiently large $\mathcal{R}$. We will return on this aspect of our model later on. When constructing the model we were driven by the following ideas: seeds need resources (nutrients, water) from the soil and produce plants which, in turn, produce seeds and release and absorb nutrients to and from the soil. Moreover, seeds might also move all throughout the lattice, being transported by animals or wind, for instance. However, plants cannot move along $\mathcal{R}$, but they may suffer from *overpopulation*: When the cells surrounding the plants in the cell α are full of vegetation, then those in α might die. All these mechanisms are modeled by the following Hamiltonian,

$$
\begin{cases}
H = H_0 + H_1 + H_2 + H_3, \\[2mm]
H_0 = \displaystyle\sum_{\beta=1}^{L^2} \left(\omega_{1,\beta} a_{1,\beta}^\dagger a_{1,\beta} + \omega_{2,\beta} a_{2,\beta}^\dagger a_{2,\beta} + \omega_{3,\beta} a_{3,\beta}^\dagger a_{3,\beta} \right), \\[4mm]
H_1 = \displaystyle\sum_{\beta=1}^{L^2} \Big[\lambda_{1,\beta} \left(a_{1,\beta} a_{2,\beta}^\dagger + a_{2,\beta} a_{1,\beta}^\dagger \right) + \lambda_{2,\beta} \left(a_{2,\beta} a_{3,\beta}^\dagger + a_{3,\beta} a_{2,\beta}^\dagger \right) \\[2mm]
\qquad\quad + \lambda_{3,\beta} \left(a_{3,\beta} a_{1,\beta}^\dagger + a_{1,\beta} a_{3,\beta}^\dagger \right) \Big], \\[4mm]
H_2 = \displaystyle\sum_{\beta=1}^{L^2} \left[\mu_\beta \sum_{\gamma=1}^{L^2} p_{\beta,\gamma} \left(a_{2,\beta} a_{2,\gamma}^\dagger + a_{2,\gamma} a_{2,\beta}^\dagger \right) \right], \\[4mm]
H_3 = \displaystyle\sum_{\beta=1}^{L^2} \left[\nu_\beta \sum_{\gamma=1}^{L^2} c_{\beta,\gamma} \left(a_{3,\beta} a_{3,\gamma}^\dagger + a_{3,\gamma} a_{3,\beta}^\dagger \right) \right].
\end{cases}
\tag{4.1}
$$

which is constructed following the general strategy and ideas proposed and analyzed in [1] and already adopted in Chapter 3. In particular, for the moment we assume that $\omega_{k,\beta}$, $\lambda_{k,\beta}$, μ_β, $p_{\beta,\gamma}$, ν_β and $c_{\beta,\gamma}$ are real constants. This is needed for H to be self-adjoint: $H = H^\dagger$. However, in Section 4.3, we will relax this assumption, since this will be a very efficient way to describe stress factors and positive effects for the system.

The operators $a_{i,\alpha}$, and their adjoints $a_{i,\alpha}^\dagger$, are fermionic ladder operators satisfying the following CAR:

$$
\left\{ a_{i,\alpha}, a_{j,\beta}^\dagger \right\} = \delta_{i,j} \delta_{\alpha,\beta} \, \mathbb{1}, \qquad \left\{ a_{i,\alpha}, a_{j,\beta} \right\} = 0,
\tag{4.2}
$$

where $i,j = 1,2,3$, and $\alpha, \beta = 1,2,3,\dots,L^2$. Our notation is the following: we adopt Latin indexes for the three agents (soil, seeds and plants), and Greek indexes for the different lattice cells.

The choice of fermionic operators is mainly based upon two reasons. The first is technical: the Hilbert space $\mathcal{H}$ of our model, see below, is finite dimensional, so that all the observables are bounded operators. More than this, they are all finite matrices. The second reason is related to the biological interpretation of our model; in fact, we are interested in measuring the quality of the soil (a sort of density of the health of the soil) and the densities of seeds and plants in the various cells, or in certain regions of $\mathcal{R}$. As it is well known, and it is discussed in a settings that are interesting for us in [7,8] and in [1], fermionic creation and annihilation operators define number operators whose mean values are automatically restricted in the range $[0,1]$, so that they can describe normalized densities quite naturally.

> *Remark:* Compared with the Hamiltonians we have introduced in Chapter 3 we
> notice that the operator in Equation (4.1) involves no reservoir. The system is made
> by the three agents, which are only interacting among them: in particular, no *external*
> *action* is considered in H. This makes the mathematics of the model, in principle, not
> so complicated, since it seems we just have three degrees of freedom, as in the polit-
> ical system considered in Section 3.5. However, this is not really so simple. In fact,
> the seeds in cell α and those in cell β, if $\alpha \neq \beta$, are considered as two independent
> agents of the model. And the same is true for plants and soil. This is reflected by the
> CAR in Equation (4.2), which show how the ladder operators of the same agent in
> different cells anticommute. So the total number of degrees of freedom of the system
> is not just three, but is $3N = 3L^2$. And this increases quite a bit the (technical) com-
> plexity of the problem for increasing values of L: each $a_{i,\alpha}$ is a finite matrix, but the
> size of the matrix increases very fast with L. This is the reason why we fix $L = 11$ in
> our computations in Section 4.3.2.

Going back to H, and to the meaning of its various terms, we start observing that $\omega_{k,\beta}$ represents the inertia of the k-th agent in the cell β: the larger its value, the smaller the amplitude of the oscillations of their related densities, see [1]. $\lambda_{1,\beta}$, $\lambda_{2,\beta}$ and $\lambda_{3,\beta}$ describe the strength of the interaction between soil and seeds, seeds and plants, plants and soil in the cell β, respectively. μ_β describes the mobility of the seeds, while $p_{\beta,\gamma}$ can be zero or not depending on whether a movement of the seeds from cell β to cell γ is forbidden or not. Finally, ν_β and $c_{\beta,\gamma}$ in H_3 take into account the effect of plant overpopulation, as we will discuss in a moment. With this in mind, it is natural to assume that $p_{\beta,\beta} = 0$ and $p_{\beta,\gamma} = p_{\gamma,\beta}$: if seeds can move from cell α to cell β, then they can also go in the opposite direction. Moreover, it makes no sense to consider as a movement the motion of seeds inside a cell; this motivates the choice $p_{\beta,\beta} = 0$. Similar reasons explain the following choices for $c_{\alpha,\beta}$: $c_{\beta,\beta} = 0$ and $c_{\beta,\gamma} = c_{\gamma,\beta}$. In particular $c_{\beta,\beta} = 0$ implies that the overpopulation of the plants in β is not affected by other plants in the same cell.

The various terms in H, following the general analysis in [1], can be interpreted as follows. H_0 is the free Hamiltonian: if $H \equiv H_0$, *i.e.*, if $\lambda_{k,\beta} = \mu_\beta = \nu_\beta = 0$ for all k and β, then all the site-dependent density operators $\hat{n}_{i,\alpha} := a_{i,\alpha}^{\dagger} a_{i,\alpha}$ are constant in time. This follows from the fact that in this case $[H, \hat{n}_{i,\alpha}] = [H_0, \hat{n}_{i,\alpha}] = 0$. The interpretation is clear: since there are no interactions between agents, and considering that H_0 does not contain any *birth* or *death* terms, the densities of each agent in each cell have no reason to change in time. Then, at a first sight, H_0 contains no interesting dynamics: in fact, this is true only if, at $t = 0$, the system $\mathcal{S}$ is in an eigenstate of the various operators $\hat{n}_{i,\alpha}$ and if H coincides with H_0, but it is not really so if $\mathcal{S}$, at $t = 0$, is in a nontrivial linear combination of these eigenstates: in this case H_0 could give rise to an interesting time evolution of the system.[1] Moreover, and more relevant for what will be discussed later, the explicit choice of the parameters involved in H_0 plays an essential role in producing some damping effect, which will be useful in the description of stress factors. We will return on this aspect later on.

If we now look at H_1 in (4.1), we see terms like $a_{2,\beta} a_{3,\beta}^{\dagger}$. This describes the fact that, in order to have a plant (agent 3), we need to destroy a seed (agent 2). In other words, this is a *germination term* in the cell β: $a_{2,\beta}$ is an annihilation operator (destroying a seed in cell β or, better, lowering the density of the seeds in that cell), while $a_{3,\beta}^{\dagger}$ is a creation operator (creating a plant, or increasing the the density of the plants, in the same cell). Of course, this contribution describes a sort of instantaneous process, which is not entirely realistic. But still we prefer to work with this H_1 since it is *not so unrealistic* and since it gives rise to the system of differential Equations (4.3) which can be solved exactly. Hence, we will not have to introduce any further approximation later, as we possibly have to do with other choices of H_1. The adjoint contribution of $a_{2,\beta} a_{3,\beta}^{\dagger}$, $a_{3,\beta} a_{2,\beta}^{\dagger}$ models seasonal plants (e.g., grass, small bushes) which die after producing seeds (again, instantaneously).[2] H_1 also contains the contribution $a_{1,\beta} a_{3,\beta}^{\dagger}$ (when the density of the plants increases, the soil deteriorates) and its adjoint $a_{3,\beta} a_{1,\beta}^{\dagger}$, which describes the fact that plants release nutrients to the soil when dying. Finally, the first term in H_1 describes the interaction between soil and seeds, and takes into account the fact that the nutrients in the soil can be used to support production of seeds. This is due to the term $a_{1,\beta} a_{2,\beta}^{\dagger}$. Its adjoint, $a_{2,\beta} a_{1,\beta}^{\dagger}$, can be understood by noticing that the quality improves when there are not so many seeds in cell β.

Let us discuss now the interpretation of H_2 in Equation (4.1). This Hamiltonian describes the dispersal of seeds along $\mathcal{R}$: in fact, because of the term $a_{2,\beta} a_{2,\gamma}^{\dagger}$, seeds disappear from cell β and appear in cell γ, at least if $p_{\beta,\gamma} \neq 0$. The opposite movement is described by $a_{2,\gamma} a_{2,\beta}^{\dagger}$. The presence of $p_{\beta,\gamma}$ is essential, since it describes

[1] This will never be the case in this chapter, where $\mathcal{S}$ is always assumed, for $t = 0$, to be in an eigenstate of the different $\hat{n}_{i,\alpha}$, see formula (4.5) below.

[2] In this case, the adjoint of a reasonable term in H is still reasonable, meaning with this that it *makes sense* for the system we are considering. Sometimes, however, this is not so and the adjoint have not a real *raison d'etre*. Most of the times, however, these *strange terms* do not contribute much to the dynamics, since initial conditions create a sort of *preferred direction of the time evolution* [1].

the fact that seeds cannot just disappear from cell β and appear in another cell γ, if β is not sufficiently close to γ: this kind of movements are just forbidden, in our model, or at least quite unlikely (see Equation [4.8]). The coefficients μ_β in H_2, as already observed, give a measure of the speed of dispersion of the seeds along $\mathcal{R}$.

The last term, H_3, contains a term of the form $\sum_\gamma c_{\beta,\gamma} a_{3,\gamma}^\dagger a_{3,\beta}$. This contribution models competition among plants: When the density of plants surrounding the cell β grows, those in β suffer for lack of resources, and this is the reason why the lowering operator $a_{3,\beta}$ appears. The adjoint contribution in H_3, $\sum_\gamma c_{\beta,\gamma} a_{3,\beta}^\dagger a_{3,\gamma}$, describes the opposite phenomenon: When the number of plants surrounding the cell β decreases, those in β can rely on more resources and, for this reason, their densities may increase.

The differential equations ruling the dynamics are deduced by adopting the Heisenberg scheme in which $\dot{X} = i[H, X]$, for each observable X of the system. We obtain the following system of differential equations:

$$\begin{cases} \dot{a}_{1,\alpha} = i\left(-\omega_{1,\alpha} a_{1,\alpha} + \lambda_{1,\alpha} a_{2,\alpha} + \lambda_{3,\alpha} a_{3,\alpha}\right), \\[2mm] \dot{a}_{2,\alpha} = i\left(-\omega_{2,\alpha} a_{2,\alpha} + \lambda_{1,\alpha} a_{1,\alpha} + \lambda_{2,\alpha} a_{3,\alpha} + \sum_{\beta=1}^{L^2} \left(\mu_\alpha + \mu_\beta\right) q_{\alpha,\beta} a_{2,\beta}\right), \\[2mm] \dot{a}_{3,\alpha} = i\left(-\omega_{3,\alpha} a_{3,\alpha} + \lambda_{2,\alpha} a_{2,\alpha} + \lambda_{3,\alpha} a_{1,\alpha} + \sum_{\beta=1}^{L^2} \left(\nu_\alpha + \nu_\beta\right) c_{\alpha,\beta} a_{3,\beta}\right). \end{cases} \quad (4.3)$$

Since H is a quadratic operator, these equations turn out to be linear. Hence, we can search for an exact analytic solution of Equation (4.3). As we have already discussed in Chapter 3, this can be seen as a good (technical) motivation not to include cubic interactions in H since these would lead to a nonlinear set of equations, for which the solution can be much harder to find. This choice leads to a possibly simplified model, which nevertheless gives nontrivial results, as we will discuss later.

It is now convenient to introduce a $3L^2$-dimensional vector $\mathbf{X}$, whose transpose is

$$\mathbf{X}^T = \left(a_{1,1}, a_{1,2}, \dots, a_{1,L^2}, a_{2,1}, a_{2,2}, \dots, a_{2,L^2}, a_{3,1}, \dots, a_{3,L^2}\right),$$

so that Equation (4.3) can be rewritten as $\dot{\mathbf{X}} = T\mathbf{X}$, where T is the $3L^2 \times 3L^2$ matrix deduced from the equation. The solution of this matrix differential equation is clearly

$$\mathbf{X}(t) = e^{Tt}\mathbf{X}(0). \quad (4.4)$$

We are in the situation described first in [7], so that the computation of the densities is now simple and the only numerical difficulty consists in finding, given T, the exponential matrix e^{Tt}. What is physically meaningful for us is the mean value of the time evolution of the *local densities* $\hat{n}_{j,\alpha}(t) = a_{j,\alpha}^\dagger(t) a_{j,\alpha}(t)$, for all j and α, on some suitable state over $\mathcal{S}$. More in detail, we want to compute the following quantities, which can be understood as the densities of the seeds and plants in cell α if $j = 2, 3$, and as the quality of soil in the same cell if $j = 1$:

$$n_{j,\alpha}(t) = \left\langle \varphi_{\mathbf{n}_1, \mathbf{n}_2, \mathbf{n}_3}, a_{j,\alpha}^\dagger(t) a_{j,\alpha}(t) \varphi_{\mathbf{n}_1, \mathbf{n}_2, \mathbf{n}_3} \right\rangle. \tag{4.5}$$

Here $\varphi_{\mathbf{n}_1, \mathbf{n}_2, \mathbf{n}_3}$ is an eigenstate of all the number operators $\hat{n}_{j,\alpha} = a_{j,\alpha}^\dagger a_{j,\alpha}$ with eigenvalue $n_{j,\alpha}$, and where $\mathbf{n}_1 = (n_{1,1}, n_{1,2}, \dots, n_{1,L^2})$, $\mathbf{n}_2 = (n_{2,1}, n_{2,2}, \dots, n_{2,L^2})$ and $\mathbf{n}_3 = (n_{3,1}, n_{3,2}, \dots, n_{3,L^2})$. They describe the initial (i.e., at $t = 0$) densities of the three agents in all the different cells.[3] The vector $\varphi_{\mathbf{n}_1, \mathbf{n}_2, \mathbf{n}_3}$ can be constructed using and adapting Equation (2.2) to the present situation. We know that the vectors $\varphi_{\mathbf{n}_1, \mathbf{n}_2, \mathbf{n}_3}$'s are mutually orthogonal, and the set of all these vectors is an o.n. basis of the Hilbert space $\mathcal{H}$ where the fermionic operators are defined.

Calling now $E_{i,j}(t)$ the generic time-dependent matrix entry of e^{Tt} we have, for instance

$$n_{1,1}(t) = \sum_{\gamma=1}^{L^2} \left| E_{1,\gamma}(t) \right|^2 n_{1,\gamma} + \left| E_{1,L^2+\gamma}(t) \right|^2 n_{2,\gamma} + \left| E_{1,2L^2+\gamma}(t) \right|^2 n_{3,\gamma},$$

$$n_{2,1}(t) = \sum_{\gamma=1}^{L^2} \left| E_{L^2+1,\gamma}(t) \right|^2 n_{1,\gamma} + \left| E_{L^2+1,L^2+\gamma}(t) \right|^2 n_{2,\gamma} + \left| E_{L^2+1,2L^2+\gamma}(t) \right|^2 n_{3,\gamma},$$

$$n_{3,1}(t) = \sum_{\gamma=1}^{L^2} \left| E_{2L^2+1,\gamma}(t) \right|^2 n_{1,\gamma} + \left| E_{2L^2+1,L^2+\gamma}(t) \right|^2 n_{2,\gamma} + \left| E_{2L^2+1,2L^2+\gamma}(t) \right|^2 n_{3,\gamma},$$

...

or, more in general,

$$n_{i,\alpha}(t) = \sum_{\gamma=1}^{L^2} \left(\left| E_{(i-1)L^2+\alpha,\gamma}(t) \right|^2 n_{1,\gamma} + \left| E_{(i-1)L^2+\alpha,L^2+\gamma}(t) \right|^2 n_{2,\gamma} \right.$$
$$\left. + \left| E_{(i-1)L^2+\alpha,2L^2+\gamma}(t) \right|^2 n_{3,\gamma} \right), \tag{4.6}$$

$i = 1, 2, 3$, and $\alpha = 1, \dots, L^2$. This formula will be the starting point of our analysis in Section 4.3.

> *Remark:* It was already observed that, in order to deduce the functions $n_{i,\alpha}(t)$ in Equation (4.6), it is not really essential to solve first the equations in Equation (4.3) and then use this solution in Equation (4.5). In fact, a possible, and apparently more direct, alternative would be to compute

---

[3] This is not very different from what was done in Section 3.2, where the various $\varphi_{j,k,l}$ were associated with different attitudes of the parties. Notice also that, if $L = 1$, the two sets of vectors here and there mathematically coincide.

$$\hat{n}_{i,\alpha}(t) = e^{iHt}\hat{n}_{i,\alpha}e^{-iHt}, \qquad (4.7)$$

and then to use this operator in Equation (4.5). However, while this procedure could be efficient for *small* systems, here it is not, since H is a $8L^2 \times 8L^2$ square matrix, whose exponential becomes more and more difficult to compute as L grows. For this reason, at least for the choice of $L = 11$, formula Equation (4.6) works much better, and much faster, than the alternative recipe in Equation (4.7).

4.3 Numerical Results

Now that the model has been introduced, the Hamiltonian has been defined and the differential equations are derived and (formally) solved, we are left with the analysis of the results and with the possible interesting consequences of these results. This is exactly what we will do in this section. In particular, we divide this analysis in two parts: in the first one, we will consider the simplest, but still not entirely trivial, situation in which the lattice $\mathcal{R}$ consists of a single cell ($L = 1$). This will help us to clarify some aspects of our model and, in particular, the role of the parameters in H. In the second part, we will consider the case in which L is larger than one,[4] so that the effects of dispersions of seeds and of overpopulation of plants, both based on the two-dimensionality of the lattice $\mathcal{R}$, will be relevant. In this more interesting situation we will discuss what happens both with and without stress, stress that can be a consequence of bad weather conditions, presence of animals which eat the plants, fires and so on. Furthermore, we will also discuss what should be modified in our conclusions if the region $\mathcal{R}$ is not spatially homogeneous. This case can be easily described in our approach by taking the parameters of the soil as really (and strongly!) cell-dependent. On the opposite side, the region $\mathcal{R}$ is homogeneous if all these parameters are chosen to be independent on β.

4.3.1 One Cell

We begin with the simplest situation: the lattice $\mathcal{R}$ is made by just one cell, and therefore $\mathcal{S}_1$, $\mathcal{S}_2$ and $\mathcal{S}_3$ are forced, of course, to occupy this single cell. In other words, $L = 1$ and the sums over Greek indices are superfluous. For this reason, in this section, these indices will always be omitted. In this case, see (4.1), H_2 and H_3 are not *active*, and the Hamiltonian reduces to $H = H_0 + H_1$. This is because (i) there is no other cell where seeds can move to and (ii) there are no surrounding cells of our single cell. Hence, there are no movements of the seeds and no overpopulation effects. As a direct consequence of this particular condition, the number of independent (fermionic) modes is just three, and then H can be written as a $2^3 \times 2^3$ matrix.

---

[4] As we have already discussed, $L = 11$ will be our working choice.

Therefore, if $H = H^\dagger$, each $\hat{n}_j(t)$ is periodic or, at most, quasi-periodic, depending on the relations among the real eigenvalues of H. This aspect has been already stressed in a more abstract context in Section 2.7, and it is a characteristic behavior of any self-adjoint Hamiltonian which can be written as a finite-dimensional matrix. Then, if the parameters of H are taken to be real, there is no way to introduce a damping into the time evolution of the number operators $\hat{n}_j(t)$, and to prove, out of this time evolution, that some asymptotic steady state of the various agents of the system is achieved, for large time. This is simply impossible, except for very particular choices of the initial conditions which, however, trivialize the system. This is the case, for instance, if we assume that the mean values of each $\hat{n}_j(0)$ is zero. In fact, if $H = H_0 + H_1$, an integral of motion for the system does exist, since the operator $\hat{N} := \hat{n}_1 + \hat{n}_2 + \hat{n}_3$ commutes with H: $[H, \hat{N}] = 0$. Then the sum of the densities (which are exactly the mean values of these operators) of the three agents is preserved in time, even if each one may have a non trivial dynamics, in principle. However, since each operator $\hat{n}_j = a_j^\dagger a_j$ is positive,[5] it turns out that the only possibility is that $\langle \hat{n}_j(t) \rangle = \langle \hat{n}_j(0) \rangle = 0$ for all t, $j = 1, 2, 3$. Hence, some limiting values are found, but in a trivial way which is not particularly interesting for us. From an ecological point of view, this result can be easily understood: When the quality of soil is very poor, and we have (almost) no seeds and plants in $\mathcal{R}$, there is no reason for this (bad) equilibrium to be broken, of course.

If $n_j(0) \neq 0$ for some j, $\mathcal{S}$ is no longer in a steady state, and in fact, an oscillating behavior is evident in Figure 4.1, where we plot $n_j(t) = \langle \varphi_{n_1, n_2, n_3}, \hat{n}_j(t) \varphi_{n_1, n_2, n_3} \rangle$, ($j = 1, 2, 3$), for the following values of the parameters: $\omega_1 = 0.5$, $\omega_2 = 0.15$, $\omega_3 = 0.3$, $\lambda_1 = 0.1$, $\lambda_2 = 0.2$ and $\lambda_3 = 0.3$, and the following initial densities[6] for the three agents: $n_1 = 0.4$, $n_2 = 0.2$ and $n_3 = 0.7$.

The same qualitative results are obtained for different choices of the values of the parameters. The conclusion of this preliminary analysis is therefore what we already expected: for real parameters of H no (nontrivial) equilibrium is possible – the quality of soil goes up and down, and appears to be quite often (but not always!) out of phase with respect to the densities of the seeds or of the plants. This is similar to what we have observed in the love affair between Alice and Bob described in [1,9], but here this effect is less evident, since the agents involved in the system, even in this simple case where $H_2 = H_3 = 0$, are three and not just two. We conclude that there is no risk of desertification, in this simple (but unrealistic) situation. But

---

[5] Positivity of an operator implies positivity of its mean value: if S is a (bounded) positive operator, S can be written as $S = s^\dagger s$, for some operator s. This is clear in our situation, where $\hat{n}_j = a_j^\dagger a_j$. Now, for any non zero vector ψ, $\langle \psi, S\psi \rangle = \|s\psi\|^2$ which is surely greater than or equal to zero.

[6] Each initial value n_j should only be zero or one, in principle. This is because of the fermionic nature of the system. However, in our interpretation, $n_j(t)$ represents the density of $\mathcal{S}_j$, which can also assume intermediate values between zero and one. For this reason, here, we will often take values of $n_j(0)$ which are neither zero, nor one. This is possible since we have the analytical solution of $n_j(t)$ in terms of $n_j(0)$ (see [4.6]).

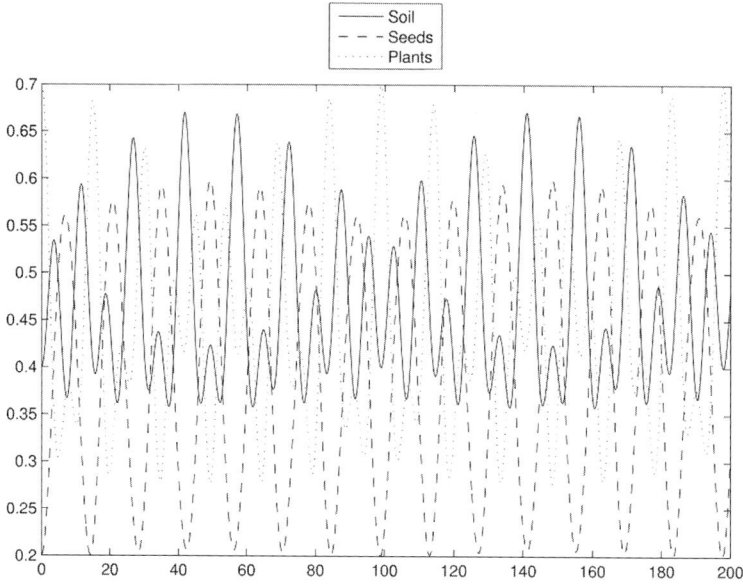

Figure 4.1 One cell: $n_1(t)$, $n_2(t)$ and $n_3(t)$ for real parameters ($\omega_1 = 0.5$, $\omega_2 = 0.15$, $\omega_3 = 0.3$, $\lambda_1 = 0.1$, $\lambda_2 = 0.2$, $\lambda_3 = 0.3$), and initial conditions $n_1(0) = 0.4$, $n_2(0) = 0.2$ and $n_3(0) = 0.7$ (Source: Bagarello et al. [66]).

the reason is clear: the model does not yet include any real stress factor: plants do not only lower the quality of the soil, by absorbing its nutrients, but they also produce nutrients which enrich the soil, when they die. This explains the plots in Figure 4.1.

There are several possibilities to deduce a somehow steady-state for the system S: the one we like most, and that we have used in many applications along the years, consists in considering S not as isolated system but as a part of a larger open system. This can be understood as if there is a region $\mathcal{R}_{out}$ surrounding $\mathcal{R}$, and other populations defining an *outer system* S_{out} interacting with S, helping the agents of S to reach some equilibrium. In this case, the evolution of the subsystem S should be studied in relation with the overall time evolution of the full system, $S_{full} = S \otimes S_{out}$.

An alternative strategy makes use of the (H, ρ)-induced dynamics introduced in Section 2.8. In this case the dynamics is driven not just by H, but also by some rule ρ that, in the present context, could be understood, for instance, as the (periodic) addition to the soil of some fertilizer which improves, from time to time, its quality. Alternatively, the rule could work *removing* the plants if they exceed a certain density in a given cell. This can be considered as a possible way to keep the over-population effect under control.

However, a much simpler and efficient possibility exists, and will be adopted here: we just give up the self-adjointness of H, assuming that some of the parameters of H can be complex, following the same general ideas discussed in Section 2.7.2.

In fact, it is sufficient to add an imaginary part to the inertia parameters, the ω_j's in H_0, to see clearly the effect of this change, and to understand that in this way we are really describing some stress acting on $\mathcal{S}$. As Figures 4.2 and 4.3 show, adding a negative imaginary part to the inertia of plants is sufficient to cause all the densities to undergo damped oscillations and disappear over time, leading in this way to desertification: this is particularly evident in Figure 4.3 where the sum of the three densities is no longer an integral of motion but converges to 0 in the time interval [0, 5000], much larger than that considered in Figures 4.1 and 4.2. After some time, the densities of plants and seeds go to zero, and the quality of soil becomes extremely poor. It is clear that, in this case, the free Hamiltonian H_0 produces already interesting consequences in the dynamics of $\mathcal{S}$. We get damping, and this is because we are considering now $H_0 \neq H_0^\dagger$, which implies, in turn, that $H \neq H^\dagger$. However, when H is not self-adjoint, we must pay attention to the fact that the ordinary rules for the Heisenberg dynamics do not apply. We have discussed some of the features occurring in this case in Section 2.7.2. In a sense, following

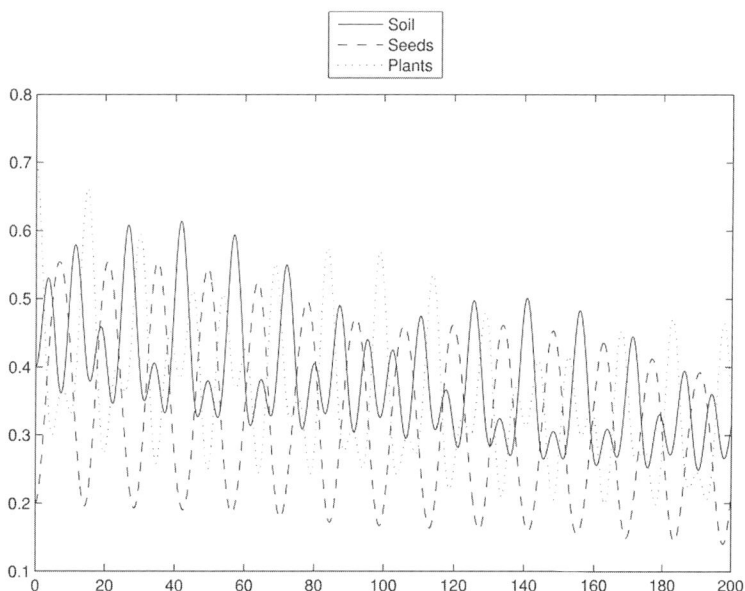

Figure 4.2 One cell: $n_1(t)$, $n_2(t)$ and $n_3(t)$ for $\omega_1 = 0.5$, $\omega_2 = 0.15$, $\omega_3 = 0.3 - 0.003\,i$, $\lambda_1 = 0.1$, $\lambda_2 = 0.2$, $\lambda_3 = 0.3$, and initial conditions $n_1(0) = 0.4$, $n_2(0) = 0.2$ and $n_3(0) = 0.7$ (Source: Bagarello et al. [66]).

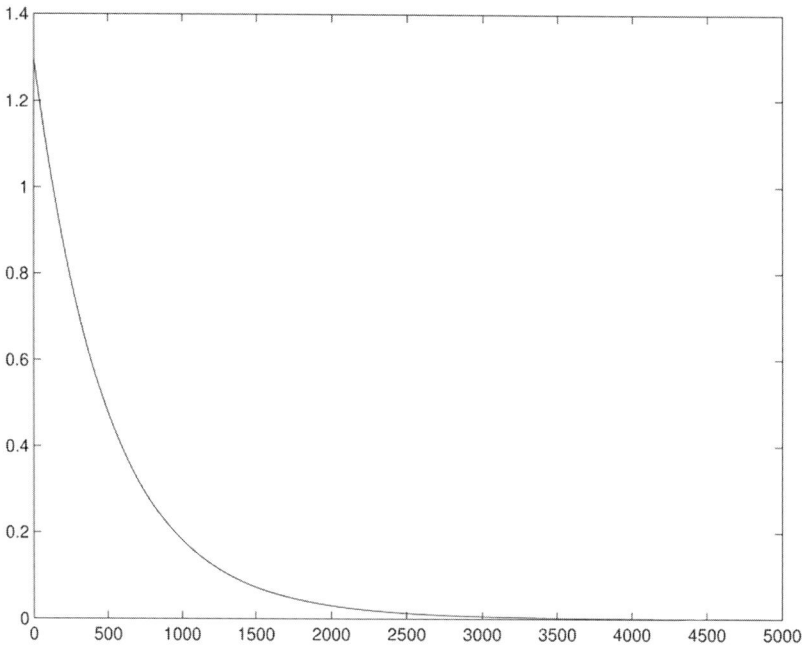

Figure 4.3 One cell: the sum of densities $n_1(t) + n_2(t) + n_3(t)$ plotted in the time interval $[0, 5000]$ for $\omega_1 = 0.5$, $\omega_2 = 0.15$, $\omega_3 = 0.3 - 0.003\,i$, $\lambda_1 = 0.1$, $\lambda_2 = 0.2$, $\lambda_3 = 0.3$, and initial conditions $n_1(0) = 0.4$, $n_2(0) = 0.2$ and $n_3(0) = 0.7$ (Source: Bagarello et al. [66]).

what it is often done in ordinary quantum mechanics, we use some non self-adjoint *effective Hamiltonian*, but we still adopt the standard rules of quantum evolution.

We can interpret the result in the following way: In our model, an external stress affecting one population (e.g., grazing for plants) is enough to trigger a desertification process, reducing drastically the densities not only of the plants, but also of the seeds, and destroying the quality of the soil as well.

What is interesting for us is that, in order to balance the external stress on one agent, it is sufficient to include a positive feedback in the system, which in our model means adding a positive imaginary part to the inertia term of another agent: it is comparable to adding seeds in the region while plants are disappearing because of grazing. The consequences are described in Figures 4.4 and 4.5: adding a small positive imaginary contribution, $0.0012\,i$, to ω_2 is enough to lead the system away from desertification. It is important to notice that the absolute value of the positive feedback is smaller than the absolute value of the negative one (less than one half in this case). We can read the result in this way: it is enough to add a *small* quantity of seeds to the system to balance the effect of the (stronger) external stress on plants. This is particularly evident from a comparison between Figures 4.5 and 4.3: in the

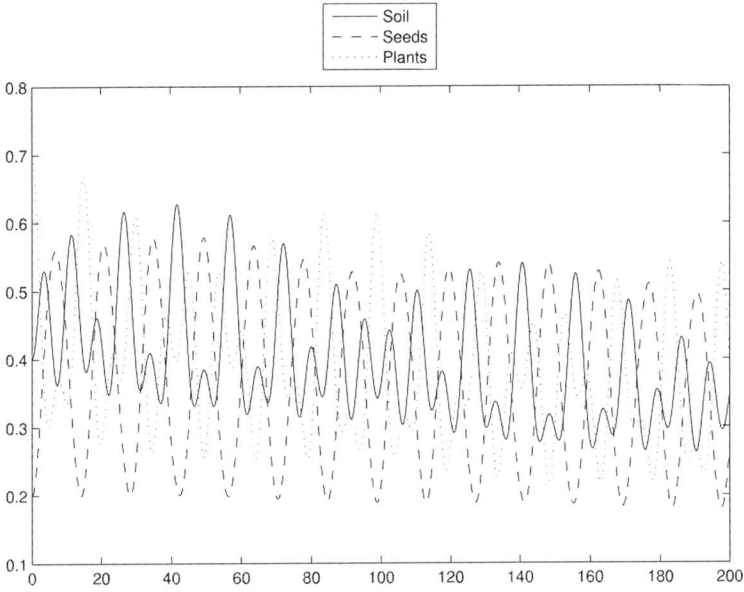

Figure 4.4 One cell: $n_1(t)$, $n_2(t)$ and $n_3(t)$ for $\omega_1 = 0.5$, $\omega_2 = 0.15 + 0.0012\,i$, $\omega_3 = 0.3 - 0.003\,i$, $\lambda_1 = 0.1$, $\lambda_2 = 0.2$, $\lambda_3 = 0.3$, and initial conditions $n_1(0) = 0.4$, $n_2(0) = 0.2$ and $n_3(0) = 0.7$ (Source: Bagarello et al. [66]).

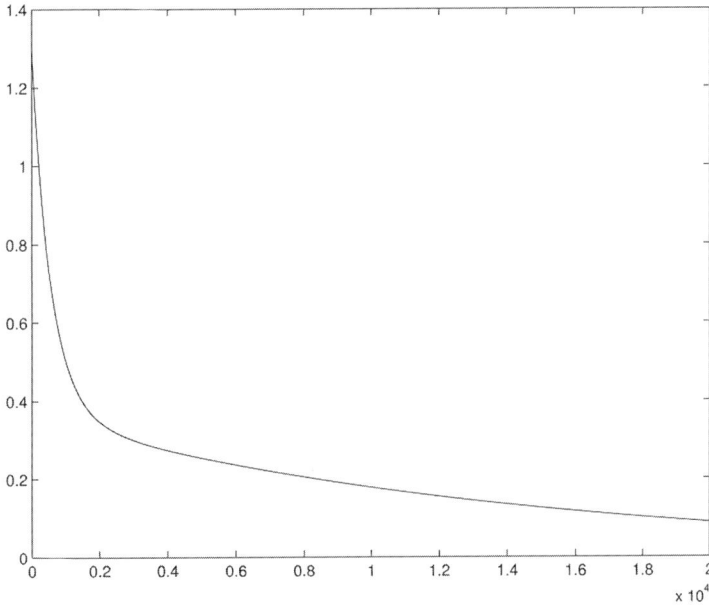

Figure 4.5 One cell: $n_1(t) + n_2(t) + n_3(t)$ in the time interval $[0, 2 \cdot 10^4]$ for $\omega_1 = 0.5$, $\omega_2 = 0.15 + 0.0012\,i$, $\omega_3 = 0.3 - 0.003\,i$, $\lambda_1 = 0.1$, $\lambda_2 = 0.2$, $\lambda_3 = 0.3$, and initial conditions $n_1(0) = 0.4$, $n_2(0) = 0.2$ and $n_3(0) = 0.7$ (Source: Bagarello et al. [66]).

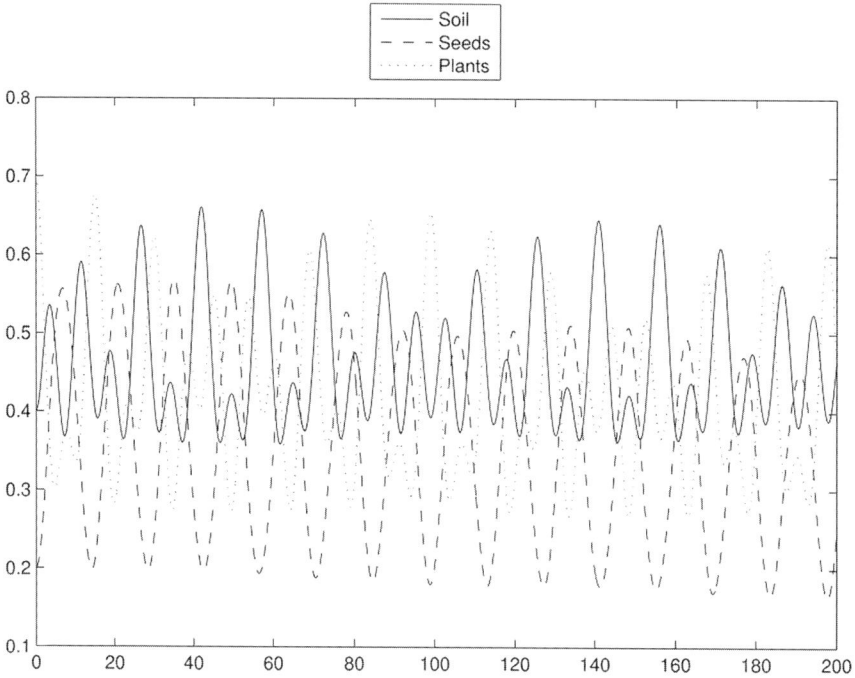

Figure 4.6 One cell: $n_1(t)$, $n_2(t)$ and $n_3(t)$ for $\omega_1 = 0.5 + 0.002\,i$, $\omega_2 = 0.15$, $\omega_3 = 0.3 - 0.003\,i$, $\lambda_1 = 0.1$, $\lambda_2 = 0.2$, $\lambda_3 = 0.3$, and initial conditions $n_1(0) = 0.4$, $n_2(0) = 0.2$ and $n_3(0) = 0.7$ (Source: Bagarello et al. [66]).

first the sum of the three densities, plotted in a very large time interval, decreases in time but, apparently, does not go to zero, as we rather observe in Figure 4.3.

The same effect can be achieved by adding a small positive term to the imaginary part of the inertia of soil, ω_1, rather than to ω_2. In this case, the densities are given in Figure 4.6, showing again that, in the considered time interval, the densities do not go to zero: a sort of nontrivial *oscillating equilibrium* is reached, quite far from desertification. Hence increasing the quality of the soil reduces also the impact of the stress factor on plants: the damping of the sum of the three densities is less accentuated than before (see Figure 4.7). Actually, Figure 4.7 suggests further that the sum increases after some time. So the system reaches a *global minimum*, but then, apparently, the ecological situation seems to improve. Notice that this effect can only be observed in the longer time interval considered in Figure 4.7; the one in Figure 4.6 is not sufficient: it takes time to recover.

Then, the conclusion of our analysis for the single cell is the following: the effect of stress factors on the plants can be safely balanced by a *mild* positive feedback, e.g., by adding very few seeds in $\mathcal{R}$, or, even better, by improving the quality of the soil, for instance adding some fertilizer from time to time, or improving the

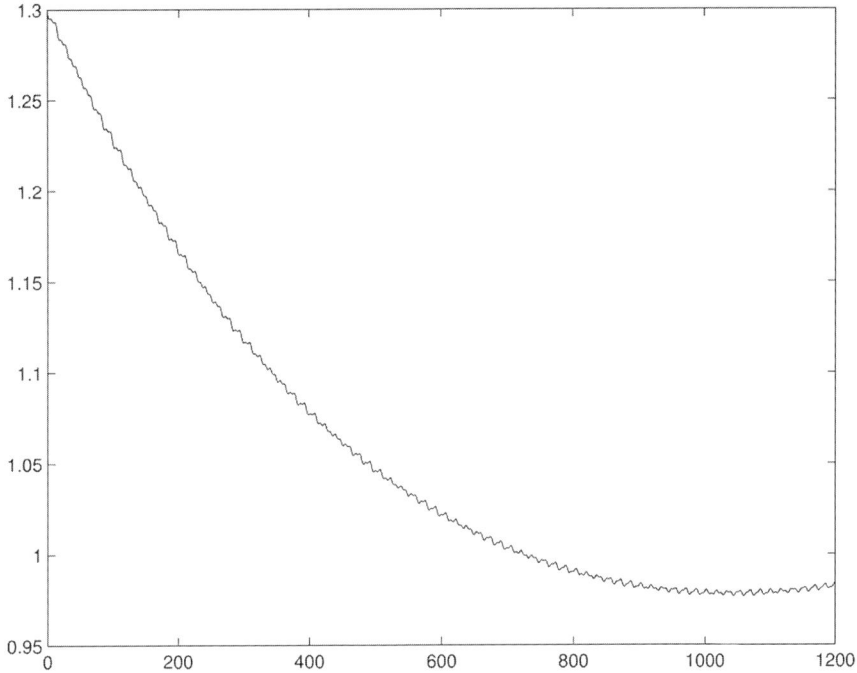

Figure 4.7 One cell: $n_1(t) + n_2(t) + n_3(t)$ in the time interval $[0, 1200]$ for $\omega_1 = 0.5 + 0.002\,i$, $\omega_2 = 0.15$, $\omega_3 = 0.3 - 0.003\,i$, $\lambda_1 = 0.1$, $\lambda_2 = 0.2$, $\lambda_3 = 0.3$, and initial conditions $n_1(0) = 0.4$, $n_2(0) = 0.2$ and $n_3(0) = 0.7$ (Source: Bagarello et al. [66]).

irrigation system. The magnitude of the *positive feedback* (positive imaginary parts) has not to be high, compared with the stress. In fact, both $|\Im(\omega_1)|$ and $|\Im(\omega_2)|$ are smaller than $|\Im(\omega_3)|$: a strong stress factor can be efficiently balanced by a little positive reaction. This aspect of the model is particularly interesting for us, since it suggests that desertification can be fought, and won, with a relatively small effort.

4.3.2 Two-Dimensional Square Lattice

We now consider the more interesting situation in which $L > 1$. In particular, to make the situation realistic but not particularly complicated from a computational point of view, from now on we fix $L = 11$. Thus, the lattice $\mathcal{R}$ is composed by 121 cells[7]: differently from the dynamics described in Section 4.3.1 we must now take into account other factors, such as mobility and overpopulation. As we have seen,

---

[7] Computations performed on larger lattices lead to similar results, with a huge increase of the time needed to compute e^{Tt} in Equation (4.4): for this reason we believe that the choice $L = 11$ is optimal, in terms of computational time and quality of results.

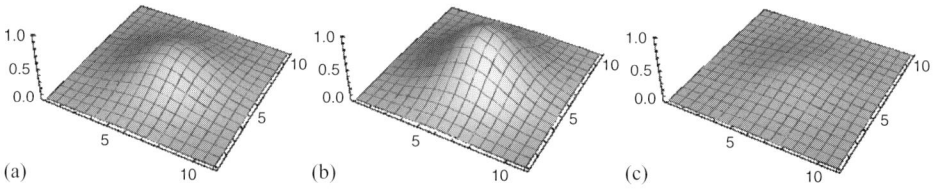

Figure 4.8 Gaussian initial data for $n_1(t)$ (a), $n_2(t)$ (b) and $n_3(t)$ (c) (Source: Bagarello et al. [66]).

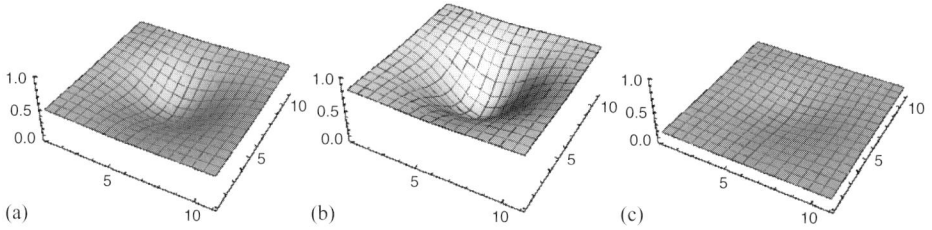

Figure 4.9 Antigaussian initial data for $n_1(t)$ (a), $n_2(t)$ (b) and $n_3(t)$ (c) (Source: Bagarello et al. [66]).

these effects are described by H_2 and H_3 respectively. In particular, we recall that H_2 describes the dispersion of the seeds, while H_3 describes competition among plants.

We considered two different sets of initial conditions for S_1, S_2 and S_3, which we call, for obvious reasons, *gaussian* and *antigaussian*. They are shown respectively in Figures 4.8 and 4.9, where we plot the spatial distributions at $t = 0$ of $n_1(0)$ (a), $n_2(0)$ (b) and $n_3(0)$ (c). In the gaussian case we see that the three functions $n_j(0)$ assume their maxima in the center of $\mathcal{R}$, while in the antigaussian situation, these maxima are assumed at the border of $\mathcal{R}$, while the center of $\mathcal{R}$ corresponds to the minima of each $n_j(0)$.

To produce some quantities related to the whole lattice $\mathcal{R}$, and not just to a single cell, we introduce the mean values $\bar{n}_1(t)$, $\bar{n}_2(t)$, $\bar{n}_3(t)$, and the variances $\sigma_1^2(t)$, $\sigma_2^2(t)$, $\sigma_3^2(t)$, of $n_{1,\alpha}(t)$, $n_{2,\alpha}(t)$ and $n_{3,\alpha}(t)$, as follows:

$$\bar{n}_i(t) = \frac{1}{L^2} \sum_{\alpha=1}^{L^2} n_{i,\alpha}(t), \quad \sigma_i^2(t) = \frac{1}{L^2} \sum_{\alpha=1}^{L^2} \left(n_{i,\alpha}(t) - \bar{n}_i(t)\right)^2, \quad i = 1, 2, 3.$$

We see that each $\bar{n}_i(t)$ describes how the densities of the agents S_j are distributed, in average, along $\mathcal{R}$. The variance $\sigma_i^2(t)$ measures how much the value $\bar{n}_i(t)$ differs from the various $n_{i,\alpha}(t)$, in each cell, and then takes a mean over all the cells.

In Figures 4.10–4.13 we plot such mean values and variances for both the gaussian and the antigaussian initial conditions and for two different sets of real inertia

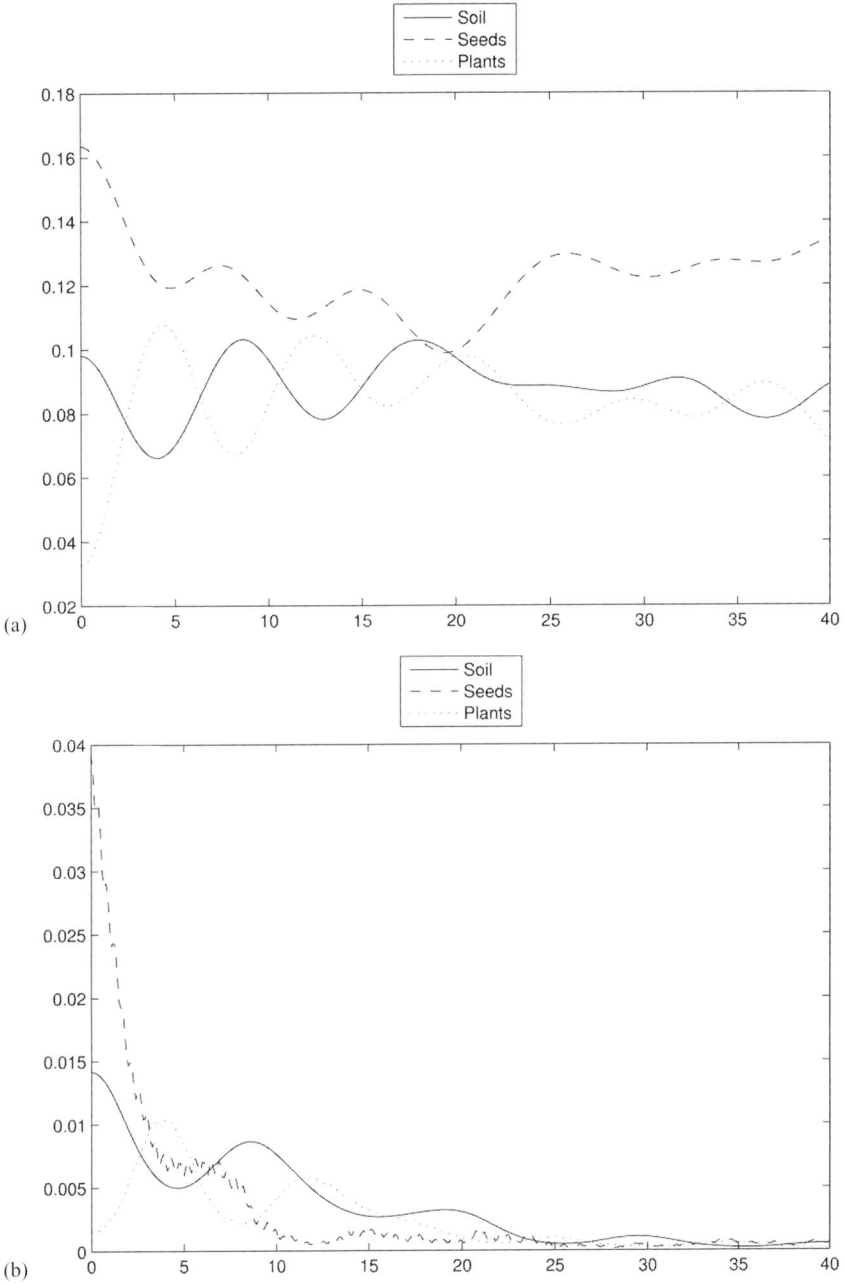

Figure 4.10 Mean values (a) and variances (b) for the first set of $\omega_{i,\alpha}$ and gaussian initial data (Source: Bagarello et al. [66]).

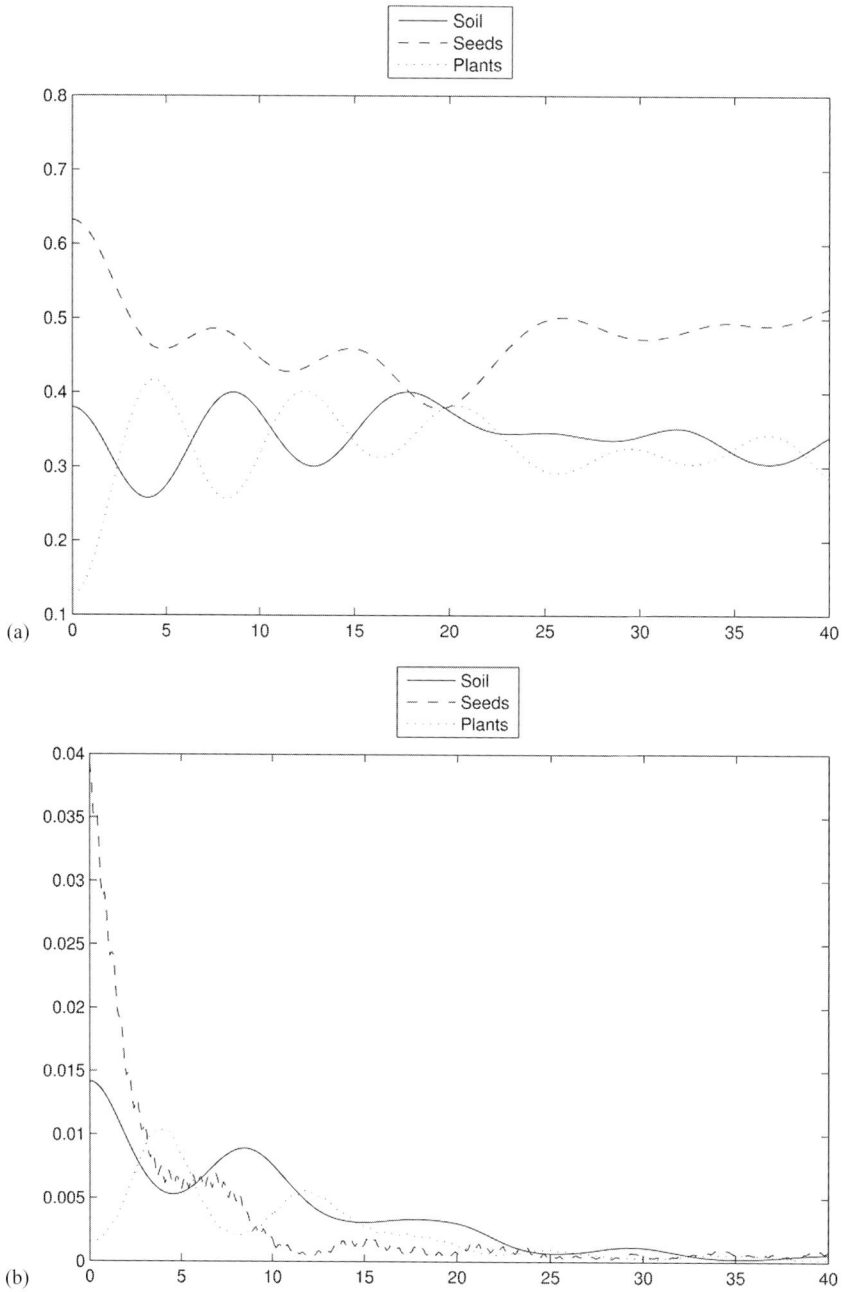

Figure 4.11 Mean values (a) and variances (b) for the first set of $\omega_{i,\alpha}$ and anti-gaussian initial data (Source: Bagarello et al. [66]).

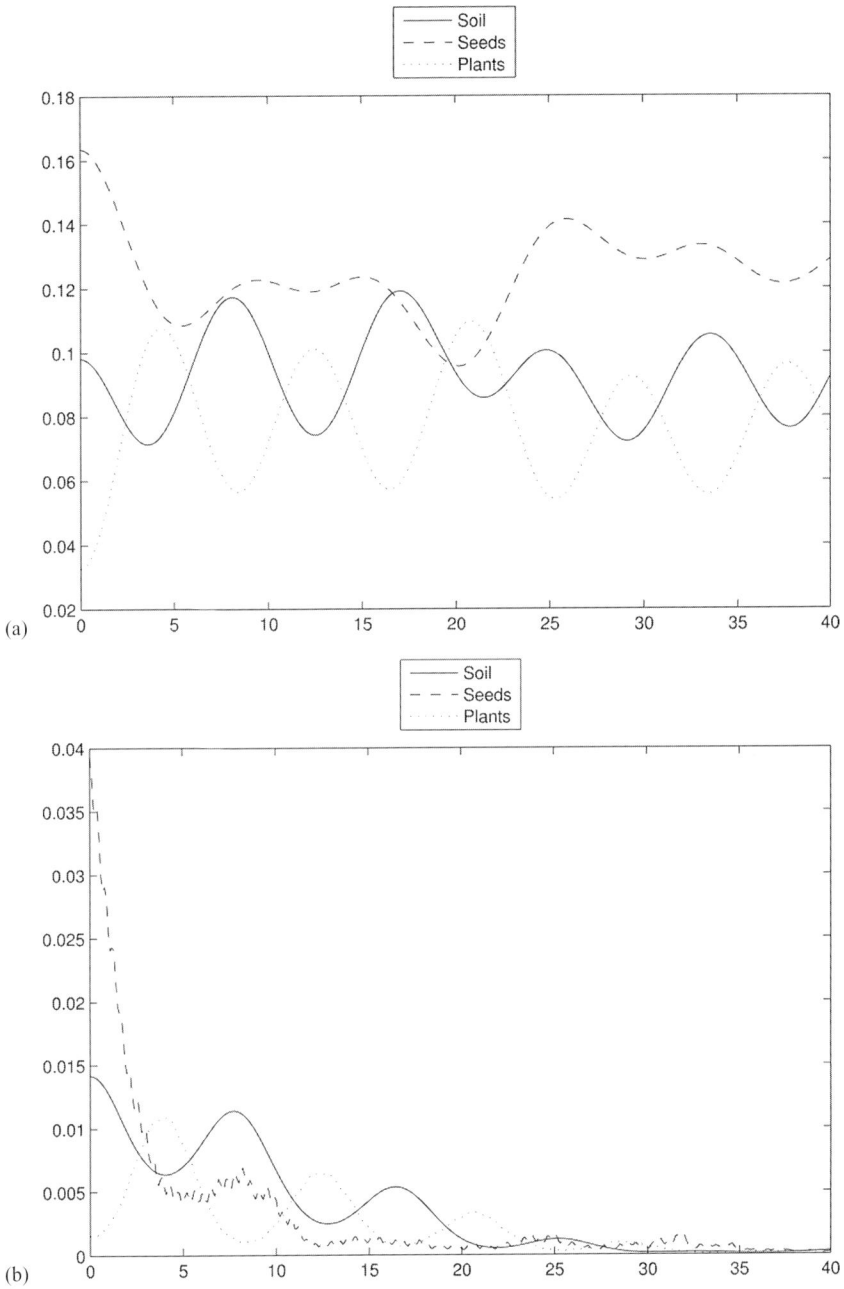

Figure 4.12 Mean values (a) and variances (b) for the second set of $\omega_{i,\alpha}$ and gaussian initial data (Source: Bagarello et al. [66]).

(a)

(b)

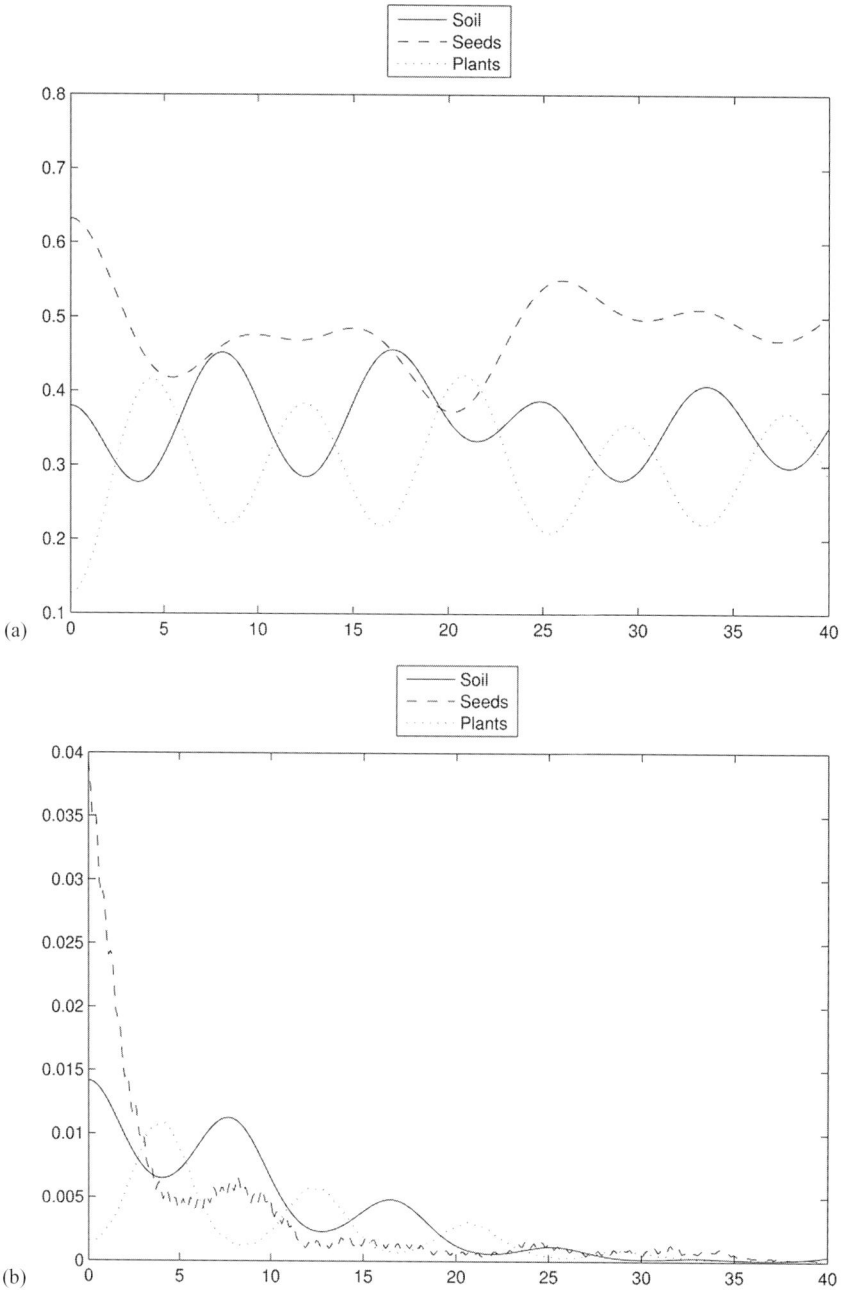

Figure 4.13 Mean values (a) and variances (b) for the second set of $\omega_{i,\alpha}$ and antigaussian initial data (Source: Bagarello et al. [66]).

parameters in H, which makes of H a self-adjoint operator: for the moment, we are not inserting any stress (or any positive effect) in the whole system. In particular, we consider the following sets:

$$\{\omega_{1,\alpha} = 0.5, \omega_{2,\alpha} = 0.15, \omega_{3,\alpha} = 0.3\} \quad \text{for all } \alpha = 1,\dots,L^2 \quad \text{(set 1)};$$

$$\{\omega_{1,\alpha} = 0.7, \omega_{2,\alpha} = 0.1, \omega_{3,\alpha} = 0.5\} \quad \text{for all } \alpha = 1,\dots,L^2 \quad \text{(set 2)}.$$

The other parameters are, for both sets, $\lambda_{1,\alpha} = 0.1$, $\lambda_{2,\alpha} = 0.2$, $\lambda_{3,\alpha} = 0.3$, $\mu_\alpha = 1$, $\nu_\alpha = 4$, for all α. We see that these parameters are all cell-independent. Hence, we are considering $\mathcal{R}$ as a homogeneous region, and the spatial differences in the system are only due to the initial distributions of the S_j, which differ from cell to cell (with some sort of rotational invariance, as Figures 4.8 and 4.9 clearly show). Of course, this is not the only possibility, and in fact we will consider a different choice later.

The parameters $p_{\alpha,\beta}$ and $c_{\alpha,\beta}$ in H_3 are fixed in the following way:

$$p_{\alpha,\beta} = \frac{0.3}{d\,(\alpha,\beta)}, \qquad c_{\alpha,\beta} = \frac{0.03}{d\,(\alpha,\beta)}, \tag{4.8}$$

for all $\alpha \neq \beta$, where $d(\alpha,\beta)$ is the Euclidean distance between the cells α and β. The choice of $p_{\alpha,\beta}$ is justified because movements of the seeds are less probable the higher the distances between the two cells involved. Analogously, a competitive term between plants is more plausible for plants which are close to each other. For these reasons, both $p_{\alpha,\beta}$ and $c_{\alpha,\beta}$ are expected to decrease with the distance between α and β, and this effect is achieved with the definitions above. Notice also that we are assuming that the effect of dispersion of the seeds is stronger than the competitive effect of overpopulation for plants. This is reflected by the numbers appearing in the formulas for $p_{\alpha,\beta}$ and $c_{\alpha,\beta}$.

In the case of $L = 11$, the system $\mathcal{S}$ can apparently stay away from desertification, while keeping self-adjointness of the Hamiltonian. In other words, in absence of stress for the plants, each $n_j(t)$ presents, in the time interval considered here, oscillations which, in most cases, are around some strictly positive value n_j^∞ and all the variances tend to zero: $n_j(t) \rightsquigarrow n_j^\infty \neq 0$, when $t \to \infty$, for $j = 1, 2, 3$. We say that the system *thermalizes*: this is what we can see from Figures 4.10–4.13, which are deduced adopting sets 1 and 2 of the parameters of H, and gaussian or antigaussian initial data for the agents. However, we should stress that what we observe for large t is not really some asymptotic value, but some (possibly small) oscillation far from zero. This is good, since otherwise we would get in contradiction with what we have discussed in Sections 2.7 and 4.3.1. In fact, even if very high-dimensional, the Hamiltonian H in Equation (4.1) can be written as an Hermitian matrix. Hence, only periodic or quasi-periodic motions are possible, strictly speaking.

This result is independent from the initial conditions and from the choice of parameters: in the examples here it holds for both gaussian or antigaussian initial values and both sets of parameters. The self-adjoint Hamiltonian (4.1), which does not allow any equilibrium if $L = 1$, exhibits a sort of global equilibrium[8] if L increases. This is true if $L = 11$, but a similar behavior has been observed already for smaller values of L, i.e., for $L = 4$ or $L = 5$, and it is expected, a fortiori, also for higher values of L.

It is worth noticing that there exists a mathematical reason why S does never go to desertification, at least, if the system is not *desertified* already at $t = 0$. The reason is the following: if $H = H^\dagger$, H commutes with the total density operator $\hat{N}_\mathcal{R} = \sum_{\beta=1}^{L^2} \sum_{k=1}^{3} \hat{n}_{k,\beta}$: $[H, \hat{N}_\mathcal{R}] = 0$. Therefore, $\hat{N}_\mathcal{R}$ is an integral of motion and during the time evolution the (densities of the) three compartments spread all along $\mathcal{R}$, approaching a common mean value of the densities in each lattice cell, which cannot be zero if S was not already desertified at $t = 0$: this is deduced by the fact that all the variances $\sigma_i^2(t)$ decay to zero after some time. Of course, such a spread cannot occur if $L = 1$, since there are no other cells where to move. However, also in this case $\hat{N}_\mathcal{R}$ is an integral of motion. The main conclusion here is that, as in the case $L = 1$, in absence of stress we do not observe desertification for the system: the agents just interact among them but these interactions do not cause the complete destruction of any of the agents. They are only responsible of a periodic or quasi-periodic change in the densities of the seeds and plants, and in the quality of the soil.

4.3.3 The Non-Conservative Case

In more realistic examples, one (or more) of the agents of the system can experience some stress. We will model such a stress as in Section 4.3.1, by assuming that some of the inertia parameters have a negative imaginary term.

In Figure 4.14 we show what happens to the densities $n_j(t)$ when a spatially homogeneous stress factor is introduced for the plants. More explicitly, we take $\omega_{1,\alpha} = 0.5$, $\omega_{2,\alpha} = 0.15$ and $\omega_{3,\alpha} = 0.3 - 0.06i$, for all $\alpha = 1, 2, \ldots, L^2$. This means that the stress if uniformly distributed all along the lattice $\mathcal{R}$, even if it only affects the plants. However, as we have already observed for the single cell lattice, this does not imply that the stress does not also, indirectly, have consequences on the seeds and on the quality of the soil.

In fact, as in the one–cell case, Figure 4.14 shows that, despite of the fact that the stress is introduced only in S_3, each $n_j(t)$ goes to zero. This is easy to understand: if plants disappear, they cannot contribute to improve the quality of the soil and to increase the number of the seeds, and so we get complete desertification.

---

[8] Once again, this only means small oscillations away from zero around some sort limiting value.

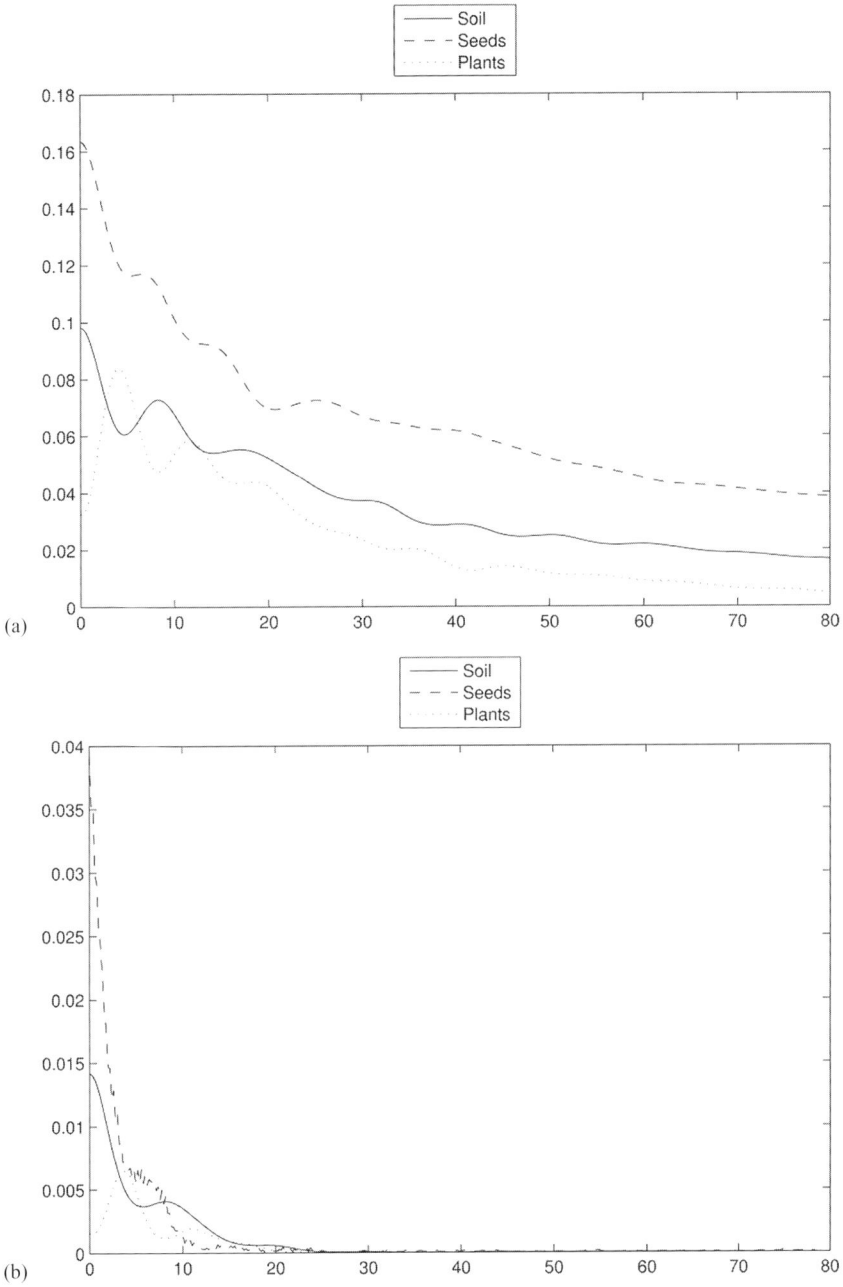

Figure 4.14 Mean values (a) and variances (b) for the model with a stress in the plants (Source: Bagarello et al. [66]).

To balance the external stress, as in Section 4.3.1, we introduce a positive effect for S_1 and S_2 by adding a spatially homogeneous, positive imaginary part in $\omega_{1,\alpha}$ and $\omega_{2,\alpha}$. In this way we improve the quality of the soil by spreading some fertilizer along $\mathcal{R}$ and we add seeds to the original population, or we replace the original seeds with other, more resistant (e.g., genetically modified), ones. Figure 4.15 shows the case in which $\omega_{1,\alpha} = 0.5 + 0.015i$, $\omega_{2,\alpha} = 0.15 + 0.015i$ and $\omega_{3,\alpha} = 0.3 - 0.06i$ for all α: the densities $n_j(t)$ start decreasing, reach a minimum and increase again; as a result, the system stays away from desertification. Once more we observe that, according to our model, it is enough to introduce a little positive effect, smaller than the stress factor itself, to avoid *bad consequences*.

4.3.4 The Non-Homogeneous Case

The last situation we want to consider here is when the region $\mathcal{R}$ is not homogeneous. We have already discussed that a natural way to describe this situation consists in taking the parameters $\omega_{k,\alpha}$ as really cell-dependent. In particular, we are interested to understand what changes when some $\omega_{k,\alpha}$ are significantly different from others. Of course, other possibilities could be considered as well, but these will not be discussed here.

Let us take $\omega_{1,\alpha} = 0.7$ and $\omega_{2,\alpha} = 0.15$ for all $\alpha = 1, \dots, L^2$. Hence, from the point of view of soil and seeds, $\mathcal{R}$ is still homogeneous. But, as far as the plants, we set $\omega_{3,\alpha} = 0.3$ everywhere except in the subregion $\mathcal{R}_{str}$ made by 16 cells located in the left lower corner of $\mathcal{R}$, where we take $\omega_{3,\alpha} = 0.3 - i$. In other words, we are considering a situation in which plants experience some stress, but only those which are localized in $\mathcal{R}_{str}$. Those distributed outside this region do not feel any *direct* stress.

The effect of such a stress on the plants in $\mathcal{R}_{str}$ is shown in Figure 4.16, where we plot the spatial distributions of the three functions $n_{j,\alpha}(t)$ in $\mathcal{R}$ at four different times, $t = 0$ (first row), $t = 10$ (second row), $t = 50$ (third row) and $t = 100$ (last row). We see that in $\mathcal{R}_{str}$ the three functions decrease after a very short time (already at $t = 10$). On the contrary, in $\mathcal{R} \setminus \mathcal{R}_{str}$ the functions *thermalize*, but on the whole lattice the average value decreases slowly to zero. In fact, even if the stress is localized in $\mathcal{R}_{str}$, its effect spreads all over $\mathcal{R}$, because of the presence of H_2 and H_3 in the Hamiltonian in (4.1).

Figure 4.17 confirms how a positive effect on S_2 can balance the stress factor on S_3, also in presence of inhomogeneities. In particular, the parameters and the initial conditions in Figure 4.17 coincide with those in Figure 4.16, except for $\omega_{2,\alpha}$. We set $\omega_{2,\alpha} = 0.15 + 0.2i$ in the central square region identified by the following cells

$$\mathcal{R}_{pos} = \{49, 50, 51, 52, 60, 61, 62, 63, 71, 72, 73, 74, 82, 83, 84, 85\},$$

while, as before, $\omega_{2,\alpha} = 0.15$ elsewhere.

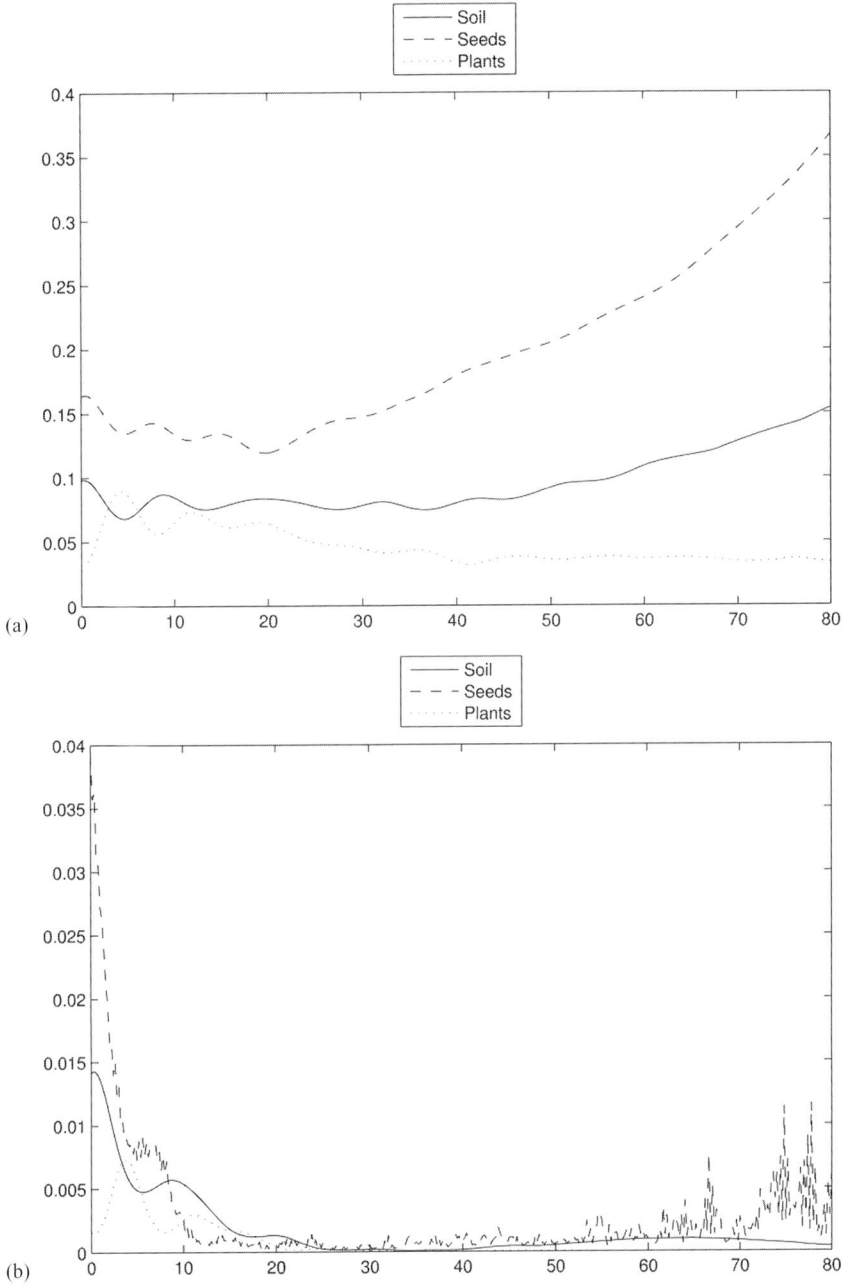

Figure 4.15 Mean values (a) and variances (b) for the model with stress and positive effects (Source: Bagarello et al. [66]).

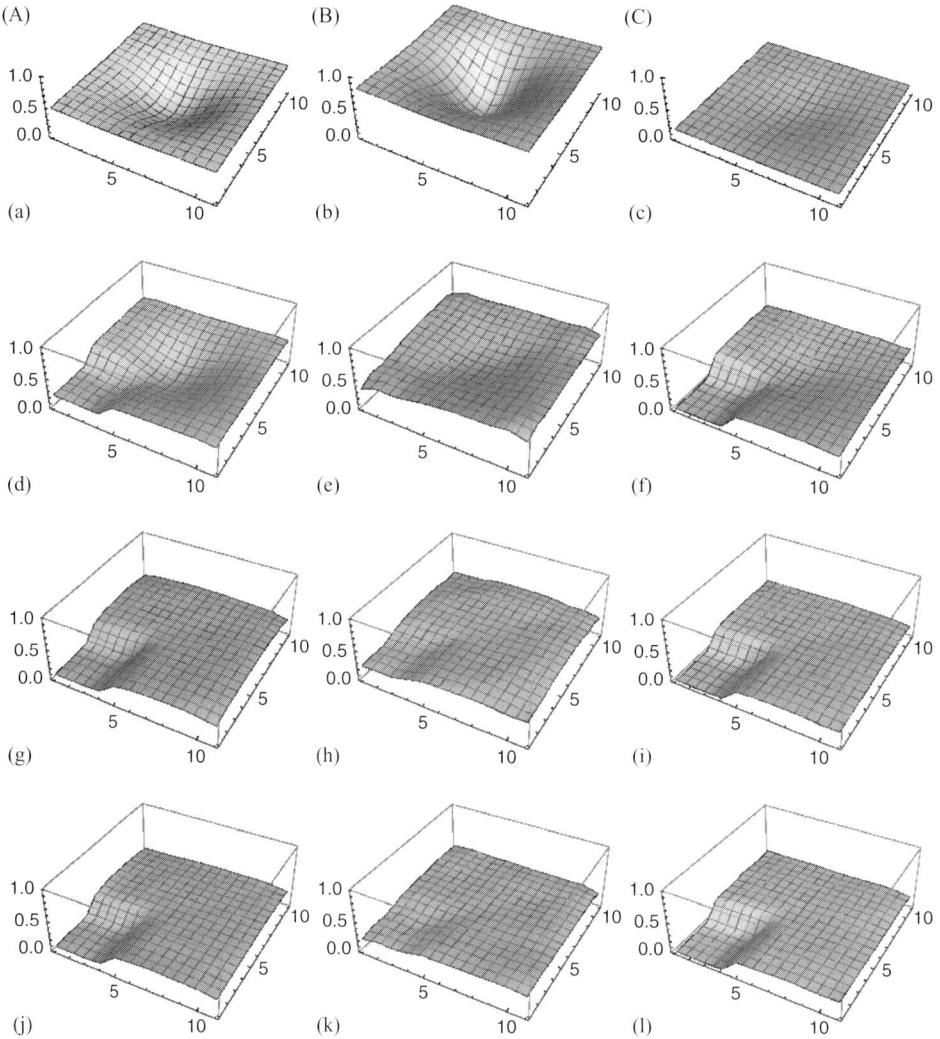

Figure 4.16 The densities $n_1(t)$ (A), $n_2(t)$ (B) and $n_3(t)$ (C) for $t = 0$ (a–c), $t = 10$ (d–f), $t = 50$ (g–i), $t = 100$ (j–l), for antigaussian initial data and stress (Source: Bagarello et al. [66]).

We observe that the positive effect is just one fifth of the stress factor ($|\mathfrak{I}\{\omega_{2,\alpha}\}| = 0.2$ while $|\mathfrak{I}\{\omega_{3,\alpha}\}| = 1$), but it is strong enough to avoid general desertification in $\mathcal{R}$ though not in $\mathcal{R}_{str}$, at least for intermediate times. This means that the stressed region still goes toward desertification, while the remaining part of $\mathcal{R}$ does not. Desertification in $\mathcal{R}_{str}$ could be avoided by adding the positive effects not in $\mathcal{R}_{pos}$, but in $\mathcal{R}_{str}$ too. This is not very surprising, in view of what we have seen before, where the stress and the positive effects were homogeneously distributed all along $\mathcal{R}$, and no desertification was observed at all (see Figure 4.15).

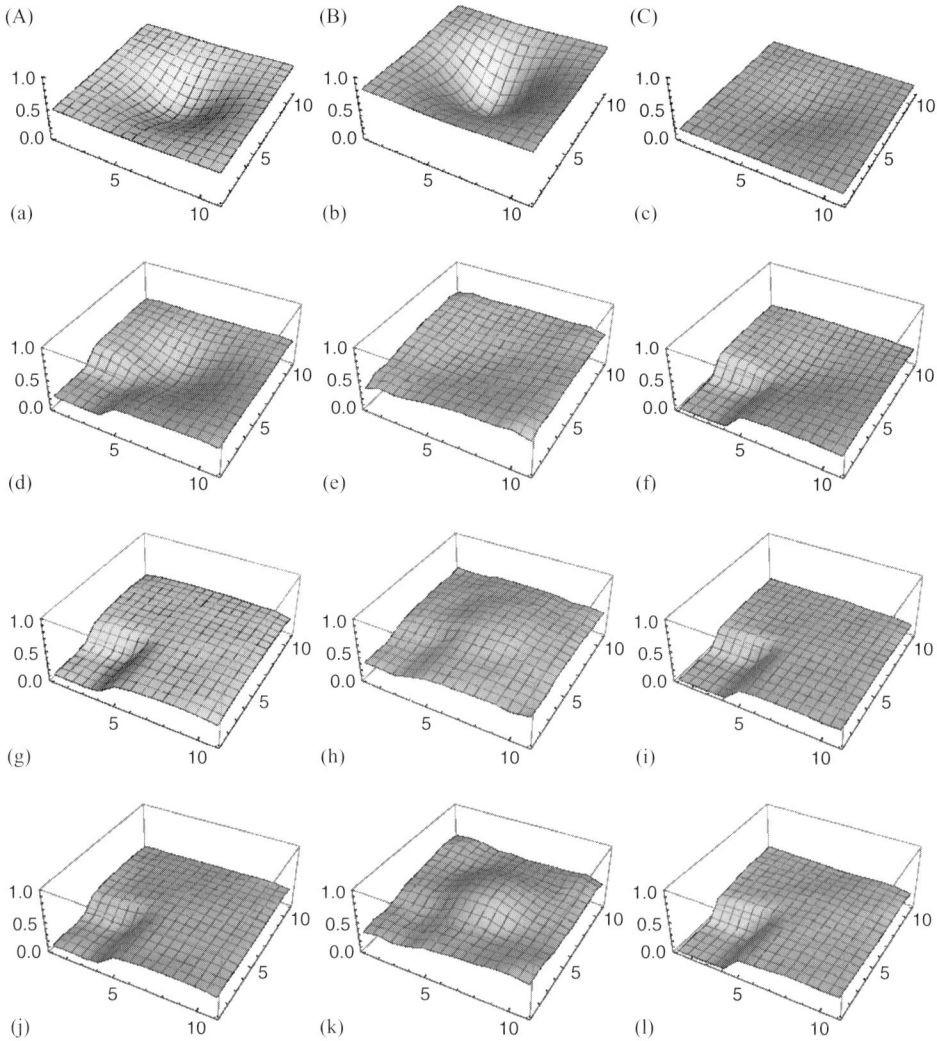

Figure 4.17 The densities $n_1(t)$ (A), $n_2(t)$ (B) and $n_3(t)$ (C) for $t = 0$ (a–c), $t = 10$ (d–f), $t = 50$ (g–i), $t = 100$ (j–l), for antigaussian initial data and stress and positive effects (Source: Bagarello et al. [66]).

4.4 Conclusions

The analysis undertaken in this chapter, based on the Hamiltonian in Equation (4.1) and on its possible, non self-adjoint extension, can be summarized as follows:

- In a nontrivial two-dimensional homogeneous lattice, if the system is purely conservative (i.e., if $H = H^\dagger$), the system *stays away* from desertification, at least if the system was not already *desertified* at $t = 0$. This is easy to understand in terms of the integral of motion $\hat{N}_\mathcal{R}$.

- Any stress factor, even when acting on just one agent, drives the whole system to desertification.
- Adding a small positive feedback may quite well balance stronger external stresses, and makes it possible to avoid desertification.
- For non-homogenous lattices the same results can be deduced, with some differences: a positive effect avoids global desertification on the lattice but not in the region where the external stress is localized, at least in the short period. The obvious alternative is to put the positive effect exactly in the same part of the lattice where the stress is localized.

It would be quite interesting to test the model in real experiments. For example, the parameters for the conservative case could be calibrated first on a real ecosystem S localized in $\mathcal{R}$, kept free of external stress and positive effects. After that, a stress factor can be introduced by removing some plants from the system. Then $\omega_{3,\alpha}$ should be changed accordingly, by adding a suitable negative imaginary part, and the evolution of the model and of the real ecosystem compared to fix this value. Finally, to check how the deterioration of the ecosystem can be reversed, a positive external effect could be calibrated by adding seeds or improving the quality of the soil in the real ecosystem: then $\omega_{1,\alpha}$ and $\omega_{2,\alpha}$ change, adding this time certain positive imaginary parts. Once the parameters of H are fixed according to the experimental data in a certain time interval, the next step would be to check if this choice allows us to predict the time evolution also for longer times, and use this prediction to keep transition to desertification under control. This is quite a fascinating problem (other than being quite useful for its applications) and we hope to be able to work on that in the close future.

5

Escape Strategies

5.1 Introduction

As we have seen in Chapters 3 and 4, ladder operators can be efficiently used in the analysis of systems that are based on some exchange between two different agents of the system: if a raising operator of agent τ_1 is linked to a lowering operator of agent τ_2, as in $a_1^\dagger a_2$, this term in a Hamiltonian mimics the fact that, for instance, τ_1 is giving a *unit of something* to τ_2. Or, that what is lost by τ_1 is gained by τ_2. As we have already observed in the previous chapter, ladder operators can be used to describe movements of some agent in a (two-dimensional) lattice. For instance, the contribution $a_{2,\beta} a_{2,\gamma}^\dagger$ in (4.1) has been used to describe a movement of agent 2 from cell β to cell γ. A similar approach will be used also in this chapter, where the agents will not be seeds, but people. The main idea of our model was first proposed in [7], and then in [1], to describe the interaction between two different species that move (i.e., migrate) along a two-dimensional lattice. In particular, in [7], our primary interest was to describe a migration of one of the two populations, from some poor to some rich area of, say, the Mediterranean Sea. As in [7] we will introduce two different families of operators for the two populations and, similarly to what we have done in the previous chapter, we label these operators with an index of cell. This is a possible way to describe a situation in which each cell can be occupied, or not, by one or two populations. The interaction between these populations, and their movements along the lattice, are described in terms of creation and annihilation operators similarly to what is done in (4.1) in the context of desertification to describe the spreading of the seeds, see H_2. The lattice considered in [1, 7] is *topologically simple*: it is just a rectangular (or a square) lattice, with all the cells essentially equivalent from the point of view of the movements: all the cell can be occupied by both populations, and there is no *forbidden region* anywhere in $\mathcal{R}$.

What we want to describe in this chapter is a similar situation, but with some relevant differences: similar from the point of view of the main dynamical ingredient, the Hamiltonian of the system. But this situation is different because of the topology of the lattice, which is rich in forbidden areas (some obstacles) and is not even closed: while in [1, 7] the two populations have to stay inside $\mathcal{R}$, in what we are going to discuss here, leaving $\mathcal{R}$ is exactly the main goal of the two populations. Moreover, they are both interested in leaving $\mathcal{R}$ as fast as possible. In other words, we are now going to discuss a possible *escape strategy*: we have two populations, $\mathcal{P}_a$ and $\mathcal{P}_b$, distributed somewhere in a room $\mathcal{R}$. $\mathcal{P}_a$ can be the set of young people moving in $\mathcal{R}$, while $\mathcal{P}_b$ could be that of the aged. Also, $\mathcal{R}$ could be a shop, a football stadium, an airport or something else. $\mathcal{R}$ has some obstacles, which can be some internal pylons or some showcase inside the shop, for instance, and has some exits, the doors of the room, which can be close to or far from the members of the two populations. We imagine that, at $t = 0$, some alarms ring and the two populations try to run away out of $\mathcal{R}$ as fast as they can. Of course, the best situation is when $\mathcal{P}_a$ and $\mathcal{P}_b$ do not disturb each other. Our main effort here will be to propose a model, constructed in terms of ladder operators, which describes this situation and suggests how the escape from the room of both populations can be optimized.

5.2 The Model

Let us consider a 2D-region $\mathcal{R}$ in which, in principle, the two populations $\mathcal{P}_a$ and $\mathcal{P}_b$ are distributed. The region $\mathcal{R}$ (e.g., rectangular or square) is divided into N cells (see Figure 5.1), labeled by $\alpha = 1, 2, \ldots, L_x L_y =: N$; to simplify the notation we will always assume that $L_x = L_y = L$, so that $N = L^2$. As we have already stressed, here, contrary to what we have done in [7], not all the cells can be occupied, in principle, since there are obstacles in $\mathcal{R}$ where the populations cannot go. Moreover, $\mathcal{R}$ is not

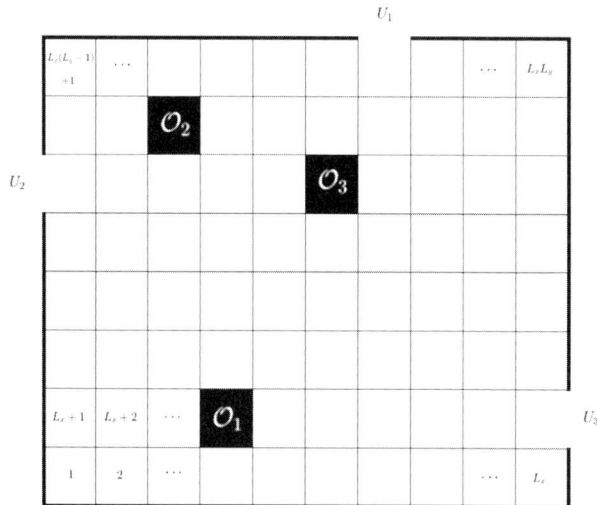

Figure 5.1 The two-dimensional lattice with three obstacles and three exits.

a closed region as it was in [7], but there are exits somewhere on the borders, exits that $\mathcal{P}_a$ and $\mathcal{P}_b$ want to reach, as fast as they can, to leave $\mathcal{R}$ under some emergency. Figure 5.1 describes one such possibility, a room with three obstacles (the cells in black O_1, O_2 and O_3) and three exits in three different sides of the region, U_1, U_2 and U_3.

Following our general scheme, the dynamics of the system $\mathcal{S} = \mathcal{P}_a \cup \mathcal{P}_b$ is given in terms of a self-adjoint Hamiltonian operator that contains the main mechanisms we expect could take place in $\mathcal{S}$. The two populations in each cell are described by the following ladder operators: a_α, $a_\alpha^\dagger$ and $\hat{n}_\alpha^{(a)} = a_\alpha^\dagger a_\alpha$ for what concerns $\mathcal{P}_a$, and b_α, $b_\alpha^\dagger$ and $\hat{n}_\alpha^{(b)} = b_\alpha^\dagger b_\alpha$ for $\mathcal{P}_b$. These operators are assumed to satisfy the following CAR:

$$\left\{ a_\alpha, a_\beta^\dagger \right\} = \left\{ b_\alpha, b_\beta^\dagger \right\} = \delta_{\alpha,\beta} \mathbb{1}, \qquad \left\{ a_\alpha^\sharp, b_\beta^\sharp \right\} = 0. \tag{5.1}$$

The reason for this choice is that, in analogy with what we have done in Chapter 4, we are interested here in deducing the time dependence of some densities of the two species, and densities are well described, in our language, when fermionic operators are used. In fact, it is natural to interpret the mean values of the operators $\hat{n}_\alpha^{(a)}$ and $\hat{n}_\alpha^{(b)}$ as *local density operators* of the two populations in the cell α: if the mean value of, say, $\hat{n}_\alpha^{(a)}$ in the state describing the system is equal to one, this means that the density of $\mathcal{P}_a$ in the cell α, at that particular time, is very high. On the other hand, if the mean value of, say, $\hat{n}_\beta^{(b)}$, in the state of the system is equal to zero, we interpret this as saying that, in cell β, we can find only very few members of $\mathcal{P}_b$, or none. Of course, this interpretation is time-dependent, meaning with this that these mean values can change significantly while the system evolves, i.e., when the people move along the room.

The description of the dynamics of the two populations inside each cell α is given by the Hamiltonian

$$H_\alpha = H_\alpha^0 + \lambda_\alpha H_\alpha^I, \quad H_\alpha^0 = \omega_\alpha^a a_\alpha^\dagger a_\alpha + \omega_\alpha^b b_\alpha^\dagger b_\alpha, \quad H_\alpha^I = a_\alpha^\dagger b_\alpha + b_\alpha^\dagger a_\alpha. \tag{5.2}$$

We have a free Hamiltonian H_α^0, which does not change the densities of $\mathcal{P}_a$ and $\mathcal{P}_b$ if the system is prepared in an eigenstate of both $\hat{n}_\alpha^{(a)}$ and $\hat{n}_\alpha^{(b)}$. Hence H_α^0 just describes the simplest situation in which the densities of the two populations in the cell α do not change in time. The interaction between $\mathcal{P}_a$ and $\mathcal{P}_b$ in α is described by H_α^I, which can be understood as follows: once the density of, say, $\mathcal{P}_a$ increases in the cell α, the density of $\mathcal{P}_b$ in the same cell must decrease, and vice versa. This means that one species tends to exclude the other. This is a sort of predator-prey term: the two populations do not like to share the same position. This is particularly reasonable if the size of the cell is small enough. When this is so, it is hard that two people occupy the same cell, of course. Notice that, assuming that all the (cell-depending) parameters in (5.2) are real, $H_\alpha = H_\alpha^\dagger$. The fact that the parameters can depend on α is useful to describe anisotropic situations, which will be particularly relevant in our analysis.

The full Hamiltonian H must consist of a sum of all the different H_α, one for each cell, plus another contribution, h, responsible for the diffusion of the populations all around the lattice: $H = \sum_\alpha H_\alpha + h$. A natural choice for this diffusion Hamiltonian h is the following one:

$$h = \sum_{\alpha,\beta} \left\{ p^{(a)}_{\alpha,\beta} \left(a_\alpha a^\dagger_\beta + a_\beta a^\dagger_\alpha \right) + p^{(b)}_{\alpha,\beta} \left(b_\alpha b^\dagger_\beta + b_\beta b^\dagger_\alpha \right) \right\}, \tag{5.3}$$

where $p^{(a)}_{\alpha,\beta}$ and $p^{(b)}_{\alpha,\beta}$ are real quantities that are rather important in our model, as we will see soon. The various terms in h can be easily understood: for instance, the term $a_\alpha a^\dagger_\beta$ describes the movement of $\mathcal{P}_a$ from the cell α to the cell β,[1] while its adjoint, $a_\beta a^\dagger_\alpha$, describes an opposite movement. Going back to the $p^{(a,b)}_{\alpha,\beta}$, their analytic expressions are based on the following minimal requirements: (i) they must be equal to zero whenever the populations cannot move from cell α to cell β; (ii) in general, we expect that $p^{(a)}_{\alpha,\beta} \neq p^{(b)}_{\alpha,\beta}$, since the two populations may have different behaviors: $\mathcal{P}_a$ can move faster than $\mathcal{P}_b$, or they can tend to move toward different exits, just to consider two obvious differences; (iii) in our approach they are used to *suggest to the populations* the fastest paths toward the exits. We will discuss this particular aspect in some detail later on.

With these considerations in mind, it is clear first that $p^{(a,b)}_{\alpha,\beta} = 0$ when β is a cell in which an obstacle (or part of it) is located. In this case, in fact, neither $\mathcal{P}_a$ nor $\mathcal{P}_b$ can occupy that cell. Also, $p^{(a,b)}_{\alpha,\beta} = 0$ whenever α and β are not nearest neighbors: in this case, to move from α to β each population must cross all the cells between the two. We will allow here also for the *diagonal* movements. For instance, $\mathcal{P}_a$ can move from cell 1 to cell $L_x + 2$ (see Figure 5.1) and not just from cell 1 to cell 2 or to cell $L_x + 1$.

A more delicate point is that we assume here that $p^{(a,b)}_{\alpha,\beta} = p^{(a,b)}_{\beta,\alpha}$: the two sets of parameters are taken to be symmetric under the exchange $\alpha \leftrightarrow \beta$. This is not really what we would like, since we are more interested in suggesting a direction to the two agents, while this choice implies, in principle, that people can go back and forth without any particular constraint. However, even if we are interested in introducing some *preferred* direction, requiring for instance that $p^{(a,b)}_{\alpha,\beta} > p^{(a,b)}_{\beta,\alpha}$, it is possible to check that, doing so, the Heisenberg equations of motion would depend not just on $p^{(a,b)}_{\alpha,\beta}$, or on $p^{(a,b)}_{\beta,\alpha}$, but on their sum $p^{(a,b)}_{\alpha,\beta} + p^{(a,b)}_{\beta,\alpha}$. Therefore, the resulting coefficients in the differential equations would be automatically symmetrical under the exchange $\alpha \leftrightarrow \beta$ anyway, even if the original coefficients $p^{(a,b)}_{\alpha,\beta}$ in (5.3) were not.

---

[1] This is because the presence of a_α causes a lowering of the density of $\mathcal{P}_\alpha$ in the cell α, density that increases in the cell β because of $a^\dagger_\beta$.

This is the main reason why, to simplify the treatment, we adopt the symmetric choice $p_{\alpha,\beta}^{(a,b)} = p_{\beta,\alpha}^{(a,b)}$ from the very beginning.

Remark: A different way to force movement from cell α to cell β could be to replace h in (5.3) with the following, manifestly non-self-adjoint (if $p_{\alpha,\beta}^{(a,b)} \neq p_{\beta,\alpha}^{(a,b)}$), operator

$$\hat{h} = \sum_{\alpha,\beta} \left\{ p_{\alpha,\beta}^{(a)} \, a_\alpha a_\beta^\dagger + p_{\alpha,\beta}^{(b)} \, b_\alpha b_\beta^\dagger \right\}.$$

However, this choice will not be discussed here. We postpone the analysis of a similar Hamiltonian in a different context to Chapter 7, where we will discuss that, sometimes, this kind of Hamiltonians can still be efficiently used in concrete applications, despite the many complications, described in Section 2.7.2 arising from the fact that $\hat{h} \neq \hat{h}^\dagger$.

Another interesting aspect of our general framework, and of the use of the CARs (5.1) in particular, is that they automatically implement the impossibility of having too many elements of a single population in a given cell. This is because $a_\alpha^{\dagger\,2} = b_\alpha^{\dagger\,2}$, for any α. So the formulas in Equation (5.1) are motivated not only because those operators can be used to describe densities, but also because of their *physical* consequences.

Now we are ready to compute the time evolution of the densities of both $\mathcal{P}_a$ and $\mathcal{P}_b$ inside $\mathcal{R}$. This is what we need to follow the movements of the two populations in $\mathcal{R}$ and to check if they are leaving $\mathcal{R}$ sufficiently fast, or not, and then to optimize the escape procedure. To achieve this result we need first to compute the time evolution $\hat{n}_\alpha^{(a)}(t)$ and $\hat{n}_\alpha^{(b)}(t)$ for each $\alpha = 1, 2, \ldots, N$, and then take their expectation values on a vector state describing the initial status of $\mathcal{S}$. This vector describes the densities, at $t = 0$, of $\mathcal{P}_a$ and $\mathcal{P}_b$ in each cell of $\mathcal{R}$. The idea reflects what we have done before, for instance in (3.35) and in (4.5), in different contexts.

Remark: We notice that, as in Chapter 4 and contrary to what was proposed in Chapter 3, the model we are considering here is *closed*: we are not considering any reservoir for the two populations. This is reasonable here, at least when the movements of $\mathcal{P}_a$ and $\mathcal{P}_b$ are not driven by any external factor, as it is assumed in the situation we are interested in now.

In order to deduce $\hat{n}_\alpha^{(a)}(t)$ and $\hat{n}_\alpha^{(b)}(t)$, it is convenient first to look for the time evolution of both a_α and b_α, by writing the Heisenberg differential equation $\dot{a}_\alpha = i[H, a_\alpha]$ and $\dot{b}_\alpha = i[H, b_\alpha]$. We get

$$\begin{cases} \dot{a}_\alpha = -i\omega_\alpha^a a_\alpha - i\lambda_\alpha b_\alpha + 2i \sum_{\beta=1}^{L^2} p_{\alpha,\beta}^{(a)} a_\beta, \\[2em] \dot{b}_\alpha = -i\omega_\alpha^b b_\alpha - i\lambda_\alpha a_\alpha + 2i \sum_{\beta=1}^{L^2} p_{\alpha,\beta}^{(b)} b_\beta. \end{cases} \tag{5.4}$$

Recall that, since $p_{\alpha,\alpha}^{(a,b)} = 0$, the sums in the right-hand sides of (5.4) are really restricted to $\beta \neq \alpha$. Equation (5.4) is linear and can be rewritten as

$$\dot{X}_{L^2} = i\mathcal{K}_{L^2} X_{L^2}, \qquad (5.5)$$

where $\mathcal{K}_{L^2} = 2T_{L^2} - P_{L^2}$, T_{L^2} and P_{L^2} being two $L^2 \times L^2$ matrices defined as follows:

$$T_{L^2} = \begin{pmatrix} V_{L^2}^{(a)} & 0 \\ 0 & V_{L^2}^{(b)} \end{pmatrix}, \qquad P_{L^2} = \begin{pmatrix} \Omega^{(a)} & \Lambda \\ \Lambda & \Omega^{(b)} \end{pmatrix}.$$

We have introduced here the following diagonal matrices:

$$\Omega^{(a)} = \mathrm{diag}\left\{\omega_1^a, \omega_2^a, \dots, \omega_{L^2}^a\right\},$$

$$\Omega^{(b)} = \mathrm{diag}\left\{\omega_1^b, \omega_2^b, \dots, \omega_{L^2}^b\right\},$$

and

$$\Lambda = \mathrm{diag}\left\{\lambda_1, \lambda_2, \dots, \lambda_{L^2}\right\},$$

while $V_{L^2}^{(a)}$ is the $L^2 \times L^2$ matrix with entries different from zero, and equal to $p_{\alpha,\beta}^{(a)}$ only for those matrix elements corresponding to the allowed movements in $\mathcal{R}$ (e.g., in the positions $(1,2)$, $(1, L+1)$, $(1, L+2)$, $(2,1)$, $(2,3)$, $(2, L+1)$, $(2, L+2)$, $(2, L+3)$). Similarly, the matrix $V_{L^2}^{(b)}$ has entries zero or equal to $p_{\alpha,\beta}^{(b)}$ in the same positions as for $V_{L^2}^{(a)}$. Finally, the transpose of the unknown vector, $X_{L^2}^T$, is defined as

$$X_{L^2}^T = \begin{pmatrix} A_1(t) & A_2(t) & \cdots & \cdots & A_{L^2}(t) & B_1(t) & B_2(t) & \cdots & \cdots & B_{L^2}(t) \end{pmatrix},$$

where $A_j(t) = a_j(t)e^{i\omega_j^a t}$ and $B_j(t) = b_j(t)e^{i\omega_j^b t}$.

The solution of Equation (5.5) is

$$X_{L^2}(t) = e^{i\mathcal{K}_{L^2} t} X_{L^2}(0).$$

Let us now call $V_{\alpha,\beta}(t)$ the generic entry of the matrix $e^{i\mathcal{K}_{L^2} t}$, and let us assume that, at $t = 0$, the system is described by the vector $\varphi_{\mathbf{n}^a, \mathbf{n}^b}$, where $\mathbf{n}^a = (n_1^a, n_2^a, \dots, n_{L^2}^a)$ and $\mathbf{n}^b = (n_1^b, n_2^b, \dots, n_{L^2}^b)$. This means that the densities of $\mathcal{P}_a$ and $\mathcal{P}_b$ at $t = 0$ are (n_1^a, n_1^b) in cell 1, (n_2^a, n_2^b) in cell 2 and so on. Hence, the mean values of the time evolution of the number operators in the cell α, assuming these initial conditions, are

$$\begin{aligned} N_\alpha^a(t) &= \left\langle \varphi_{\mathbf{n}^a, \mathbf{n}^b}, a_\alpha^\dagger(t) a_\alpha(t) \varphi_{\mathbf{n}^a, \mathbf{n}^b} \right\rangle = \left\langle \varphi_{\mathbf{n}^a, \mathbf{n}^b}, A_\alpha^\dagger(t) A_\alpha(t) \varphi_{\mathbf{n}^a, \mathbf{n}^b} \right\rangle, \\ N_\alpha^b(t) &= \left\langle \varphi_{\mathbf{n}^a, \mathbf{n}^b}, b_\alpha^\dagger(t) b_\alpha(t) \varphi_{\mathbf{n}^a, \mathbf{n}^b} \right\rangle = \left\langle \varphi_{\mathbf{n}^a, \mathbf{n}^b}, B_\alpha^\dagger(t) B_\alpha(t) \varphi_{\mathbf{n}^a, \mathbf{n}^b} \right\rangle, \end{aligned} \qquad (5.6)$$

which can be rewritten as

$$N_\alpha^a(t) = \sum_{\theta=1}^{L^2} \left| V_{\alpha,\theta}(t) \right|^2 n_\theta^a + \sum_{\theta=1}^{L^2} \left| V_{\alpha,L^2+\theta}(t) \right|^2 n_\theta^b,$$

(5.7)

$$N_\alpha^b(t) = \sum_{\theta=1}^{L^2} \left| V_{L^2+\alpha,\theta}(t) \right|^2 n_\theta^a + \sum_{\theta=1}^{L^2} \left| V_{L^2+\alpha,L^2+\theta}(t) \right|^2 n_\theta^b,$$

in terms of the matrix elements $V_{\alpha,\beta}(t)$. These formulas, and their global counterparts

$$N^a(t) = \sum_{\alpha=1}^{L^2} N_\alpha^a(t), \qquad N^b(t) = \sum_{\alpha=1}^{L^2} N_\alpha^b(t),$$

will be the starting point for our numerical simulations, which we consider in the next section.

5.3 Numerical Simulations

Before looking for consequences of Equation (5.7), we need to discuss how we should describe the fact that people is going out of the room. And, even more relevant, which is the most efficient way to do this? In fact, what is interesting for us is to describe escape strategies; or, stated differently, we want to model people leaving $\mathcal{R}$. For this reason, even if the main dynamical features of the two populations in $\mathcal{R}$ are not expected to be particularly different from those of a migration process, see [7], the two cases differ for an essential feature: while in the model for migration the global densities of the populations were kept constant during time evolution (people move around $\mathcal{R}$ but nobody dies[2] or disappears), here we are interested in densities of the populations that decrease (hopefully) to zero, and the faster they decrease, the better for us. Of course, nothing like this could be possible if $H = H^\dagger$ and if the system is closed (i.e., if it has no exit): in these cases, we still deduce (for our particular form of H in Equation [5.2]), that the sum of the densities should be constant in time. This is because the Hamiltonian H commutes with the total density operator, $\hat{N} = \sum_\alpha \hat{n}_\alpha^{(a)} + \sum_\alpha \hat{n}_\alpha^{(b)}$, so that $\hat{N}(t)$, and its mean value as a consequence, remains constant in time. So people move along $\mathcal{R}$, but always stay inside! This is reasonable, since $\mathcal{R}$ has no exit at all, being closed. But, also in presence of exits, we need to understand how to model this decay of densities, and how to describe first, and to produce after, this decay in the most efficient way, from the point of view of the numerical implementation.

A possible, natural way to deal with this problem would be to consider $\mathcal{R}$ as a sub-lattice of a bigger lattice, $\mathcal{R}_{tot}$, which differs from $\mathcal{R}$ because it consists also of

---

[2] Which, unfortunately, we all know is not a realistic assumption.

more cells surrounding $\mathcal{R}$, the *outside of the room*, $\mathcal{R}_{out}$: people leave the room $\mathcal{R}$, but stay in $\mathcal{R}_{out}$. So, even if the sum of the densities of $\mathcal{P}_a$ and $\mathcal{P}_b$ stays constant in $\mathcal{R}_{tot} = \mathcal{R} \cup \mathcal{R}_{out}$, this sum can decrease when restricted to $\mathcal{R}$. This idea is surely reasonable but it is not very efficient, mainly for technical reasons: we need $\mathcal{R}_{tot}$ to be much larger than $\mathcal{R}$ for the mechanisms to work properly. But this creates serious difficulties because the computational time increases rapidly, even if the size of $\mathcal{R}$ is not so large. Moreover, it is not completely impossible that some of the people who leave the room go back inside after some time! This is, of course, something we want to avoid but, in principle, it can happen: our Hamiltonian contains terms that describe movements from a cell α to a cell β, and terms describing the reverse motion. As we have already seen, this is connected to our requirement on H, $H = H^\dagger$.

A second possibility to model some decaying densities could be to insert some negative imaginary part in the inertia of some parameters of H_0. We have seen in Chapter 4 how this mechanism describes, in a quite efficient way, a decrease in the mean values of some observables of the model. But, again, this idea cannot be implemented *sic et simpliciter*, here. In fact, if we just add a negative imaginary part to, say, ω_1^a or to $\omega_{L_x+2}^b$, the effect that we obtain is that the density of $\mathcal{P}_a$ decays in cell 1, while the density of $\mathcal{P}_b$ decays in cell $L_x + 2$, see Figure 5.1. And, since we do not observe an increment of the density of $\mathcal{P}_a$ in the cells close to 1, and of the density of $\mathcal{P}_b$ in those close to $L_x + 2$, and since cells 1 and $L_x + 2$ are far from the exits, these decays cannot be understood as *movements* of the populations. They would just describe the fact that some population is disappearing, which makes not much sense in our context, of course.

We can combine the two ideas above by adding first a region outside $\mathcal{R}$, $\mathcal{R}_{out}$, and using imaginary parts in some parameters of the Hamiltonian, but only for the cells in $\mathcal{R}_{out}$. All the parameters of H inside $\mathcal{R}$ are kept real. $\mathcal{R}_{out}$ can only be reached through the exit U. In this approach, the Hamiltonian H in (5.2) refers not only to $\mathcal{R}$, but also to $\mathcal{R}_{out}$. In other words, the sums over α and β are extended to all the lattice cells in $\mathcal{R}_{tot}$. The main difference between $\mathcal{R}$ and $\mathcal{R}_{out}$ is that, as already stated, while all the parameters of H *inside* $\mathcal{R}$ are real, those in $\mathcal{R}_{out}$, and the inertia in particular, could be complex valued, with a small negative imaginary part. In this way, when some elements of $\mathcal{P}_a$ or $\mathcal{P}_b$ reach $\mathcal{R}_{out}$, they begin to disappear and, as a consequence, they cannot return back to $\mathcal{R}$: the densities of the populations decrease simply because they have reached the courtyard $\mathcal{R}_{out}$, and they have no reason to go back to $\mathcal{R}$. Despite the fact that the idea looks reasonable, our numerical attempts show that, for this procedure to be efficient, $\mathcal{R}_{out}$ must again be sufficiently large, whereas the imaginary parts can be rather small. Otherwise we do not get what we are interested in. And, as we have seen before, a large $\mathcal{R}_{out}$ implies that the computational time and the need for memory increase quite a bit.

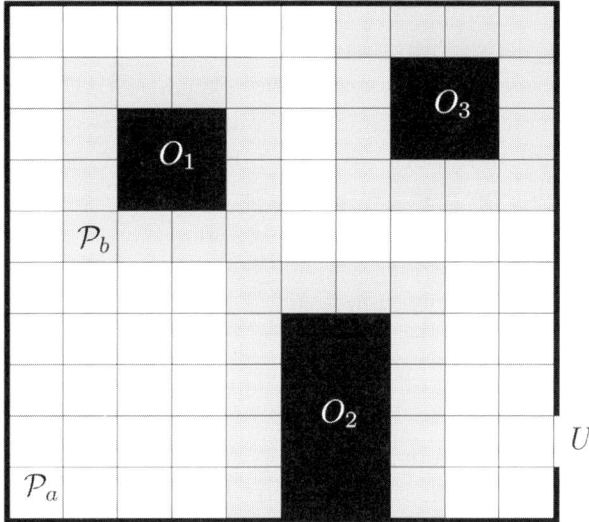

Figure 5.2 Setting S_2: The region $\mathcal{R}$ is a square of $L_x \cdot L_y = 10 \cdot 10$ cells. The exit cell U is located at $(11,2)$. At $t = 0$ the populations $\mathcal{P}_a$ and $\mathcal{P}_b$ are located in the cells $(1,1)$ and $(2,6)$ respectively. The black cells in O_1, O_2, O_3 represents the obstacles, surrounded by a region, ∂O, which might be slightly different from the rest of $\mathcal{R}$ (Source: Bagarello et al. [8]).

A different strategy, which has proven to be much more efficient for our purposes, is the following: we fix a (small) time interval ΔT and two *threshold density values* N_{thr}^a and N_{thr}^b between zero and one. Then we compute $N_\alpha^a(t)$ and $N_\alpha^b(t)$ as in Equation (5.7) in each lattice cell α. In particular, we compute these functions in the exit cells. Let us call $N_U^a(t)$ and $N_U^b(t)$ these particular densities, and let us assume, for the moment, that there is only one exit, U, as in Figure 5.2. Incidentally we observe that U is not a cell in $\mathcal{R}$; U is a sort of very small courtyard $\mathcal{R}_{out}$, made of a singe cell. During the computations, we check the values of the densities at some time intervals of length ΔT. Hence we compute $N_U^a(k\Delta T)$ and $N_U^b(k\Delta T)$, where $k = 1, 2, 3, \dots$. If, for some k, $N_U^a(k\Delta T)$, $N_U^b(k\Delta T)$, or both, they do exceed their threshold values N_{thr}^a and N_{thr}^b, we stop the computation. Otherwise we consider the step $k + 1$, and we let the system evolve as before. Now, suppose that $N_U^a(k\Delta T) > N_{thr}^a$ or $N_U^b(k\Delta T) > N_{thr}^b$. Then we go back to the solution Equation (5.7), but considering new initial conditions, i.e., those in which, at the *new* initial time $t_0 = k\Delta T$, there is no population $\mathcal{P}_a$ (if $N_U^a(k\Delta T) > N_{thr}^a$) and $\mathcal{P}_b$ (if $N_U^b(k\Delta T) > N_{thr}^b$) at all in U: those that have reached U have just left $\mathcal{R}$, and for this reason they will no longer contribute to the global densities of $\mathcal{P}_a$, or of $\mathcal{P}_b$! This choice, besides being natural, is also faster than our previous suggestions, since in that case the numerical computations involve all of $\mathcal{R}_{tot}$ and not only $\mathcal{R}$, while here we have to add a single cell to the original lattice $\mathcal{R}$. This is convenient, since a larger domain involves more degrees of freedom and, consequently, a larger dimension of the Hilbert space. The related matrix $\mathcal{K}_{L^2}$ in (5.5) becomes much larger, and therefore the numerical computations slow down significantly. Here, on the contrary, the size of $\mathcal{R}_{tot}$ is just the size of $\mathcal{R}$, N, plus one (or two or three if $\mathcal{R}$ has two or three exits), $N+1$.

> *Remark:* It is interesting to observe that the strategy discussed above fits well our definition of *rule*, as discussed in Section 2.8. In fact, what we are proposing here is to perform a periodic check on some aspects of the system (here we check the values of $N_U^a(k\Delta T)$ and $N_U^b(k\Delta T)$) and, depending on whether they are larger than the thresholds N_{thr}^a and N_{thr}^b, the time evolution starts again, but with different modalities: nothing really happens if $N_U^a(k\Delta T) \leq N_{thr}^a$ or $N_U^b(k\Delta T) \leq N_{thr}^b$ while we produce discontinuities in the global densities, since part of the population ($\mathcal{P}_a$ or $\mathcal{P}_b$, or both), the part in cell U, is removed from the system if the related threshold is exceeded. Hence we are operating letting the system evolve from $t=0$ to $t=\Delta T$ using the Heisenberg equations of motion produced by H. Then we perform a check that can modify, or not, the state of the system. And then we let the system evolve again with H for another time interval of length ΔT, until the next check and so on. This is exactly how the (H, ρ)-induced dynamics works.

A typical situation of what we have in mind is showed in Figure 5.2, which describes what is the topology of the lattice, and how the populations are distributed at $t=0$. This specific topology of the room and initial distributions of the populations is what we class *Setting S_2*, since it involves both $\mathcal{P}_a$ and $\mathcal{P}_b$. Setting S_1, which will be considered in Section 5.3.1, refers to a single population. It is convenient sometimes, and here in particular, to adopt a double-indexes (referring to the x and the y axes) notation to label the cells of the lattice. We see that population $\mathcal{P}_a$ occupies, at $t=0$, the cell $\alpha=1$ or, in this different notation, cell $(1,1)$. Analogously, $\mathcal{P}_b$ is all concentrated in cell 52 or in cell $(2,6)$. We have three obstacles in the room, O_1, O_2 and O_3. The presence of O_1, in particular, implies that cells $(3,7)$, $(3,8)$, $(4,7)$ and $(4,8)$ cannot be occupied by any of the agents of the model. The room has a single exit U, which we label as $(11,2)$, which is outside the room, and which is the cell $\mathcal{P}_a$ and $\mathcal{P}_b$ are trying to reach. So the check on the densities $N_U^a(k\Delta T)$ and $N_U^b(k\Delta T)$ refers to this cell. It is not so relevant, at least to implement our rule, what is happening in the other cells. Of course, densities of the two populations inside $\mathcal{R}$ are essential to understand if $\mathcal{P}_a$ and $\mathcal{P}_b$ are leaving the room or not.

Before considering some concrete situations, it is worth pointing out that, in order to work always (almost) in the same conditions, we have considered the same *internal* topology in $\mathcal{R}$ in any of the choices discussed in this chapter by putting three obstacles in $\mathcal{R}$, always located in the same places, except in Figure 5.3, where

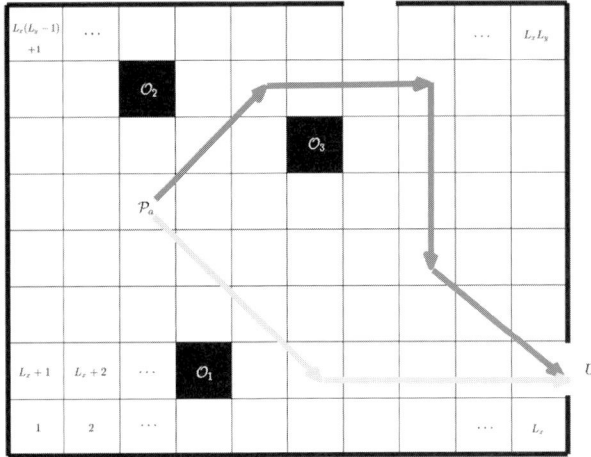

Figure 5.3 An optimal path γ_{OP} (pale gray arrows) and a non-optimal path (dark gray arrows).

a slightly different choice is shown to put in evidence the role of what we call *optimal paths*. In this way our results will not be affected much *also* by the topology. Concerning the exits in the analysis presented here, we will consider only rooms with a single exit and rooms with two exits. As for the distribution of the two populations at $t = 0$, we consider two opposite cases. In the first one, they are located in just two different cells, always chosen to be the same. In the second situation, $\mathcal{P}_a$ and $\mathcal{P}_b$ are rather distributed all along $\mathcal{R}$. Our idea is that in the first case the interactions between the two groups of people are not expected to play an essential role since there are few people in the room, while their interactions become more and more relevant when the densities of $\mathcal{P}_a$ and $\mathcal{P}_b$ increase.

5.3.1 One Population, One Exit (Setting S_1)

In order to propose an optimal escape strategy, our first goal is to define $p_{\alpha,\beta}^{(a,b)}$ in a convenient way. This is because $p_{\alpha,\beta}^{(a,b)}$ is exactly what contains, in Equation (5.3) of the diffusion Hamiltonian h, the information on which movements are possible and which are not. This was, in fact, all what was included in [7]: when discussing migration, movements from one cell to any neighboring one were all considered equivalent. Here we want more: we need $p_{\alpha,\beta}^{(a,b)}$ to speed up the movement of each population in some direction, but not in others. In other words, we need $p_{\alpha,\beta}^{(a,b)}$ to define, in some clever way, some *preferable paths* across $\mathcal{R}$. This is because we want people to reach the exits and not just to move around the room. It is clear, for example, that if part of the population is initially located in the cell (1,1) as in Figure 5.2, a convenient escape path to reach the exit cell U should go around (and close to) the obstacle O_2, while escape paths going around O_1 or close to O_3 would not be good choices, for obvious reasons. Hence, the general procedure we consider to determine, for instance, $p_{\alpha,\beta}^{(a)}$, consists in requiring that $\mathcal{P}_a$ take the shortest path possible to reach the exit cell U. To implement this requirement, and to clarify what is going on, we suppose first that, at $t = 0$, $\mathcal{P}_a$ is located in a single cell, α_0. We will extend our treatment to a more general situation after. We put

$$p_{\alpha,\beta}^{(a)} = \rho_a \hat{\delta}_{\alpha,\beta} \left(\gamma_\alpha^{(a)} + \gamma_\beta^{(a)} \right) / 2, \tag{5.8}$$

where ρ_a is a positive real parameter, whose meaning will be discussed soon, while $\hat{\delta}_{\alpha,\beta}$ is a symmetric tensor that is equal to 1 if the population can move from α to β and 0 otherwise. Hence $p_{\alpha,\beta}^{(a)}$ can be different from 0 only if α and β are neighboring cells and if neither α nor β are cells where some obstacle is localized. As we have already discussed before, we assume that $p_{\alpha,\beta}^{(a)} = p_{\beta,\alpha}^{(a)}$ from the beginning.

The quantities $\gamma_\alpha^{(a)}$ and $\gamma_\beta^{(a)}$ are defined as follows: first we introduce a *function* $f_d(\alpha, \beta)$, called the *Dijkstra function*, named from Edsger Dijkstra, see [73], which returns the length of the minimal path among all the paths going from cell α to cell β, subjected to the constraints given by the presence of $\hat{\delta}_{\alpha,\beta}$ and by the presence of the obstacles in $\mathcal{R}$ (with the additional assumption that all the paths compatible with the constraints have the same weight, see [73]).

Now, the sum $f_d(\alpha_0, \alpha) + f_d(\alpha, U)$ can be understood as the length of the path from cell α_0 to cell α, and then from cell α to the exit U. Let us call γ_{OP} the optimal path from α_0 to U, that is, the path with minimal length between α_0 and U, compatible with all the constraints above. If we define

$$g(\alpha) = (f_d(\alpha_0, \alpha) + f_d(\alpha, U))^{-1},$$

it is clear that this quantity takes its maximum value if $\alpha \in \gamma_{OP}$, i.e., if the optimal path includes the cell α. Otherwise, if $\alpha \notin \gamma_{OP}, f_d(\alpha_0, \alpha) + f_d(\alpha, U)$ increases more and more when α moves away from γ_{OP}, so that $g(\alpha)$ decreases. Figure 5.3 shows two possible paths that take $\mathcal{P}_a$ out of $\mathcal{R}$. These paths differ since the one in pale gray is optimal (i.e., the shortest possible path) while the one in darker gray is not. We also see that the optimal path is not unique: it is not hard to construct a second path along which $\mathcal{P}_a$ can move to the exit having the same length as the pale gray path in Figure 5.3. Now, let us call $M = \max_{\alpha \in \mathcal{R}} g(\alpha)$. It is clear that

$$\frac{g(\alpha)}{M} \leq 1,$$

and, in particular, it is equal to 1 if $\alpha \in \gamma_{OP}$. Then, if we define

$$\gamma_\alpha^{(a)} = \left(\frac{g(\alpha)}{M}\right)^{\sigma_a},$$

for some positive σ_a, this ratio is again less or equal to 1. In particular, $\gamma_\alpha^{(a)} = 1$ if $\alpha \in \gamma_{OP}$, for any positive value of σ_a, while $\gamma_\alpha^{(a)}$ becomes smaller and smaller when σ_a increasing, if $\alpha \notin \gamma_{OP}$. In particular, in this case, $\gamma_\alpha^{(a)}$ goes to zero when σ_a diverges.

Of course, a similar construction can be repeated for the other coefficients $\gamma_\beta^{(a)}$ in Equation (5.8). Because of (5.8), this means that the parameter $p_{\alpha,\beta}^{(a)}$ becomes small if α (or β, or both) is away from the optimal path, while takes its maximum value if both α and β belong to γ_{OP}, independently of the value of σ_a. This is important for us, since the values of $p_{\alpha,\beta}^{(a)}$ are proportional to the mobilities of the two populations, and to the time needed to reach the exit [7]. This increases when the populations move far away from the optimal path, while it decreases otherwise.

> *Remark:* If $\mathcal{P}_a$, at $t = 0$, is located in a set of cells, $\alpha_1, \alpha_2, \dots, \alpha_n$ rather than in a single one, we can extend the above construction by introducing first
>
> $$g_j(\alpha) = \left(f_d\left(\alpha_j, \alpha\right) + f_d(\alpha, U)\right)^{-1}, \quad \text{and} \quad M_j = \max_{\alpha \in \mathcal{R}} g_j(\alpha) \quad \forall j = 1, \dots, n, \qquad (5.9)$$
>
> and then
>
> $$\gamma_\alpha^{(a)} = \max_{j=1,\dots,n} \left[\left(\frac{g_j(\alpha)}{M_j}\right)^{\sigma_a}\right], \qquad (5.10)$$
>
> where $\sigma_a \geq 0$ is, again, a fixed quantity.

By increasing or decreasing the value of the parameter ρ_a in Equation (5.8), we speed up or slow down further the population $\mathcal{P}_a$, since, as we have just commented, the magnitude of $p_{\alpha,\beta}^{(a)}$ is a measure of the mobility of the first population. This aspect was discussed in detail in [1, 7].

Summarizing, the greatest values of $p_{\alpha,\beta}^{(a)}$ are obtained if α and β are along a minimal path going from the given cell α_0 to U, while $p_{\alpha,\beta}^{(a)}$ decreases to zero when the direction from cell α to cell β is not along a minimal path. Obviously, the coefficients $p_{\alpha,\beta}^{(b)}$, relevant in Setting S_2, are defined in a similar way and share the same properties of the $p_{\alpha,\beta}^{(a)}$.

> *Remark:* The construction of $p_{\alpha,\beta}^{(a)}$ proposed here is just one among all the possible choices: for example, one could modify the function $g_j(\alpha)$ in Equation (5.9). However, our choice is reasonable and it seems to work very well, as the numerical results discussed in the rest of this section show.

After this long preamble on the definition and the role of $p_{\alpha,\beta}^{(a)}$, we can move to the analysis of $\mathcal{R}$, where the single population $\mathcal{P}_a$ is distributed. The presence of a single population can be easily described using our Hamiltonian H, assuming that,

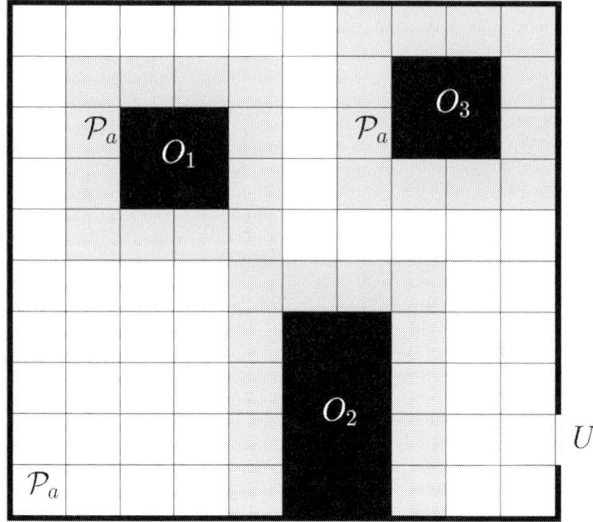

Figure 5.4 Setting S_1: At $t = 0$ the population $\mathcal{P}_a$ is located in the cells $(1,1), (2,8)$ and $(7,8)$. The black cells O_1, O_2, O_3 represents the obstacles, and the exit cell U is located at $(11,2)$ (Source: Bagarello et al. [8]).

at $t = 0$, the vector $\mathbf{n}^b = (n^b_1, n^b_2, \ldots, n^b_{L^2})$ introduced before is all made by zeros. This is because this vector corresponds to a density zero, for the second population, in each cell of the lattice. In other words, there is no member of $\mathcal{P}_b$ in the room at $t = 0$. We further fix $\lambda_\alpha = 0$, $\forall\, \alpha$, in Equation (5.2). Therefore, since people are not created out of nothing, the density of $\mathcal{P}_b$ will stay zero in all cells for all $t \geq 0$. So we don't need to worry about $\mathcal{P}_b$. As for $\mathcal{P}_a$, we assume, for the moment, that $\mathcal{P}_a$ is originally located as shown in Figure 5.4: some are in cell $(1,1)$, others in cells $(2,8)$ and $(7,8)$. They all want to reach cell $(11,2)$, our courtyard $\mathcal{R}_{out}$.

As discussed before, the analysis of the dynamics of $\mathcal{P}_a$ requires the knowledge of the threshold value N^a_{trh} and of the time interval ΔT defining the periodicity of the control on $N^a_U(t)$, that is, on the density of $\mathcal{P}_a$ in the exit cell. These are the main ingredients of the rule. To fix their values, we have performed several simulations, some of them reported in Figures 5.5 and 5.6.

In Figure 5.5 we plot the different densities of $\mathcal{P}_a$, $N^a(t)$, inside $\mathcal{R}$ for different values of N^a_{trh} with fixed $\Delta T = 0.08$. Also, here and in the rest of this section, we put $\omega^a_\alpha = 1$ for all α, if not explicitly stated. We see that for values of the threshold not so small, $N^a_{trh} = 10^{-1}$ or $N^a_{trh} = 10^{-2}$, the plots of the function $N^a(t)$ change significantly. On the other hand, for smaller values of N^a_{trh}, the difference between the various plots becomes smaller and smaller and, in fact, the numerical results for $N^a_{trh} = 10^{-4}$ and $N^a_{trh} = 10^{-5}$ are almost indistinguishable. Hence, at least for

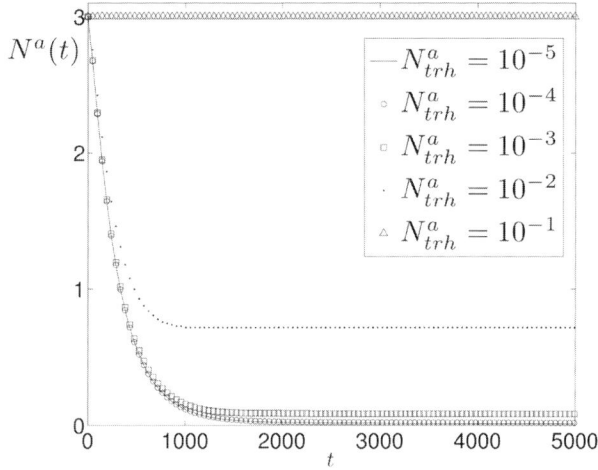

Figure 5.5 Setting S_1: The total density $N^a(t)$ is plotted for $\Delta T = 0.08$ and different values of N^a_{trh} (Source: Bagarello et al. [8]).

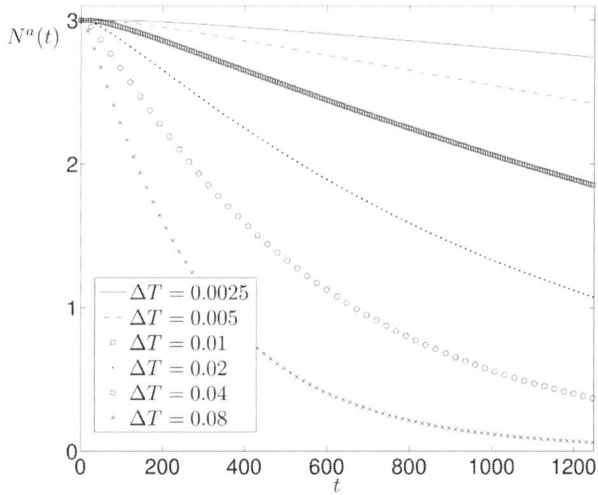

Figure 5.6 Setting S_1: The total density $N^a(t)$ is plotted for $N^a_{trh} = 10^{-5}$ and different values of ΔT (Source: Bagarello et al. [8]).

the value of $\Delta T = 0.08$ and for this setting, $N^a_{trh} = 10^{-5}$ can be considered a very good threshold value. In Figure 5.6 we keep fixed the *optimal value* of N^a_{trh} just found, $N^a_{trh} = 10^{-5}$, and we play with the value of ΔT. We see that increasing the value of ΔT improves convergence to zero of the density of $\mathcal{P}_a$. This can be easily understood: when the value of ΔT increases, a larger amount of population can accumulate in U and, because of our exit strategy, this larger amount quite likely exceeds N_{trh}^a and, therefore, it is removed from $\mathcal{R}$ after ΔT. On the other hand, if ΔT is small, it is not as easy for $\mathcal{P}_a$ to accumulate in U in such a way to exceed the threshold value. For this reason, we do not want ΔT to be too large, to prevent all the populations from disappearing in a few *time steps*, which is not really plausible. However, we cannot even take ΔT to be too small, also because the numerical computations would become very slow, since in this case the numerical check on the densities of $\mathcal{P}_a$ (and $\mathcal{P}_b$, see Setting S_2) in U should be repeated quite often. After some tests, we have found a good compromise by fixing $\Delta T = 0.08$.

In Figure 5.7 we plot the densities of $\mathcal{P}_a$ for different values of the parameter ρ_a in Equation (5.8), for the above optimal choices of N_{trh}^a and ΔT, $N_{trh}^a = 10^{-5}$ and $\Delta T = 0.08$. It is evident that lower values of ρ_a do not help the exit from the room, and $N^a(t)$ does not decrease as fast as it happens for larger values of ρ_a. This is in agreement with our understanding of $p_{\alpha,\beta}^{(a)}$, see also [1, 7], as related to the *mobility* of $\mathcal{P}_a$: the higher its value, and in particular the value of ρ_a, the faster $\mathcal{P}_a$ is moving. Incidentally, this suggests that, if we are interested in describing two populations $\mathcal{P}_a$ and $\mathcal{P}_b$ with different features, we could take ρ_a larger (or even much larger) than ρ_b to describe two sets of young ($\mathcal{P}_a$) and aged ($\mathcal{P}_b$) people: the first is expected to have a better mobility than the second.

We have already discussed that the parameters of H_0 are related to a sort of inertia of those degrees of freedom (or those agents) to which they refer. With this

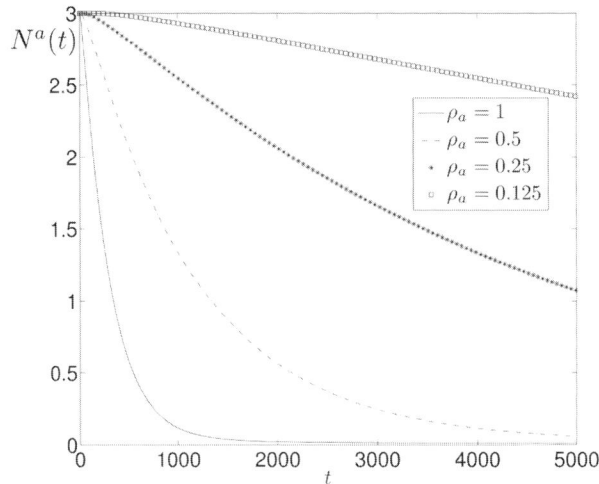

Figure 5.7 Setting S_1: The total density $N^a(t)$ of the population $\mathcal{P}_a$ is plotted for $\Delta T = 0.08, N_{trh}^a = 10^{-5}$, and different values of the parameter ρ_a (Source: Bagarello et al. [8]).

in mind, to get some extra hint on some possible optimal exit strategy, we have considered what happens if we consider two different values of ω_α^a in $\mathcal{R}$. This means that we are making $\mathcal{R}$ more and more inhomogeneous: $\mathcal{R}$ is inhomogeneous not only because of the presence of the obstacles O_j, but also because the regions around the obstacles, ∂O_j (the parts in gray in Figure 5.4, for instance), are not exactly the same as the cells in white: people try to avoid the obstacles, of course, but they also try not to run close to them, if possible.

Then we assume a high value of the ω^a in the cells in ∂O_j and a smaller value in the rest of $\mathcal{R}$. The results of this analysis are given in Figure 5.8, where we plot the total density $N^a(t)$ for $\Delta T = 0.08$, $N_{trh}^a = 10^{-5}$ and $\rho_a = 1$. The various curves correspond to different choices of $\omega_{\partial O}^a$, where $\partial O = \cup_{j=1}^3 \partial O_j$. We see that $\mathcal{P}_a$ slows down significantly if $\omega_{\partial O}^a$ becomes much larger than $\omega_{\mathcal{R}\backslash\partial O}^a$, the value of ω_α^a outside ∂O: the worst performance is observed when $\omega_{\mathcal{R}\backslash\partial O}^a = 0.01$ and $\omega_{\partial O}^a = 200$. In this case the difference between the inertia in the gray and that in the white region of $\mathcal{R}$ is very large. On the opposite side, if $\omega_{\mathcal{R}\backslash\partial O}^a = \omega_{\partial O}^a = 0.01$, i.e., when there is no difference at all, we get the best exit performance. It is like, when people approach an obstacle (one of the parts in gray), they get stuck in there, and it becomes difficult for them to run away: stated differently, *obstacles extend their (negative) effects to the surrounding regions*. From a mathematical side, this is easily understood: when someone reaches a cell in gray, its inertia becomes much larger, and then it tends to stay in that cell.

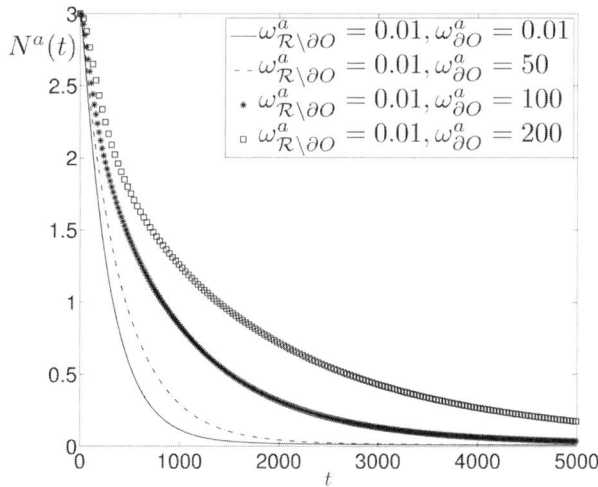

Figure 5.8 Setting S_1: The total density $N^a(t)$ of the population $\mathcal{P}_a$ is plotted for $\Delta T = 0.08$, $N_{trh}^a = 10^{-5}$, $\rho_a = 1$ and different values of the parameter ω_α^a inside and outside ∂O (Source: Bagarello et al. [8]).

Remark: Of course, the gray areas could also act to improve the escape strategy for $\mathcal{P}_a$, if $\omega^a_{\mathcal{R}\backslash\partial O}$ is now much larger than $\omega^a_{\partial O}$. This is because those agents of $\mathcal{P}_a$ that arrive in the gray region have no reason to stay there (the inertia is rather low, and they are free to move). This can correspond, in real life, to the presence of several signs indicating the exits close to the obstacles.

We conclude that, to decrease the time needed by $\mathcal{P}_a$ to leave $\mathcal{R}$, it is convenient to adopt the following strategy: (i) clarify, from each possible cell in $\mathcal{R}$, which direction goes straight to the exit (this can be done, for instance, by clearly indicating the exit); (ii) try to increase the mobility of $\mathcal{P}_a$ (for instance, removing all the unnecessary obstacles in $\mathcal{R}$) and (iii) try to keep a certain homogeneity in the accessible part of $\mathcal{R}$. Of course, while (i) and (ii) are quite expected suggestions, (iii) is not evident a priori, and its implementation could really help improve the exit strategies.

It is useful to stress that changing the topology of the room and the initial distribution of $\mathcal{P}_a$ in $\mathcal{R}$ would change the numerical outputs, but not the main conclusions deduced in the particular situation considered above.

5.3.2 Two Populations, One Exit

So far, even if the Hamiltonian (5.2) involves two populations, $\mathcal{P}_a$ and $\mathcal{P}_b$, we have not considered $\mathcal{P}_b$ at all. This was meant to analyze the essential aspects of the model, and to understand in detail some of its features. Now, we are ready to consider the case in which neither $\mathbf{n}^a$ nor $\mathbf{n}^b$ are $\mathbf{0}$: this means that, at $t = 0$, at least one cell of $\mathcal{R}$ is surely occupied by at least one population. We will consider separately two different situations: in the first one $\mathcal{P}_a$ and $\mathcal{P}_b$ do not mutually interact. This is compatible with the fact that the densities of the two populations are small in $\mathcal{R}$ already at $t = 0$. The densities will stay small also for $t > 0$, since the people are leaving the room. In the second situation, relevant when the densities are high, $\mathcal{P}_a$ and $\mathcal{P}_b$ do interact, since elements of $\mathcal{P}_a$ and $\mathcal{P}_b$ are more likely to meet while they are trying to reach the exit. We will see that the two cases share some similarities, but also that some interesting differences arise.

5.3.2.1 Without Interaction (Setting S_2^{ld})

We begin our analysis by considering the case in which the two populations $\mathcal{P}_a$ and $\mathcal{P}_b$ are originally located as shown in Figure 5.2 and they have the same initial density, $N^a(0) = N^b(0) = 1$. The suffix *ld* in S_2^{ld} stands for *low density*, while the subscript 2 indicates, as we have already seen, that we are considering now two populations. This is in contrast with the single population considered before, corresponding to a setting S_1. As the total densities of both $\mathcal{P}_a$ and $\mathcal{P}_b$ are small, we neglect here the effect of the interaction between the two populations, since $\mathcal{P}_a$ and

$\mathcal{P}_b$ quite likely do not meet during the time evolution. Hence we can take $\lambda_\alpha = 0$, $\forall \alpha \in \mathcal{R}$, in H, since λ_α is exactly the strength of this interaction in the cell α. To fix the values of the coefficients $p_{\alpha,\beta}^{(a)}$ and $p_{\alpha,\beta}^{(b)}$ in the Hamiltonian, we apply the procedure outlined in the previous section for each population: of course, since the two populations have different initial conditions, the coefficients $p_{\alpha,\beta}^{(a)}$ and $p_{\alpha,\beta}^{(b)}$ are not identical. This difference is also a consequence of the different values of the mobilities we want to consider, in principle, for $\mathcal{P}_a$ and $\mathcal{P}_b$. As in the previous section, we fix $\Delta T = 0.08$ and $N_{trh}^a = N_{trh}^b = 10^{-5}$. We further take $\omega_\alpha^a = \omega_\alpha^b = 1$, for all α, $\rho_a = 1$, and we consider different situations corresponding to the following values of ρ_b: $\rho_b = 1, 0.75, 0.5, 0.25$, in order to describe different mobilities for (the aged population) $\mathcal{P}_b$. The first choice, $\rho_b = 1$, corresponds to having (essentially) the same mobilities for $\mathcal{P}_a$ and $\mathcal{P}_b$. Hence, even if $\mathcal{P}_b$ is made of old people, these people are well trained. Smaller values of ρ_b correspond to slower and slower $\mathcal{P}_b$: they are not so well trained anymore. We should expect that if both the populations have the same inertia and the same mobilities ($\omega_\alpha^a = \omega_\alpha^b$ and $\rho_a = \rho_b$), then $N^b(t)$ should decay to zero faster than $N^a(t)$: in fact, see Figure 5.2, the optimal path going from the cell occupied by $\mathcal{P}_b$ at $t = 0$ to the exit cell U has length 9, while for $\mathcal{P}_a$ the optimal path has length 10. In Figure 5.9 this result is clearly shown, when comparing $N^a(t)$ and $N^b(t)$ for $\rho_a = \rho_b = 1$: convergence to zero of $N^b(t)$ is faster than the analogous decay of $N^a(t)$. On the other hand, if we decrease ρ_b, we slow

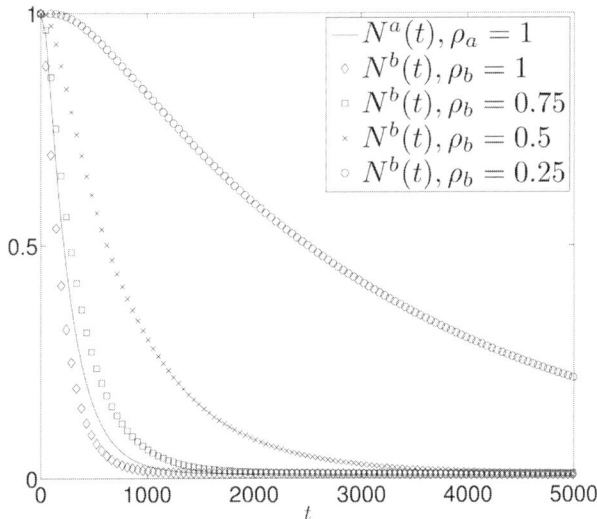

Figure 5.9 Setting S_2^{ld}: The total densities $N^a(t), N^b(t)$ are shown for $\Delta T = 0.08, N_{trh}^a = N_{trh}^b = 10^{-5}, \omega_\alpha^a = \omega_\alpha^b = 1, \forall \alpha, \rho_a = 1$, and for different values of the parameter ρ_b (Source: Bagarello et al. [8]).

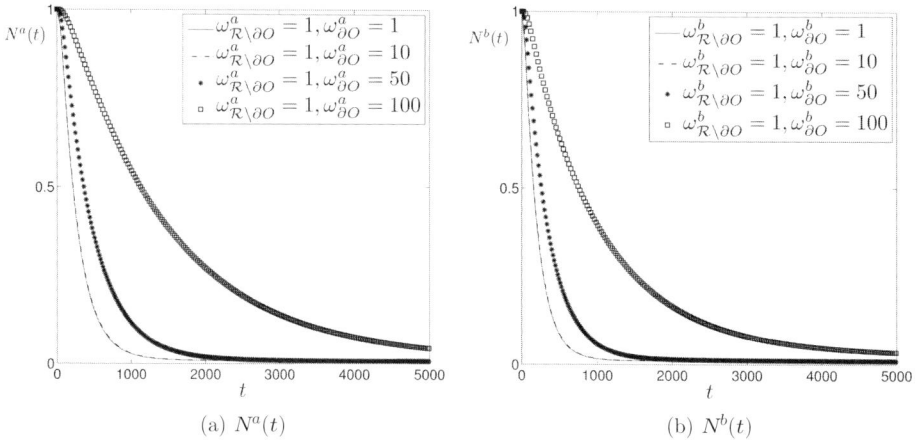

(a) $N^a(t)$ (b) $N^b(t)$

Figure 5.10 Setting S_2^{ld}: The total densities $N^a(t), N^b(t)$ are shown for $\Delta T = 0.08, N_{trh}^a = N_{trh}^b = 10^{-5}, \rho_a = \rho_b = 1$ and different values of $\omega_\alpha^a, \omega_\alpha^b$ inside and outside ∂O (Source: Bagarello et al. [8]).

down $\mathcal{P}_b$, and this allows $N^a(t)$ to go to zero faster than $N^b(t)$: not surprisingly, the fact that one population leaves the room faster than the other is not only a matter of where the populations were originally located, but also of how fast they can move.

It is interesting to study the effect of a strong inhomogeneity in $\mathcal{R}$ also for this setting S_2^{ld}, inhomogeneity represented by strongly changing the values of ω_α^a and ω_α^b in $\mathcal{R}$. For concreteness, and analogous to what we have done for the single population, we try to understand what happens when the parameters $\omega_\alpha^{a,b}$ *inside* ∂O are different (and, sometimes, much bigger) than those outside: $\omega_\alpha^{a,b} \gg \omega_\beta^{a,b}$, if $\alpha \in \partial O$ while $\beta \in \mathcal{R} \setminus \partial O$. The results are shown in Figure 5.10(a) and (b), where the densities $N^a(t), N^b(t)$ are plotted, respectively: as previously seen in the Setting S_1 (see Figure 5.8), increasing values of $\omega_\alpha^{a,b}$ inside ∂O slow down the populations in the region ∂O, and, therefore, $N^a(t), N^b(t)$ decay slower. Of course, the explanation for that coincides with that proposed in Section 5.3.1. Then again we conclude that if we are interested in an optimal escape strategy, we should avoid these kinds of inhomogeneities in $\mathcal{R}$.

5.3.2.2 With Interaction (Setting S_2^{hd})

Suppose now that the populations $\mathcal{P}_a$ and $\mathcal{P}_b$ are originally located as shown in Figure 5.11, so that they have the same initial density, $N^a(0) = N^b(0) = 7$, significantly higher than in the previous situation. Here, the suffix *hd* in S_2^{hd} stands for *high density*,[3] and the subscript 2 refers again to the number of populations. In this case

---

[3] More precisely, we should not really talk of *high* density, for these values of $N^a(0)$ and $N^b(0)$. But, of course, these densities are high when compared with those in S_2^{ld}.

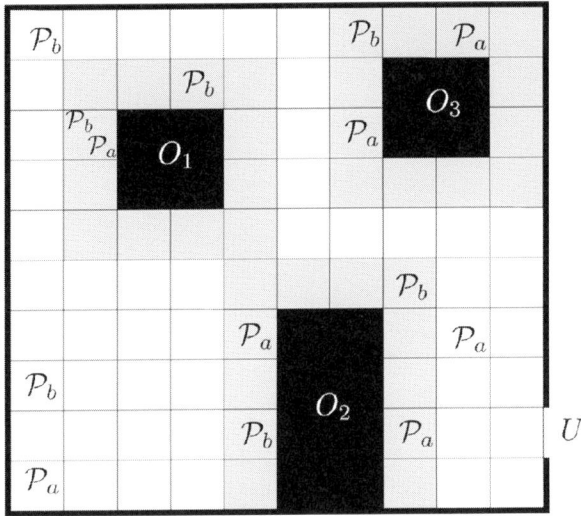

Figure 5.11 Setting S_2^{hd}: At $t = 0$ $\mathcal{P}_a$ is located in the cells $(1,1), (8,2), (5,4)$, $(9,4), (2,8), (7,8), (9,10)$, while $\mathcal{P}_b$ in the cells $(5,2), (1,3), (8,5), (2,7), (2,8)$, $(4,9), (1,10), (7,10)$. The black cells O_1, O_2, O_3 represents the obstacles, and the exit cell U is located at $(11,2)$ (Source: Bagarello et al. [8]).

interactions between the two populations can reasonably occur, during the time evolution while the two populations try to reach the exit. For this reason, contrary to what we have done before, we consider here the *local* interaction parameters λ_α to be different from zero. The population $\mathcal{P}_a$ is, in this configuration, *globally* closer to the exit cell U than $\mathcal{P}_b$. This is because the sum of all the optimal paths from the cells initially occupied by $\mathcal{P}_a$ to U is 43, while the analogous quantity for $\mathcal{P}_b$ is 55. Of course, it may happen that part of $\mathcal{P}_b$ reaches U before any agent of $\mathcal{P}_a$, but it is not expected, at least if all the other parameters for $\mathcal{P}_a$ and $\mathcal{P}_b$ coincide, that all the agents of $\mathcal{P}_b$ reach U before each agent of $\mathcal{P}_a$ has gone away. We have performed several simulations by varying the interaction parameter λ_α, and the results are shown in Figure 5.12(a)–(d). In all these plots we have taken $\rho_a = \rho_b = 1$, since the role of the mobility is already understood and there is no need to analyze it further. This implies that different decays are not due to different mobilities, but only to the difference in the initial distributions of $\mathcal{P}_a$ and $\mathcal{P}_b$ along $\mathcal{R}$. Figure 5.12(a)–(d) shows that, when the value of λ_α is small enough, for instance when $\lambda_\alpha = 0.05$ for all α, the effect of the interaction is essentially negligible: $N^a(t)$ decays faster than $N^b(t)$, as expected in view of our previous considerations, and in particular because of the different lengths of the two *global* optimal paths. On the other hand, already taking $\lambda_\alpha = 0.5$ for all α, the differences between $N^a(t)$ and $N^b(t)$ become almost negligible, and the two functions really look as a single one, even for larger

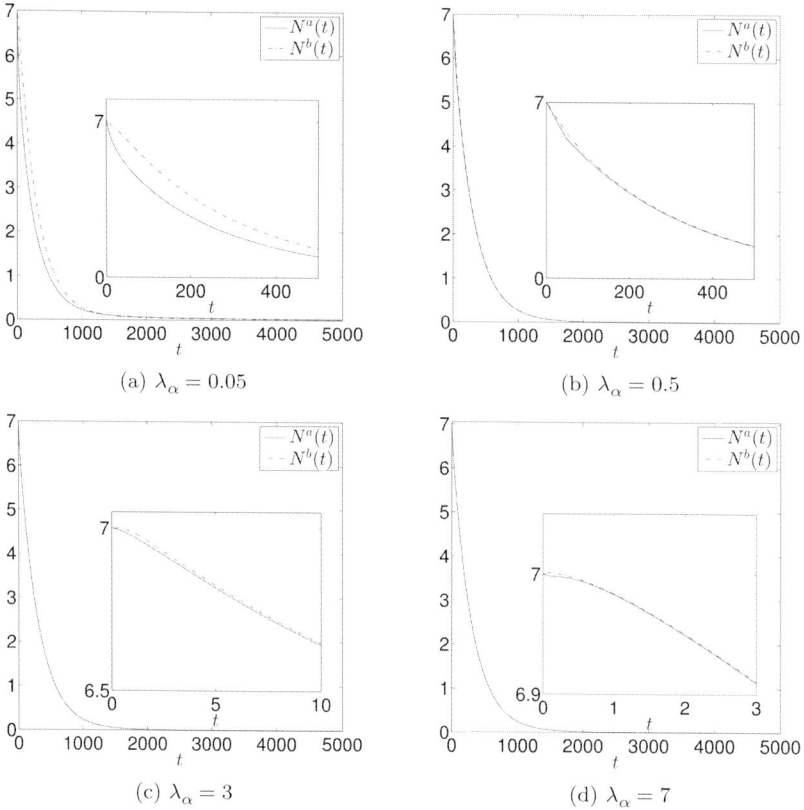

(a) $\lambda_\alpha = 0.05$

(b) $\lambda_\alpha = 0.5$

(c) $\lambda_\alpha = 3$

(d) $\lambda_\alpha = 7$

Figure 5.12 Setting S_2^{hd}: The total densities $N^a(t), N^b(t)$ are shown for $\Delta T = 0.08$, $N_{trh}^a = N_{trh}^b = 10^{-5}, \omega_\alpha^a = \omega_\alpha^b = 1\ \forall \alpha,\ \rho_a = \rho_b = 1$, and for different values of the parameter λ_α escape strategies (Source: Bagarello et al. [8]).

values of λ_α, as for $\lambda_\alpha = 3$ or $\lambda_\alpha = 7$, for all α, see Figure 5.12(b) and (d). Roughly speaking, if we increase λ_α we obtain a better efficiency of the system. In fact, we see clearly that population $\mathcal{P}_b$ leaves the room much faster than when we have $\lambda_\alpha = 0.05$. Moreover, comparing the plots of $N^a(t)$ in Figure 5.12(a) and (b), we also see that $N^a(t)$ decays faster when $\lambda_\alpha = 0.5$ than when $\lambda_\alpha = 0.05$. The escape strategy improves when the interaction parameters increase: the two populations help each other. This is a cooperative effect.

5.3.3 Two Populations, Two Exits

The next situation we are going to describe is again the case in which both $\mathcal{P}_a$ and $\mathcal{P}_b$ are, at $t = 0$, in the room $\mathcal{R}$, but the room has now two exits and not just one. So, both $\mathbf{n}^a$ and $\mathbf{n}^b$ are different from 0. As in the previous section we will consider separately the low-densities and high-densities cases, to simulate respective absence or presence of interactions between $\mathcal{P}_a$ and $\mathcal{P}_b$.

Due to the presence of two exit cells, we need to slightly modify the definitions in Equations (5.9) and (5.10) used to define $p_{\alpha,\beta}^{(a,b)}$: then, assuming that $\mathcal{P}_a$ is initially located in the cells $\alpha_1, \alpha_2, ..., \alpha_n$, and that the exit cells are $U_1, U_2, ..., U_m$, we put

$$\gamma_\alpha^{(a)} = \max_{j=1,...,n, k=1,...,m} \left[\left(\frac{g_{jk}(\alpha)}{M_{jk}} \right)^{\sigma_a} \right], \tag{5.11}$$

where

$$g_{jk}(\alpha) = (f_d(\alpha_j, \alpha) + f_d(\alpha, U_k))^{-1} \qquad \forall j = 1, ..., n, \quad k = 1, ..., m, \tag{5.12}$$

and

$$M_{jk} = \max_{\alpha \in \mathcal{R}} g_{jk}(\alpha), \qquad \forall j = 1, ..., n, \quad k = 1, ..., m. \tag{5.13}$$

A similar construction can be repeated for $\gamma_\alpha^{(b)}$. Then we construct $p_{\alpha,\beta}^a$ following Equation (5.8). The positive parameter σ_a tunes, as before, how fast $p_{\alpha,\beta}^a$ decreases to zero if α or β is not along the optimal path going from the cell α_j to some exit U_k. In general, $p_{\alpha,\beta}^{(a)}$ assumes its greatest values if α and β are along some minimal path going from the given cell α_j to some exit U_k. Otherwise, $p_{\alpha,\beta}^{(a)}$ decreases and goes to zero for very large σ_a.

5.3.3.1 Without Interaction (Setting $S_{2,2}^{ld}$)

The two populations $\mathcal{P}_a$ and $\mathcal{P}_b$ are originally located as shown in Figure 5.13, and they have the same global initial density, $N^a(0) = N^b(0) = 1$. We use $S_{2,2}^{ld}$ to indicate this settings because it has low density, two populations and two exits. As we have done for the Setting S_2^{ld}, we neglect for the moment the effects of any possible interaction between the populations because of their low initial densities. Therefore, we take $\lambda_\alpha = 0$ for all α. As before, motivated by our previous analysis, we fix $\Delta T = 0.08, N_{trh}^a = N_{trh}^b = 10^{-5}, \omega_\alpha^a = \omega_\alpha^b = 1, \rho_a = \rho_b = 1$. In this configuration the population $\mathcal{P}_b$ is globally closer to the exit cells U_1 and U_2, as the sum of all the optimal paths going from the cell initially occupied to U_1 and U_2 is 14, while for $\mathcal{P}_a$ this sum is 20: we therefore expect that, if the populations have the same mobility, that is, if $\rho_a = \rho_b$ as we are assuming here, then $N^b(t)$ decays faster than $N^a(t)$. This is, in fact, what we observe in Figure 5.14, where $\rho_a = \rho_b = 1$. In the same figure, we also plot $N^b(t)$ smaller values of ρ_b, and, as we already noticed while discussing, for instance, Setting S_2^{ld} (see Figure 5.9), if we decrease ρ_b then $\mathcal{P}_b$ slows down and $N^b(t)$ decays slower than $N^a(t)$, even if $\mathcal{P}_b$ is closer to the exits.

To confirm our understanding of the parameters $\omega_\alpha^a, \omega_\alpha^b$, that were previously interpreted as the inertia of the populations, we show in Figure 5.15(a) and (b) the densities of the populations for different values of $\omega_\alpha^a, \omega_\alpha^b$ within $\mathcal{R}$: analogous to

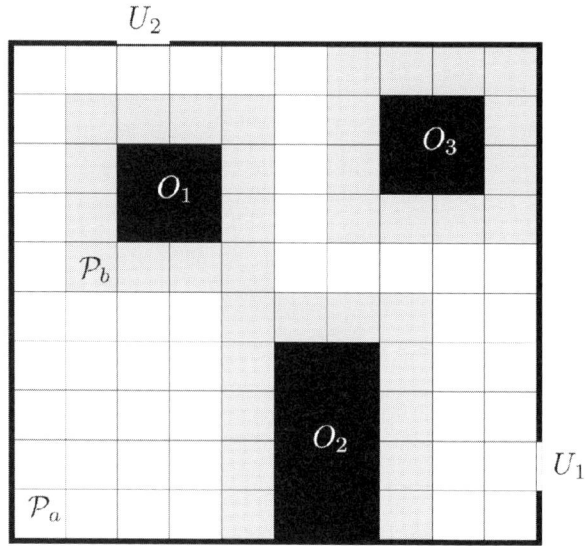

Figure 5.13 Setting $S_{2,2}^{ld}$: At $t = 0$ the populations $\mathcal{P}_a$ and $\mathcal{P}_b$ are located in the cells $(1,1)$ and $(2,6)$ respectively. The black cells represents the obstacles. The exit cells U_1, U_2 are located at $(11,2)$ and $(3,11)$ (Source: Bagarello et al. [8]).

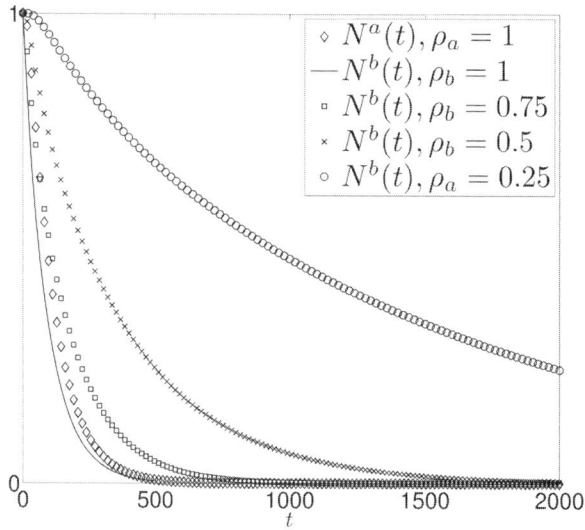

Figure 5.14 Setting $S_{2,2}^{ld}$: The total densities $N^a(t), N^b(t)$ are shown for $\Delta T = 0.08, N_{trh}^a = N_{trh}^b = 10^{-5}, \omega_\alpha^a = \omega_\alpha^b = 1 \quad \forall \alpha, \rho_a = 1$, and different values of the parameter ρ_b (Source: Bagarello et al. [8]).

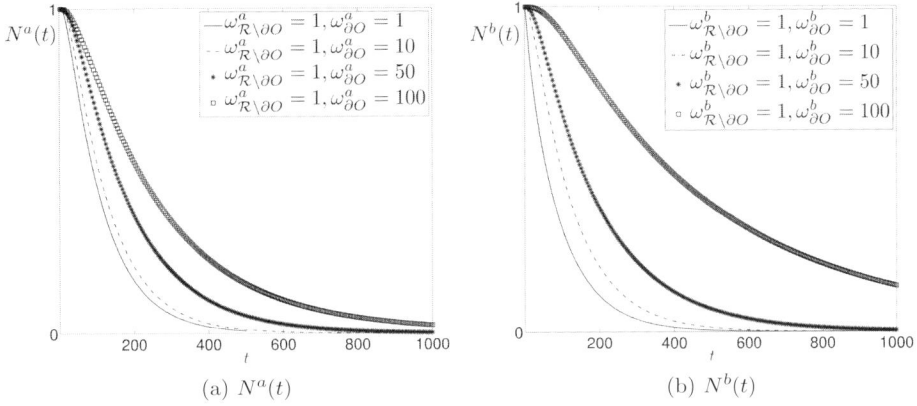

(a) $N^a(t)$ (b) $N^b(t)$

Figure 5.15 Setting $S_{2,2}^{ld}$: The total densities $N^a(t), N^b(t)$ are shown for $\Delta T = 0.08, N_{trh}^a = N_{trh}^b = 10^{-5}, \rho_a = \rho_b = 1$ and different values of ω^a, ω^b, i.e., for inhomogeneous $\mathcal{R}$ (Source: Bagarello et al. [8]).

what we have seen in Figures 5.8 and 5.10 for the Setting S_1 and S_2^{ld}, when ω^a, ω^b inside ∂O are much larger than outside, in $\mathcal{R} \setminus \partial O$, then the two populations are rather slower because they become more static in ∂O.

5.3.3.2 With Interaction (Setting $S_{2,2}^{hd}$)

Suppose now that the populations $\mathcal{P}_a$ and $\mathcal{P}_b$ are originally located as shown in Figure 5.16 and they have the same initial densities, $N^a(0) = N^b(0) = 7$, larger than in the previous case. As for Setting S_2^{hd}, we consider now the effect of the interaction between the populations due to their sufficiently high initial densities, so that we cannot take λ_α equal to zero anymore. When compared to the Setting S_2^{hd}, in which only one exit was present, here the second exit cell U_2 creates an easy way out for the population $\mathcal{P}_b$, and we observe a kind of equivalence between the optimal paths for the two populations: in fact, the sum of all the lengths of the optimal paths going from the initial cells occupied by the populations to the exit cells is 90 for $\mathcal{P}_a$ and 89 for $\mathcal{P}_b$, and therefore we should expect that $N^a(t)$ and $N^b(t)$ decay in a similar way, at least if they have a similar mobility. In Figure 5.17 the difference $N^a(t) - N^b(t)$ is shown for $\lambda_\alpha = 0.05, 1, 3, 7$: as we increase λ_α, $N^a(t) - N^b(t)$ decreases. This is the same *equalizing effect* we observed in presence of a single exit: interactions make the populations more and more similar. We can also distinguish in Figure 5.17 two different time regions: in a first time interval, which depends on λ_α, we see that $N^a(t) > N^b(t)$, while in the next interval, we have the opposite, $N^a(t) < N^b(t)$. This result can be understood as follows: when time evolution starts, $\mathcal{P}_b$ is initially closer to U_2 than $\mathcal{P}_a$ and therefore $N^b(t)$ decays faster

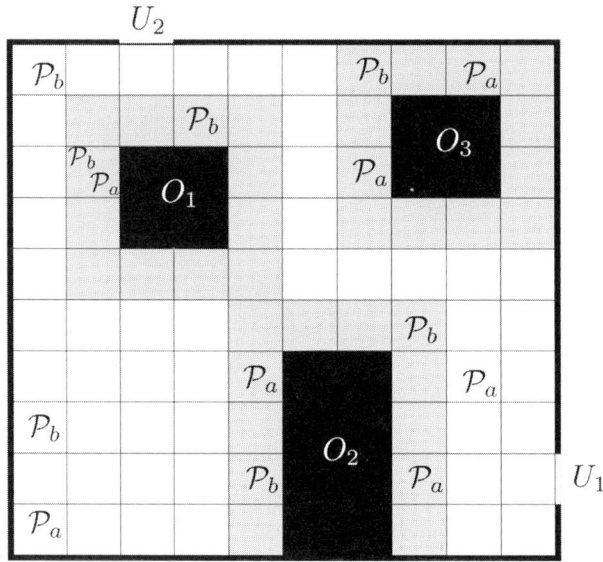

Figure 5.16 Setting $S_{2,2}^{hd}$: At $t = 0$ the population $\mathcal{P}_a$ is located in the cells $(1,1)$, $(8,2)$, $(5,4)$, $(9,4)$, $(2,8)$, $(7,8)$, $(9,10)$, while $\mathcal{P}_b$ is located in the cells $(5,2)$, $(1,3)$, $(8,5)$, $(2,7)$, $(2,8)$, $(4,9)$, $(1,10)$, $(7,10)$. The black cells O_1, O_2, O_3 represents the obstacles, and the exit cells U_1, U_2 are located at $(11,2)$ and $(3,11)$ (Source: Bagarello et al. [8]).

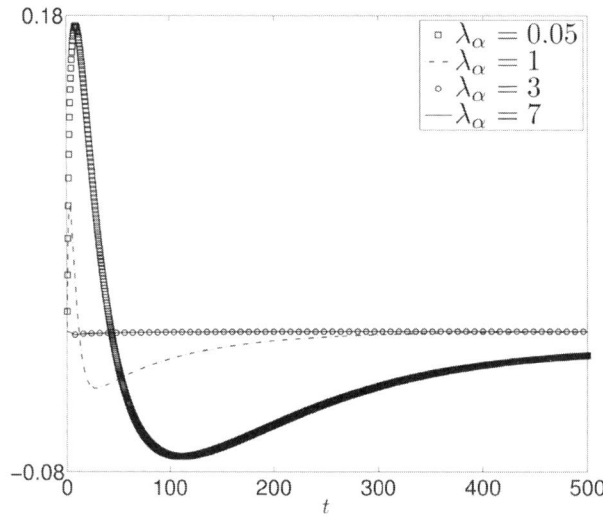

Figure 5.17 Setting $S_{2,2}^{hd}$: The difference $N^a(t) - N^b(t)$ of the densities $N^a(t)$, $N^b(t)$ for different values of λ_α (Source: Bagarello et al. [8]).

than $N^a(t)$. After some time, we get $N^a(t) < N^b(t)$, since the remaining part of $\mathcal{P}_b$ that is left in $\mathcal{R}$ and that moves toward U_1 needs longer than $\mathcal{P}_a$ to exit.

Changing the value of ρ_b, or the values of $\omega_\alpha^{a,b}$ inside ∂O, produces exactly the same phenomena we have already described in the previous settings, and the conclusions are exactly the same: to speed up the escape procedure, it is convenient to keep $\mathcal{R}$ homogeneous (or to put some *accelerating effects* in ∂O).

5.4 Conclusions

It is obvious that the problem considered in this chapter can be relevant in many concrete situations. This is proved by the large number of contributions that exist in the literature where a variety of techniques have been proposed to deal with escape strategies and related systems. We refer to [74–80], and to references therein, for a partial list of contributions in this area.

Summarizing, what we have done in this chapter was to describe an escape strategy for two populations scattered in a room (or in any restricted region) $\mathcal{R}$ with one or more exits, considering that they can have, in general, different mobilities. Apart from quite natural conclusions, as the fact that the path toward the exit(s) should be clearly identified and that a larger mobility implies that people requires less time to leave the room, we have also deduced that another improvement in the escape procedure is easily obtained when that part of $\mathcal{R}$ that can be occupied by the populations is *homogeneous*. In other words, the presence in $\mathcal{R}$ not only of obstacles (this is obvious) but also of some *points of interest* can slow down the escape procedure. We have also seen that the effect of the interaction between the two populations is to help both populations to leave the room faster, at least for moderate-high values of the interaction parameters. As for the role of the parameter σ_a in (5.10), this has also been studied in [8] by considering several choices of its value, and not many differences have been observed. This could be understood as follows: when σ_a increases, the path is narrow but direct to the exit. On the other hand, when σ_a decreases, the path is large, but not so direct. However, a larger amount of populations can use this larger path simultaneously [8].

It may be useful to observe that, despite the apparent difficulty of the model, we have been able to recover analytically the densities of the populations in each cell in a very simple way, with obvious advantages in terms of computational effort. Other fluid-dynamic models working on a macroscopic scale, for example, can work well in certain environments, but the presence of nonlinearities in these other models usually produces serious difficulties in any analytical and numerical approach. We refer to [81, 82], and to the references therein, for some results along this line.

6

Closed Ecosystems

The application considered in this chapter consists in finding the dynamical behavior of some (micro or macroscopic) *closed ecosystems*, something that can be relevant in ecology and in biology, for instance. Hence the analysis discussed here continues, in a certain sense, the study began in Chapter 4 in a slightly different context.

Closed ecological systems are quite interesting [83, 84] for some of their features: in particular, they do not exchange matter (or energy) with any part outside the system – what happens, happens inside the system. From this point of view, closed systems are the perfect dual version of the (quantum) open systems we have already met before. Closed ecological systems are often used to describe small artificial systems designed and controlled by humans, e.g., agricultural systems and activated sludge plants, or aquaria, or fish ponds [85]. Also fascinating is the possibility of using some suitably designed mathematical models of these ecosystems to optimize strategies of life support during space flights, in space stations or in space habitats [84]. What makes closed system S a sustainable system is the essential fact that any waste product produced by one species of S must be used by at least one other species and converted into nutrients.

With this in mind, we will discuss here two mathematical models of closed ecosystems based on several compartments, one for each species (i.e. the various agents) of the system S. Once again, rather than describing the time evolution of S directly in terms of differential equations, as in [86], we will introduce a Hamiltonian operator, which describes the essential interactions and the main features of S, written in terms of ladder operators, and then deduce the dynamics from it. As in Chapter 4, the ladder operators used will be assumed to satisfy the CAR.

The models we consider in this chapter all share a common structure: they are all made by N different *internal* compartments (i.e., the *levels*, made of different living organisms), interacting with a certain number of *external* compartments[1] playing the role of the *nutrients* needed to feed the organisms in level 1 (autotroph organisms), and the *garbage* produced by all the elements occupying the various levels. Part of the garbage turns into nutrients after some time. The organisms of levels 2, 3,... (heterotroph organisms) are fed by those of the immediately preceding level: the organisms of level k are food for those in level $k + 1$. The only dynamical degrees of freedom which are interesting for us are the densities of the organisms in the different levels, of the garbage(s) and of the nutrients. Of course, the simplest model is the one with a single level of heterotroph organisms: $N = 2$. This is a good starting point to describe how our general framework could be used in this context, and to check its usefulness also for more complicated situations. We will consider this particular situation later in the chapter, while we will keep N unfixed in the general description of the models.

We begin our analysis in the next section by considering a simple linear model, describing N different levels of the system that interact with two external compartments. Compartments play the role of the *nutrients* and the *garbage* for the organisms. In Section 6.2, we consider another linear model with two different garbages. The reason for doubling the garbages is because we are interested in modeling the possibility that part of the garbage turns into nutrients quite fast, while for another part this change is much slower. This is exactly what happens if we have some *soft* and *hard* garbage, like flesh and bones. In Section 6.3, we propose a nonlinear version of this last system, while, in Section 6.4, we introduce phenomenologically a damping effect to model the fact that, after a sufficiently long time, if the ecosystem is not completely efficient, i.e., if S is unable to recycle all the garbage produced during time, the densities of the species are expected to decrease significantly, and, eventually, to approach zero. This recalls what we have done in Chapter 4, for instance, to mimic the existence of some stress in the soil quality, and in fact it can be interpreted in the same way: the *recycling efficiency* of S, i.e., the capacity of S to transform garbage into nutrients, is less than 100%, because of some possible effect that we can call, again, a *stress factor*. Hence, it will be natural, and successful, to model this effect as we have done in Chapter 4, simply replacing some real parameters of the Hamiltonian with their complex valued counterparts.

6.1 A Linear Model with a Single Garbage

In Figure 6.1, we show how our first system S looks: it is made of N levels of organisms, $L_1, L_2, \ldots, L_N$, one compartment for the nutrients and a single compartment for the garbage. The arrows indicate the possible mechanisms occurring in

---

[1] These compartments are external with respect to the living species, the levels, but are still part of the system S. That's why the system is closed.

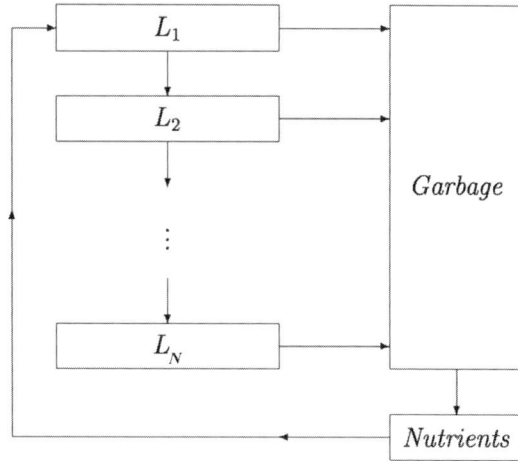

Figure 6.1 A schematic view of the single garbage ecosystem (Source: Bagarello and Oliveri [87]).

$\mathcal{S}$; all the organisms, independently of their level, contribute to the garbage, when dying. The garbage, after some time, becomes nutrient, which feeds the organisms in level L_1. These organisms contribute not only to the garbage, but they become food for those in L_2, which are food for those in L_3, and so on. Since we are interested in the description of the time dependence of the densities of these compartments, we adopt here fermionic ladder operators as the relevant dynamical variables, as we have done so far in this book, since they describe densities well.

The time evolution of the densities is described by a Hamiltonian operator containing the essential features of the system we want to model. Here we adopt the following Hamiltonian:

$$\begin{cases} H = H_0 + H_I, \qquad \text{with} \\[2mm] H_0 = \sum_{j=0}^{N+1} \omega_j\, a_j^\dagger\, a_j, \\[2mm] H_I = \sum_{j=0}^{N} \lambda_j \left(a_j\, a_{N+1}^\dagger + a_{N+1} a_j^\dagger \right) + \sum_{j=0}^{N-1} v_j \left(a_j\, a_{j+1}^\dagger + a_{j+1} a_j^\dagger \right), \end{cases} \tag{6.1}$$

where, as stated, a_j's are ladder fermionic operators satisfying the following CAR:

$$\{a_j, a_k^\dagger\} = \delta_{j,k}\, \mathbb{1}, \qquad a_j^2 = 0, \tag{6.2}$$

for all $j, k = 0, 1, \ldots, N+1$, and where ω_j, v_j and λ_j are real constants. We see that $H = H^\dagger$, and that H does not depend explicitly on time. The zeroth mode is related to the nutrients, the $(N+1)$-th mode to the garbage, while all the remaining modes describe the organisms of the various trophic levels. The Hamiltonian (6.1) contains a free *standard* part, H_0, whose parameters measure the inertia of the different compartments [1]: the higher the value of a certain ω_j, the higher the tendency of the density of the j-th agent (organism, nutrient or garbage) to stay constant in time, even in presence of interaction, or to have only *small* oscillations around some fixed value. Moreover, if $\lambda_j = \nu_j = 0$, $H = H_0$ commutes with each $\hat{n}_j = a_j^\dagger a_j$: in absence of interactions between the compartments, their densities stay constant. The next term, H_I, which is quadratic in the raising and lowering operators, can be understood as follows: $\lambda_j a_j a_{N+1}^\dagger$ describes the death of some organisms in level L_j and a related increasing of the garbage – metabolic waste and death organisms become garbage! For $j = 0$, H_I contains the contribution $\lambda_0 a_{N+1} a_0^\dagger$, which describes the fact that the garbage is recycled by decomposers and transformed into nutrients. Recall that, to make the Hamiltonian self-adjoint, we are also forced to add the adjoint contributions, even if their presence in H has no evident natural interpretation. For instance, we have to include in H the terms $\lambda_j a_{N+1} a_j^\dagger$ and $\lambda_0 a_0 a_{N+1}^\dagger$: it is as if the arrows in Figure 6.1 could also be reversed, describing fluxes in both directions. However, the effects of these terms in the Hamiltonian are to a certain extent controlled by the initial conditions on S, which in general make more favorable the flux in just one direction. We will return to this aspect of our approach in Chapter 7, exploring the role of terms like those above in H rather than the initial conditions, in another application to biology. The term $\nu_j a_j a_{j+1}^\dagger$ describes the fact that the nutrients are used by the organisms of level 1, and that the organisms of level L_j feed those of the level L_{j+1} ($j = 1, \dots, N-1$). Again, the adjoint contribution $\nu_j a_{j+1} a_j^\dagger$ needs to be inserted in H_I to keep H self-adjoint, but the initial conditions on S make the probability that organisms of level L_{j+1} feed those of level L_j less likely.

The Heisenberg equations of motion for the lowering operators of the system, deduced by $\dot{X} = i[H, X]$, follow from the CAR's in (6.2):

$$\begin{cases} \dot{a}_0 = i \left(-\omega_0 a_0 + \lambda_0 a_{N+1} + \nu_0 a_1\right), \\ \dot{a}_l = i \left(-\omega_l a_l + \lambda_l a_{N+1} + \nu_{l-1} a_{l-1} + \nu_l a_{l+1}\right), \\ \dot{a}_N = i \left(-\omega_N a_N + \lambda_N a_{N+1} + \nu_{N-1} a_{N-1}\right), \\ \dot{a}_{N+1} = i \left(-\omega_{N+1} a_{N+1} + \sum_{j=0}^{N} \lambda_j a_j\right), \end{cases} \tag{6.3}$$

where $l = 1, 2, \dots, N-1$. Of course, the equations for a_0 and a_{N+1} are different from that for a_l, since they describe compartments of the model which are not the organisms. Moreover, the equation for a_N is also slightly different from those of the a_l, $l = 1, 2, \dots, N-1$, because, as Figure 6.1 shows, level L_N has just an outgoing arrow (to the garbage), while the others L_k, $k = 1, 2, \dots, N - 1$ have 2 such arrows, one going to the garbage and the other to the next level, L_{k+1}.

System (6.3) can be rewritten as $\dot{A} = XA$, where

$$
A = \begin{pmatrix} a_0 \\ a_1 \\ \vdots \\ \vdots \\ a_N \\ a_{N+1} \end{pmatrix}, \qquad
X = i \begin{pmatrix}
-\omega_0 & \nu_0 & 0 & \cdots & \cdots & \lambda_0 \\
\nu_0 & -\omega_1 & \nu_1 & \cdots & \cdots & \lambda_1 \\
\cdots & \cdots & \cdots & \cdots & \cdots & \cdots \\
\cdots & \cdots & \cdots & \cdots & \cdots & \cdots \\
\cdots & \cdots & \cdots & \cdots & -\omega_N & \lambda_N \\
\lambda_0 & \lambda_1 & \lambda_2 & \cdots & \lambda_N & -\omega_{N+1}
\end{pmatrix},
$$

X being a symmetric matrix. The solution is $A(t) = V(t)A(0)$, with $V(t) = e^{Xt}$. Calling $V_{k,l}(t)$ the entries of the matrix $V(t)$, and defining, in analogies with formulas (4.5) and (5.6), $n_k(t) = \left\langle \varphi_{\mathbf{n}}, a_k^\dagger(t)a_k(t)\varphi_{\mathbf{n}} \right\rangle$, where $\mathbf{n} = (n_0, n_1, \dots, n_N, n_{N+1})$ is the vector describing the initial densities of the system, we find that

$$
n_k(t) = \sum_{l=0}^{N+1} |V_{k,l}(t)|^2 \, n_l. \tag{6.4}
$$

These are the densities at time t of the various compartments of the system, $k = 0, 1, 2, \dots, N + 1$, with initial conditions on $\mathcal{S}$ fixed by the vector $\varphi_{\mathbf{n}}$.

Rather than studying further what comes out from formula (6.4), we will now discuss how the model introduced in this section can be extended in order to model the possibility of having garbages of different kinds: *soft garbage*, which easily turns into nutrients, and *hard garbage*, which produces nutrients but only after a much longer period.

6.2 A Linear Model with Two Garbages

In this section, we discuss what happens if we add a second compartment for the garbage to the system. The leading idea is that, considering different coupling constants between the garbages G_1 and G_2, and the nutrients F, we will be able to model the fact that part of the waste products and dead organisms are turned into nutrients rather quickly, while other parts are converted into nutrients only after some time. This approach is suggested by many previous applications considered over the years, which show that the role of the coupling constants in H is often related to some (generalized) mobility of the agents involved. For instance, this is what we have observed in Chapter 5, where the coupling parameters $p_{\alpha,\beta}^{(a)}$ and $p_{\alpha,\beta}^{(b)}$ in formulas (5.3) and (5.8) are directly related to the speed of diffusion of the two populations in the room.

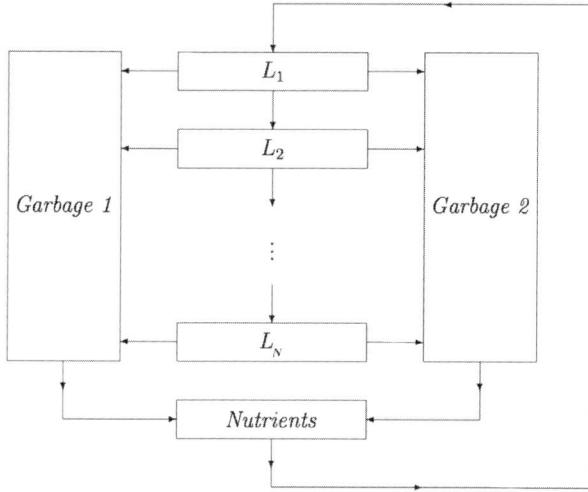

Figure 6.2 A schematic view of the two-garbages ecosystem (Source: Bagarello and Oliveri [87]).

Of course, the presence of two compartments for the garbages implies that we need to add another degree of freedom (i.e., another agent) to the model and, consequently, an extra fermionic operator, a_{N+2}. The structure of the ecosystem is given in Figure 6.2. We see that, when compared with Figure 6.1, two horizontal arrows start from each level $L_j, j = 1, 2, \ldots, N$, one going to the soft garbage G_1, and the second toward the hard garbage, G_2. Moreover, both G_1 and G_2 (with different time scales) contribute to the nutrients, and these nutrients are used to feed the organisms in L_1, as in the previous model.

The Hamiltonian now looks like

$$
\begin{cases}
H = H_0 + H_I, \qquad \text{with} \\[2mm]
H_0 = \displaystyle\sum_{j=0}^{N+2} \omega_j \, a_j^\dagger \, a_j, \\[4mm]
H_I = \displaystyle\sum_{j=0}^{N} \lambda_j^{(1)} \left(a_j \, a_{N+1}^\dagger + a_{N+1} a_j^\dagger \right) + \sum_{j=0}^{N} \lambda_j^{(2)} \left(a_j \, a_{N+2}^\dagger + a_{N+2} a_j^\dagger \right) \\[4mm]
\qquad + \displaystyle\sum_{j=0}^{N-1} \nu_j \left(a_j \, a_{j+1}^\dagger + a_{j+1} a_j^\dagger \right),
\end{cases}
\tag{6.5}
$$

where $\lambda_j^{(1)}$ describes the interaction between the organisms and G_1, while the value of $\lambda_j^{(2)}$ fixes the strength of the interaction between the organisms and G_2. The operators a_j and $a_j^\dagger$ satisfy the same CAR as in (6.2), $\{a_j, a_k^\dagger\} = \delta_{j,k} \mathbb{1}$, with $a_j^2 = 0$, for all $j, k = 0, 1, \ldots, N + 2$.

The meaning of the various contributions in H are completely analogous to those given in Section 6.1 and will not be repeated here. We just want to observe that the two terms responsible for the transformation of garbage into nutrients are $\lambda_0^{(1)} a_{N+1} a_0^\dagger$ and $\lambda_0^{(2)} a_{N+2} a_0^\dagger$: part of G_1 and G_2 turns into nutrients, with a speed that is related to the numerical values of $\lambda_0^{(1)}$ and $\lambda_0^{(2)}$, respectively. Notice also that the last term in H_I is identical to a contribution already appearing in (6.1) and describes the fact that the organisms in L_j are food for those in L_{j+1}. The equations of motion are the following:

$$
\begin{cases}
\dot{a}_0 = i\left(-\omega_0 a_0 + \lambda_0^{(1)} a_{N+1} + \lambda_0^{(2)} a_{N+2} + \nu_0 a_1\right), \\[4pt]
\dot{a}_l = i\left(-\omega_l a_l + \lambda_l^{(1)} a_{N+1} + \lambda_l^{(2)} a_{N+2} + \nu_{l-1} a_{l-1} + \nu_l a_{l+1}\right), \\[4pt]
\dot{a}_N = i\left(-\omega_N a_N + \lambda_N^{(1)} a_{N+1} + \lambda_N^{(2)} a_{N+2} + \nu_{N-1} a_{N-1}\right), \\[4pt]
\dot{a}_{N+1} = i\left(-\omega_{N+1} a_{N+1} + \sum_{l=0}^{N} \lambda_l^{(1)} a_l\right), \\[4pt]
\dot{a}_{N+2} = i\left(-\omega_{N+2} a_{N+2} + \sum_{l=0}^{N} \lambda_l^{(2)} a_l\right),
\end{cases}
\tag{6.6}
$$

$l = 1, 2, \ldots, N - 1$, and can be solved as in Section 6.1, due to the fact that the system (6.6) is linear. Setting $\hat{N}_{tot} := \sum_{l=0}^{N+2} \hat{n}_l$, where $\hat{n}_l = a_l^\dagger a_l$, we can check that $[H, \hat{N}_{tot}] = 0$, so that $\hat{N}_{tot}$ is a conserved quantity: What disappears from the organisms appears in the garbages and in the nutrients, and vice versa. This is a good reason to consider our ecological system *closed*: nothing really disappears. It is only moved, (and changed) from one compartment to another. To make the situation easy, but still not trivial, let us fix $N = 2$: such a simplifying choice corresponds to identify levels L_1 and L_2 with the autotroph and heterotroph organisms, respectively. In Figure 6.3 we show how the densities of the various compartments of the system change in time, where the parameters are set as follows: $\lambda_1^{(1)} = 0.005$, $\lambda_2^{(1)} = 0.009$, $\lambda_1^{(2)} = 0.05$, $\lambda_2^{(2)} = 0.09$, $\nu_0 = 0.1$, $\nu_1 = 0.01$, $\nu_2 = 0.1$, $\omega_0 = 0.05$, $\omega_1 = 0.1$, $\omega_2 = 0.2$, $\omega_3 = 0.3$ and $\omega_4 = 0.45$. This particular choice is motivated by the following reasons: since G_2 is the hard garbage, while G_1 is the soft one, it is clear that the inertia of G_2, measured by ω_4, must be larger than that of G_1, which is measured by ω_3. For this reason we have taken $\omega_4 > \omega_3$. Moreover, since the nutrients should be easily used by the organisms, ω_0 is taken to be rather small, representing almost no inertia. Levels L_1 and L_2 are distinguished by assuming a larger inertia for level L_2 with respect to that of level L_1: $\omega_1 < \omega_2$. We are further assuming that the strength of the interaction of the first level with G_1 and G_2 is smaller than that of the second one:

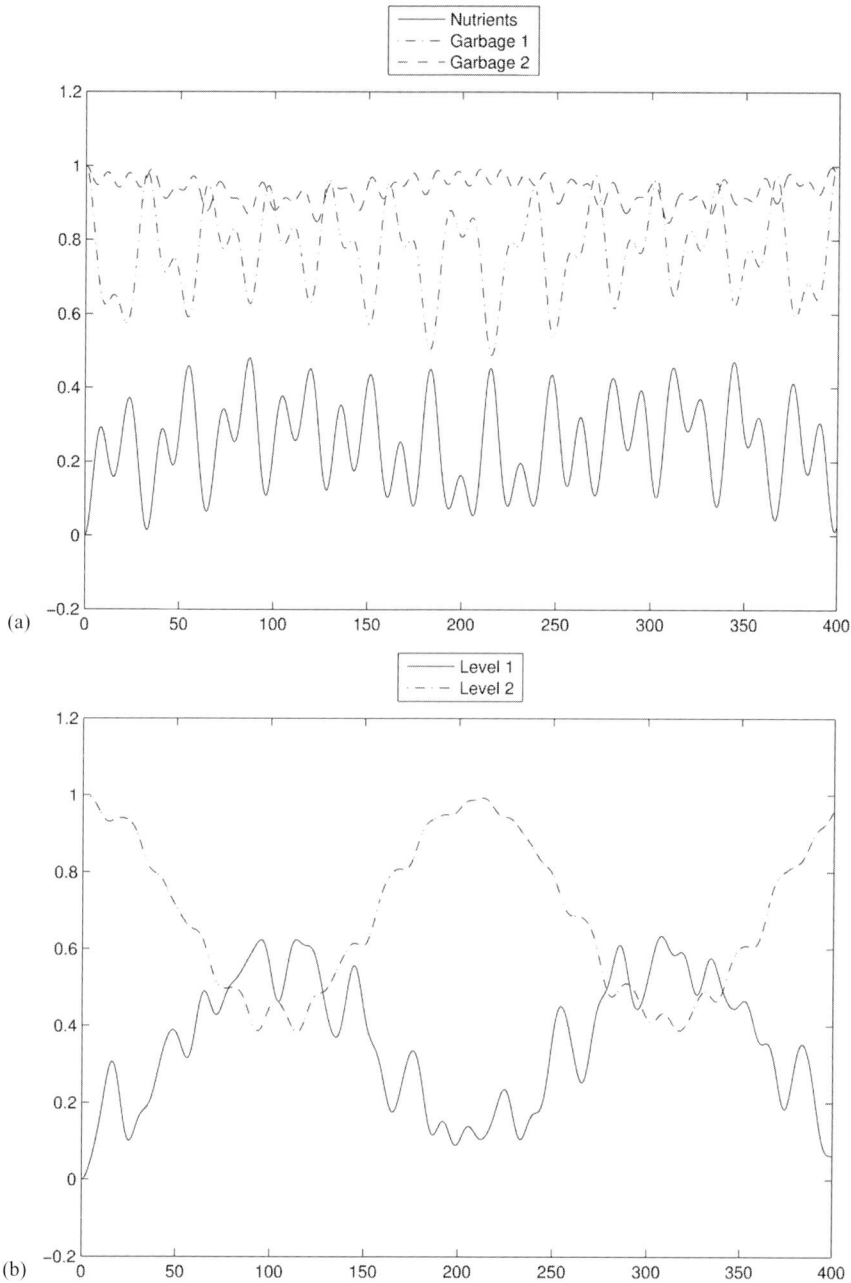

Figure 6.3 Densities of the nutrients, G_1 and G_2 (a) and of levels L_1 and L_2 (b). Initial conditions: nutrients and level L_1 empty, G_1, G_2 and level L_2 completely filled (Source: Bagarello and Oliveri [87]).

for this reason, we are taking $\lambda_1^{(k)} < \lambda_2^{(j)}$, $j, k = 1, 2$. The parameter ν_1 is very small, compared with ν_0, because organisms of level L_1 use the nutrients to increase their densities with timescales smaller than those needed by the organisms of level L_2, which are fed by the organisms of level L_1.

Figures 6.3–6.5 are all related to these values of the parameters but to different initial conditions for the various compartments. In Figure 6.3 we assume that the nutrients and level L_1 are empty (or that their densities are very low) at $t = 0$, while the two garbages and level L_2 are completely filled (or have high densities). Figure 6.4 describes the results obtained assuming that the nutrients and level L_2 are empty at $t = 0$, while the two garbages and level L_1 are completely filled. Finally, in Figure 6.5, the nutrients and the levels L_1 and L_2 are completely filled at $t = 0$, while the two garbages are empty.

Among other things, these figures show that the fluctuations of G_2 are smaller than those of the other compartments. This is expected, since the value of ω_4 is larger than the other ω_j, so that G_2 presents the largest inertia. As for the densities of levels 1 and 2, Figures 6.3 and 6.4 shows the originally empty level acquires a density that is larger than the density of the other, originally filled, level. In other words, we see that an inversion of the populations is possible. Figure 6.5 shows that, starting with *clean* initial conditions (nutrients, levels L_1 and L_2 completely filled, and no waste products at all, at $t = 0$), the densities of the garbages stay reasonably low, while the nutrients and the densities of the organisms in levels L_1 and L_2 oscillate around high values, close to one. This could be interesting for concrete applications, for which one needs to maximize the *efficiency* of the ecosystem, trying to keep low the quantity of the garbage (of any kind), and to keep high the density of the nutrients. However, in real life, some dissipation mechanism should also be added into the system: this is important to model, for instance, the fact that not all the garbage really turns into nutrient! This aspect will be discussed in Section 6.4.

> *Remark:* The fact that Figures 6.3 and 6.4 only show oscillation is in complete agreement with what we have discussed in Section 2.7. In fact, our ecological system is defined on a finite-dimensional Hilbert space, and since the Hamiltonian in (6.5) is self-adjoint, $H = H^\dagger$, the motion of any observable of S can only be periodic or quasi-periodic. In Section 6.4, we will relax this aspect of the system, and we will consider a Hamiltonian satisfying condition $H \neq H^\dagger$. As already stated, this is useful and convenient to model some dissipation. In fact, the consequence of this choice will be that the densities decay to zero after some time (showing again many small oscillations while decaying).

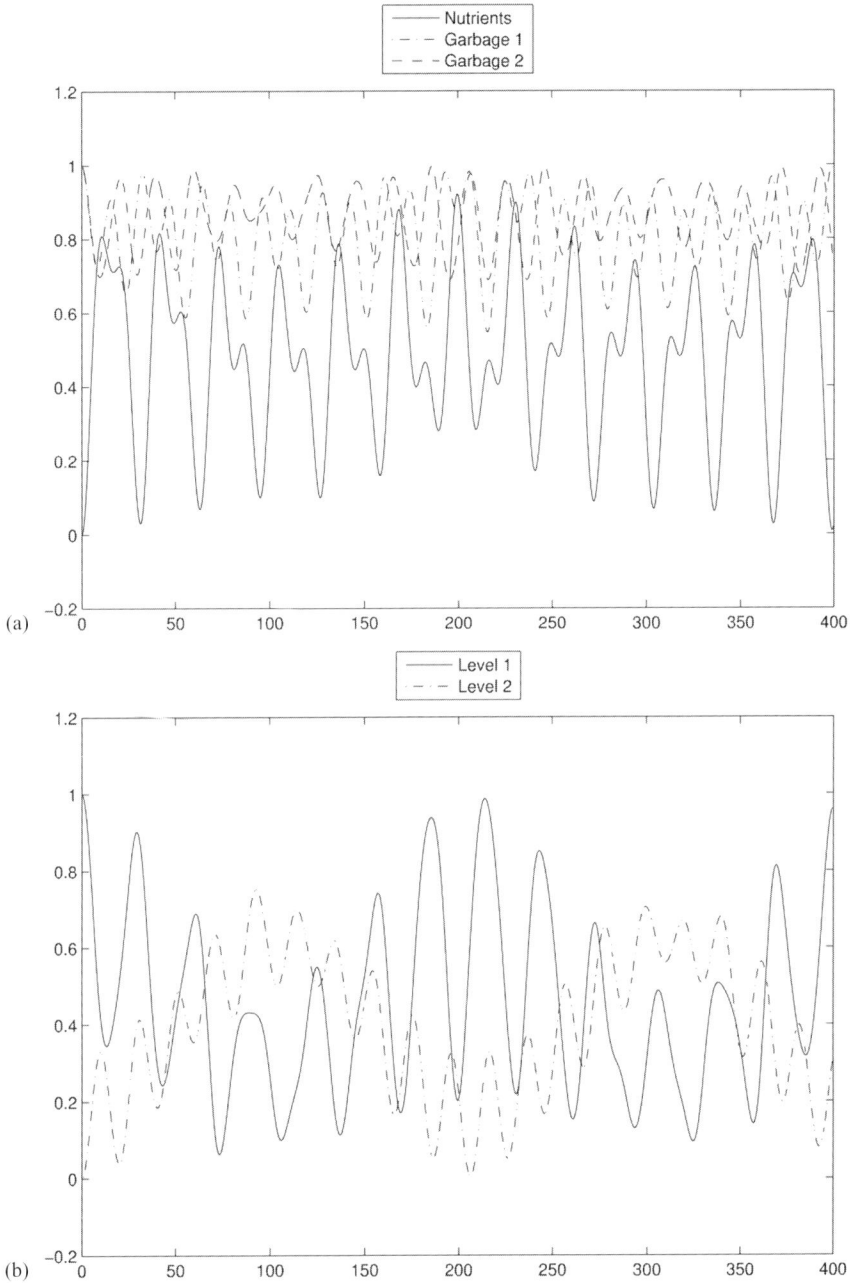

Figure 6.4 Densities of the nutrients, G_1 and G_2 (a) and of levels L_1 and L_2 (b). Initial conditions: nutrients and level L_2 empty, G_1, G_2 and level L_1 completely filled (Source: Bagarello and Oliveri [87]).

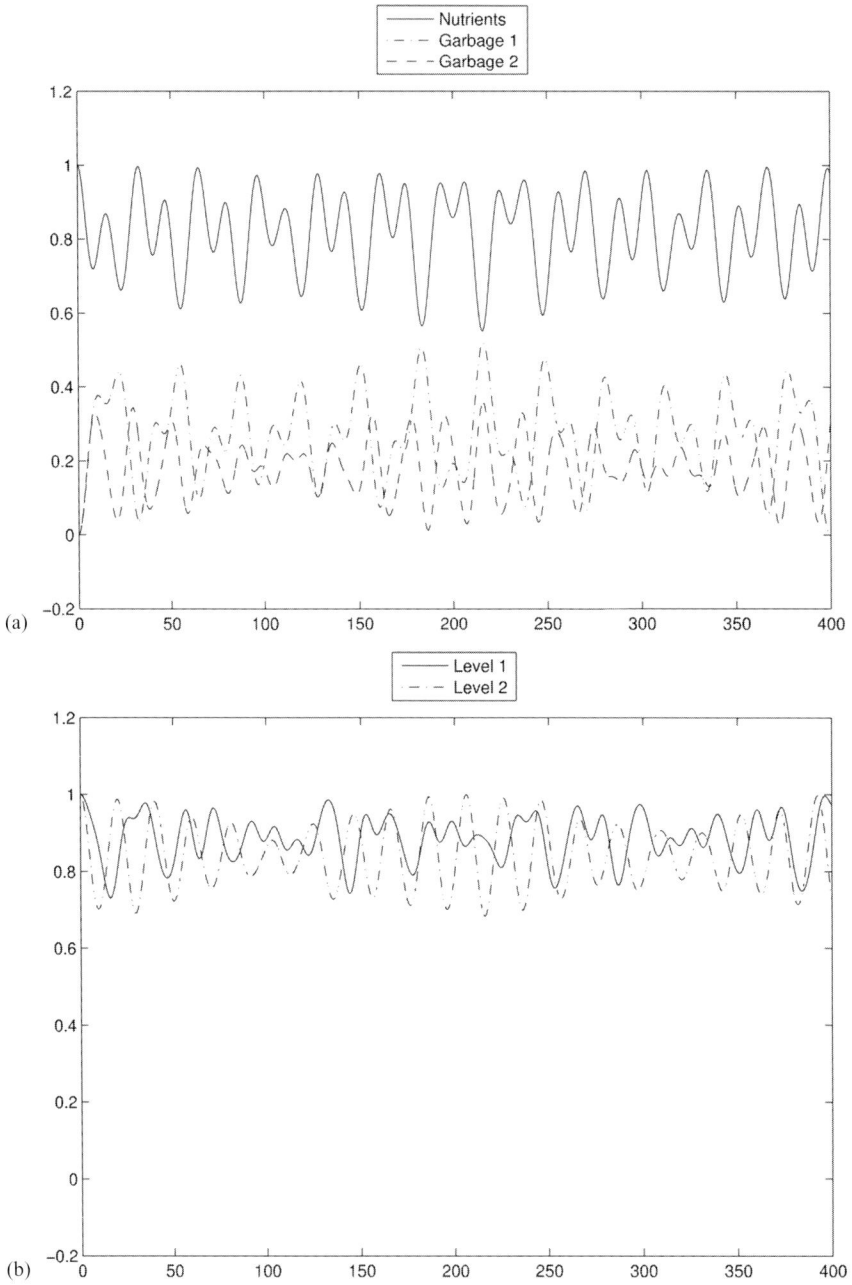

Figure 6.5 Densities of the nutrients, G_1 and G_2 (a) and of levels L_1 and L_2 (b). Initial conditions: nutrients, levels L_1 and L_2 completely filled, G_1 and G_2 empty (Source: Bagarello and Oliveri [87]).

6.3 A Nonlinear Model with Two Garbages

The same system schematically described in Figure 6.2 will be considered now in a slightly different way: Instead of adding two different, quadratic terms, in H to model the interaction of the various levels with the two garbages, as in $\lambda_j^{(1)} a_j a_{N+1}^\dagger$ and $\lambda_j^{(2)} a_j a_{N+2}^\dagger$, we could also consider a single cubic contribution (see the first term in H_I below, formula [6.7]). For instance, $a_j a_{N+1}^\dagger a_{N+2}^\dagger$ models the fact that the density of the j-th level decreases while, simultaneously, the densities of both G_1 and G_2 increase: through its metabolism, or dying, an organism produces, in a single event, garbage of two different kinds, soft and hard. A term like this (together with its hermitian conjugate) can be used to replace the two quadratic contributions in (6.5) proportional to $\lambda_j^{(1)}$ and $\lambda_j^{(2)}$. However, the consequences of having cubic terms in a Hamiltonian are not so easy to handle, in general. The reason is that the Heisenberg differential equations of motion for the lowering operators deduced out of the Hamiltonian will no longer be linear in the ladder operators. This creates (too) many technical problems when the ladder operators satisfy the CCR (which is not the case here), while the model can still be analyzed, with just a little extra effort, if the a_j's satisfy the CAR. This is because the fermionic Hilbert space, constructed as in Section 2.2, is finite-dimensional, while its bosonic counterpart is not.

The full Hamiltonian of the system is the following:

$$
\left\{
\begin{aligned}
&H = H_0 + H_I, \qquad \text{with} \\[4pt]
&H_0 = \sum_{j=0}^{N+2} \omega_j \, a_j^\dagger \, a_j, \\[4pt]
&H_I = \sum_{j=1}^{N} \lambda_j \left(a_j \, a_{N+1}^\dagger \, a_{N+2}^\dagger + a_{N+2} \, a_{N+1} \, a_j^\dagger \right) \\[4pt]
&\qquad + \sum_{j=1}^{2} \nu^{(j)} \left(a_0 \, a_{N+j}^\dagger + a_{N+j} a_0^\dagger \right) + \sum_{j=0}^{N-1} \nu_j \left(a_j \, a_{j+1}^\dagger + a_{j+1} a_j^\dagger \right).
\end{aligned}
\right.
\tag{6.7}
$$

The notation is the same as in the previous section. For instance, zero is the mode for the nutrients, while $N + 1$ and $N + 2$ are the modes for the two garbages. The physical interpretation of the Hamiltonian is easy to understand: first of all, H_0 is identical to the one in (6.5). H_I describes an interaction between the organisms and the two garbages (first contribution); the nutrients and the two garbages (second contribution); and a *hopping* term (third contribution): the nutrients are used to feed the organisms of level 1, and the organisms of level L_j feed those of level L_{j+1} ($j = 1, \ldots, N-1$). The conjugate term, $a_{j+1} a_j^\dagger$, is needed in order to make the Hamiltonian self-adjoint, since all the parameters are here supposed to be real.

The Heisenberg differential equations of motion look not so *friendly*, especially when compared with those in (6.6). Indeed, calling $X := \sum_{l=1}^{N} \lambda_l \, a_l$, we have

$$
\begin{cases}
\dot{a}_0 = i \left(-\omega_0 a_0 + v_0 a_1 + 2X a_0 a_{N+1}^{\dagger} a_{N+2}^{\dagger} + 2a_{N+2} a_{N+1} X^{\dagger} a_0 + v^{(1)} a_{N+1} \right. \\
\qquad \left. + v^{(2)} a_{N+2} \right), \\[4pt]
\dot{a}_j = i \left(-\omega_j a_j + v_j a_{j+1} + v_{j-1} a_{j-1} + 2X a_j a_{N+1}^{\dagger} a_{N+2}^{\dagger} \right. \\
\qquad \left. + a_{N+2} a_{N+1} (2X^{\dagger} a_j - \lambda_j \mathbb{1}) \right), \\[4pt]
\dot{a}_N = i \left(-\omega_N a_N + v_{N-1} a_{N-1} + 2X a_N a_{N+1}^{\dagger} a_{N+2}^{\dagger} \right. \\
\qquad \left. + a_{N+2} a_{N+1} (2X^{\dagger} a_N - \lambda_N \mathbb{1}) \right), \\[4pt]
\dot{a}_{N+1} = i \left(-\omega_{N+1} a_{N+1} + X a_{N+2}^{\dagger} \left(\mathbb{1} - 2a_{N+1}^{\dagger} a_{N+1} \right) + v^{(1)} a_0 \right), \\[4pt]
\dot{a}_{N+2} = i \left(-\omega_{N+2} a_{N+2} + X a_{N+1}^{\dagger} \left(2a_{N+2}^{\dagger} a_{N+2} - \mathbb{1} \right) + v^{(2)} a_0 \right),
\end{cases}
\tag{6.8}
$$

with $j = 1, 2, \ldots, N-1$. It is evident that this system is not closed. In order to close it, we have to consider also the hermitian conjugate of these equations. Putting all the equations together we get a nonlinear system, whose solution can be found numerically, but not analytically, at a first sight.

Remark: As already suggested, for instance in (4.7), an alternative strategy would be to compute $e^{\pm iHt}$ for our H given in (6.7), and then compute the time evolution of the various densities $\hat{n}_j = a_j^{\dagger} a_j$, as prescribed by the Heisenberg picture:

$$
\hat{n}_j(t) = e^{iHt} \hat{n}_j e^{-iHt},
$$

$j = 0, 1, \ldots, N+2$. This approach might look simpler than the one based on the solution (numerical or analytical) of the system in (6.8), but it is not always so. The reason is that, as it is not difficult to imagine, the explicit computation of the exponential of a matrix (our $\pm iHt$) can be hard. And it gets harder and harder when N increases. Already for $N = 2$, H is a 16×16 matrix! Only under some serious simplifying assumption we can hope to exponentiate a similar matrix, of course. We can conclude that, even for fermions, Hamiltonians that are not quadratic in the ladder operators produce a dynamics, which can be almost impossible to compute. And the task becomes even harder when the number of degrees of freedom of the system increases, and essentially hopeless if CAR are replaced by CCR.

Notice now that, because of the nonlinearity in H, the operator $\hat{N}_{tot}$ introduced in Section 6.2 does not commute with the Hamiltonian anymore: $[H, \hat{N}_{tot}] \neq 0$. This means that $\hat{N}_{tot}$ is not an integral of motion for this nonlinear version of the system, and it is not evident if any other integral of motion exists at all. In a certain sense, we can say that loosing the linearity looks like *opening the system* to the outer world: part of $\hat{N}_{tot}$ could be lost or created, during the time evolution. This could have interesting consequences, since we might expect that a realistic ecosystem is not really *fully closed*. However, the plots in Figures 6.6–6.8, which are deduced by choosing, as in Section 6.2, $N = 2$, do not show any evident decay for the densities and, in fact, no such a decay is really expected at all, even in this nonlinear version of the model. We will return to this aspect of the model at the end of the section. For this reason, adapting to the present situation the strategy already considered in Section 4.3 (and see also Section 6.4), we will use a phenomenological non self-adjoint Hamiltonian, constructed by adding a small imaginary part to some of the parameters of the Hamiltonian itself. This will prove to be an efficient mechanism, in the present context, to describe some *lack of efficiency* of the recycling mechanisms of S.

Figures 6.6–6.8 are deduced fixing the parameters of the Hamiltonian as follows: $\omega_0 = 0.05$, $\omega_1 = 0.1$, $\omega_2 = 0.2$, $\omega_3 = 0.3$, $\omega_4 = 0.45$, $\lambda_1 = 0.005$, $\lambda_2 = 0.009$, $\nu_0 = 0.1$, $\nu_1 = 0.01$, $\nu^{(1)} = 0.1$ and $\nu^{(2)} = 0.03$. Notice that the value of $\nu^{(1)}$ is taken larger than that of $\nu^{(2)}$ since G_1 is assumed to be faster in producing nutrients with respect to G_2.

Figure 6.6 shows, among other things, that level L_2 and G_2 change in time less than the other compartments, as expected, because of their ecological interpretation, which is mathematically reflected by our choice of the related ω's. The nutrients and G_1 appear to be exactly out of phase. This is interesting, since it suggests that the soft garbage turns into nutrients quite easily (actually, simultaneously), while the hard garbage is almost not involved into this transformation (see Figure 6.6[a]). Figure 6.6(b) shows that, during the time evolution, the density of level L_1 can change by, at most, 30% of its initial value, while the density of level L_2 can decrease, at most, by 10%. Similar features, with slightly different percentages, can be deduced from Figure 6.7, which differs from the previous figure only in the initial conditions.

Also in Figure 6.7, we see clearly the effect of the inertia, which makes the second level and G_2 almost constant in time, especially when compared with the time behavior of the other compartments. In this case, the nutrients and G_1 are no longer exactly out of phase, as it is surely more realistic: the garbage in G_1 does not turn into nutrients instantaneously, as the process takes some time.

In Figure 6.8, we consider yet another initial condition on the system, but with the same choice of parameters. In particular, at $t = 0$, we assume that there is no garbage of any kind. Figure 6.8 shows that organisms of level L_2 undergo negligible variations, nutrients and organisms of level L_1 oscillate around very high values, whereas the garbage compartments are only partially filled, with G_2 always almost empty. Hence, the presence of nonlinearities in the model does not affect what we observed in Figure 6.5, deduced with the same initial conditions considered in Figure 6.8.

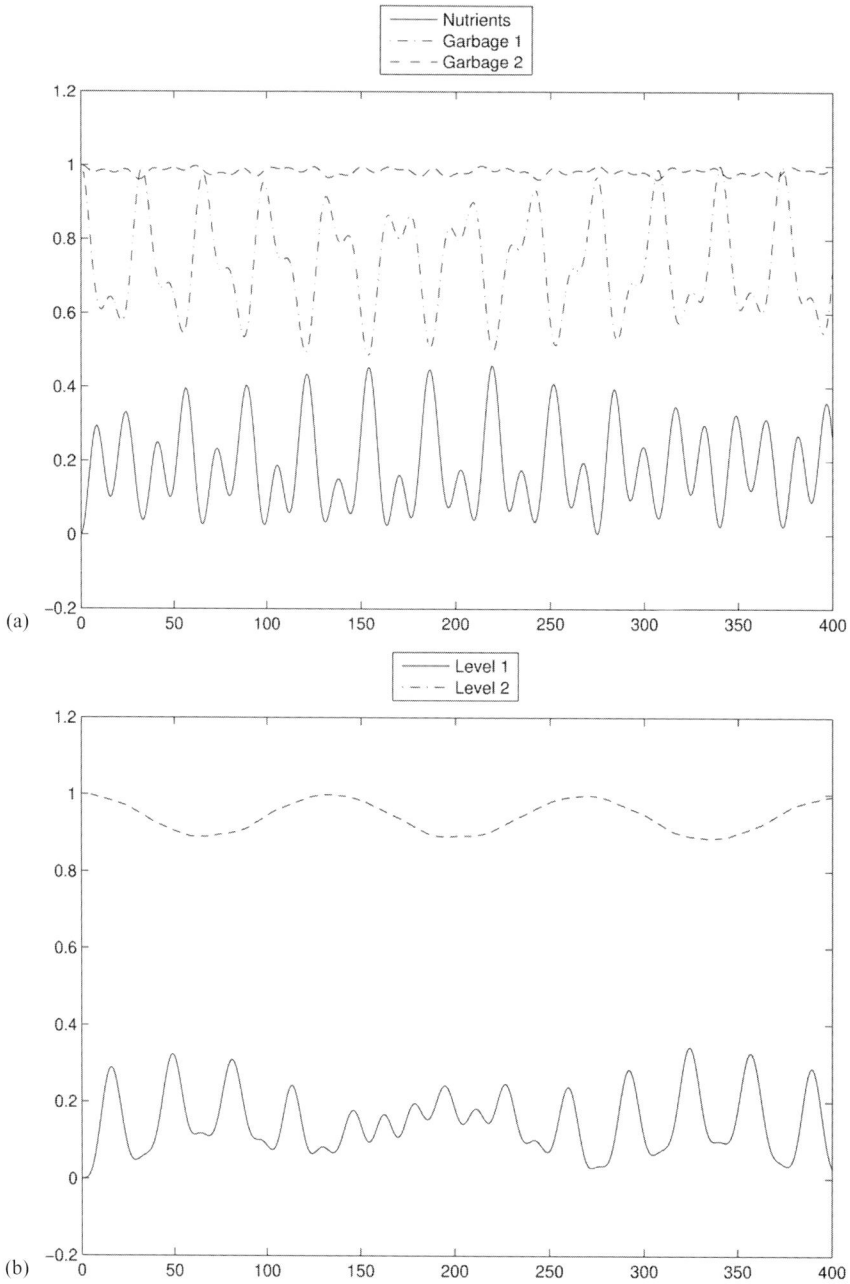

Figure 6.6 Densities of the nutrients, G_1 and G_2 (a) and of levels L_1 and L_2 (b). Initial conditions: nutrients and level L_1 empty, G_1, G_2 and level L_2 completely filled (Source: Bagarello and Oliveri [87]).

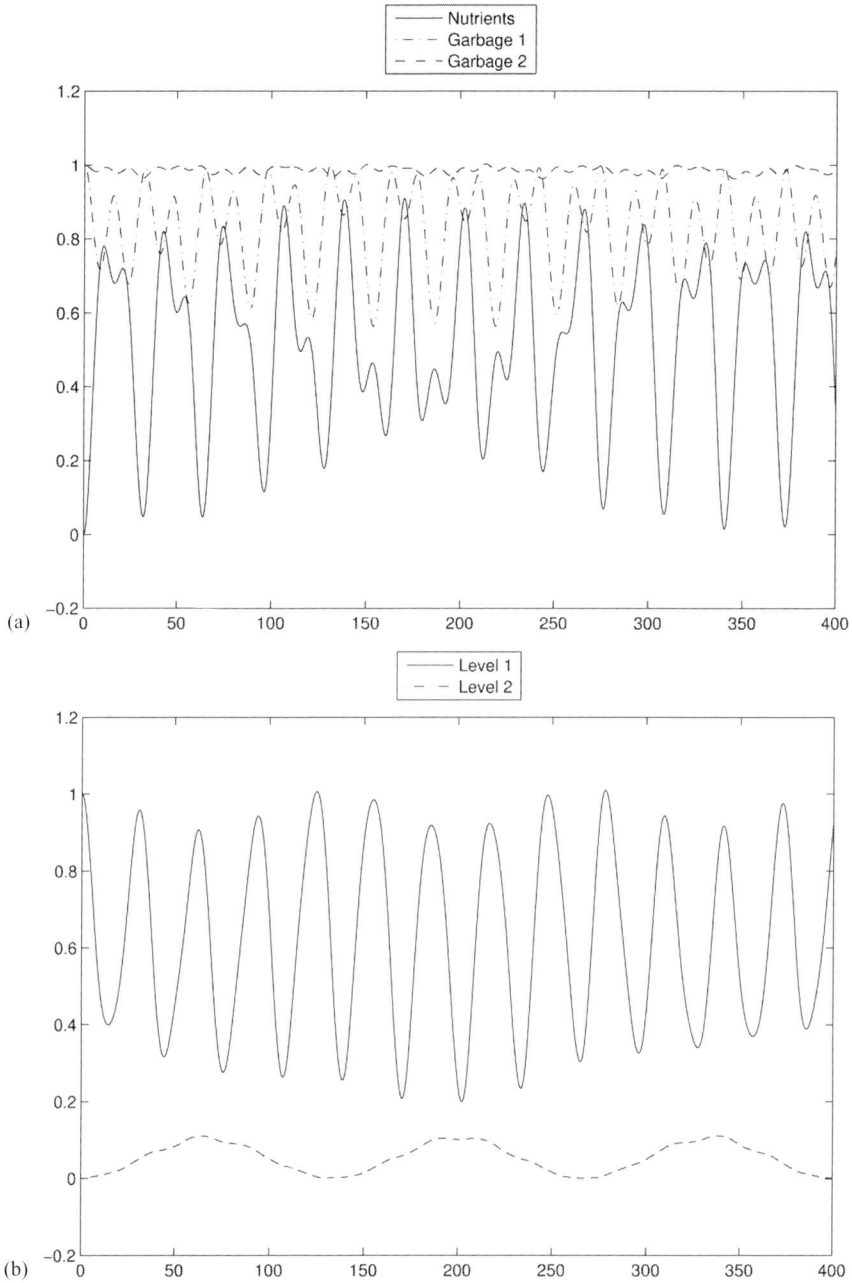

Figure 6.7 Densities of the nutrients, G_1 and G_2 (a) and of levels L_1 and L_2 (b). Initial conditions: nutrients and level L_2 empty, G_1, G_2 and level L_1 completely filled (Source: Bagarello and Oliveri [87]).

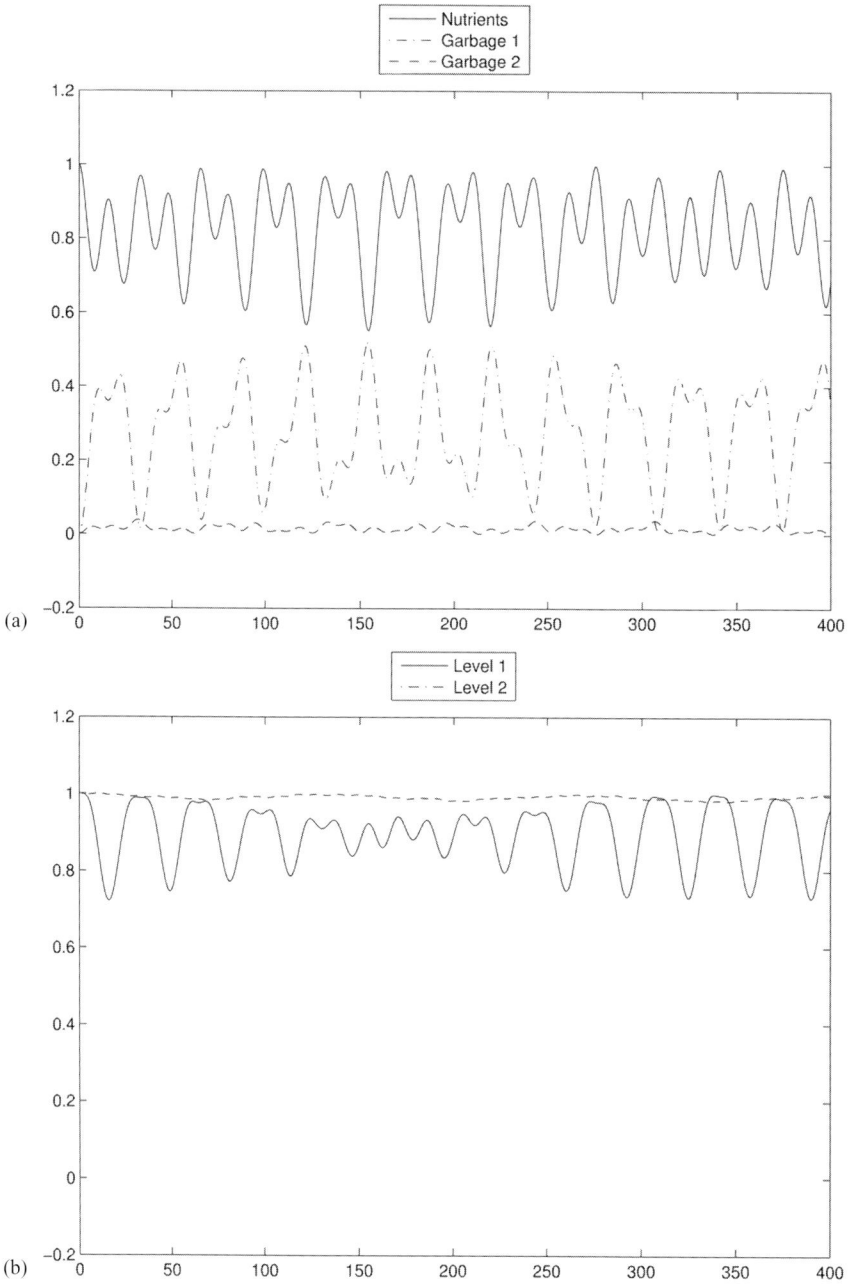

Figure 6.8 Densities of the nutrients, G_1 and G_2 (a) and of levels L_1 and L_2 (b). Initial conditions: nutrients, levels L_1 and L_2 completely filled, G_1 and G_2 empty (Source: Bagarello and Oliveri [87]).

All the figures considered here suggest that, at least with the choice of the parameters we have fixed in our simulations, the densities of the various compartments can only oscillate, but they don't appear to decay to any asymptotic value. This is expected because H, even in the presence of cubic interactions, is again a self-adjoint finite matrix, which can only produce periodic or quasi-periodic evolutions, as widely discussed several times along this manuscript (see in Section 2.7, in particular). To get some asymptotic value, we need to act on S in some *clever* and *meaningful* way: we could *open the system*, adding some external environment that interacts with S. Or we could replace the Heisenberg evolution with a (H, ρ)-induced dynamics, for some suitable rule ρ. Or, and this will be our choice in the next section, we could replace the self-adjoint Hamiltonian H with an *effective* Hamiltonian, H_{eff}, with $H_{eff} \neq H_{eff}^\dagger$, which describes well the lack of recycling efficiency of the ecosystem.

6.4 A Phenomenological Damping

As we have seen before, our interest is also in describing some *realistic damping* for the system by means of an *efficient approach* consisting in replacing the self-adjoint Hamiltonian with a suitable non self-adjoint version of it, keeping the other rules of the game unchanged. In particular, for our purposes, it is often sufficient to replace some real parameters involved in the Hamiltonian with complex numbers. This is exactly what we have done in Section 4.3.3 and what we will do here. More explicitly, we will show that it is enough to add a small negative imaginary part to just a single parameter of the free Hamiltonian, to induce damping for all the compartments of our closed ecosystem. And this is true both when we consider the quadratic Hamiltonian in (6.5) and when we work with the cubic Hamiltonian in (6.7). These two cases are considered separately in the following sections.

6.4.1 The Linear Case

The numerical values of the parameters are exactly those given in Section 6.2, except for ω_3, which instead of being real and equal to $\omega_3 = 0.3$, is now taken to be complex-valued: $\omega_3 = 0.3 - 0.01\,i$. As we see, we are adding a negative and relatively small imaginary part to ω_3. The reason for the choice of the sign of the imaginary part is suggested by what we have already done in Chapter 4, where we have seen, in particular, that the sign in the imaginary part of ω_3 is really essential: a positive imaginary part for ω_3 produces a blow up of the solution, while a negative imaginary part gives rise to a decay! From the point of view of the interpretation, we can surely say that negative and positive imaginary parts of the inertia of some compartment describe respectively stress factors and positive effects (see also Section 4.3.3). They act first on that particular compartment, but then, because of the interactions in S, they also affect the other agents of the system.

> *Remark:* The different effect of the sign of $\Im(\omega_3)$ can be easily understood as follows. Let us consider $h = \omega\, c^\dagger c$, with $[c, c^\dagger] = \mathbb{1}$ (or with $\{c, c^\dagger\} = \mathbb{1}$: our conclusions will be exactly the same). Then, the Heisenberg equation of motion for $c(t)$ is the following: $\dot{c}(t) = i[h, c] = -i\omega c(t)$. The solution is $c(t) = e^{-i\omega t} c(0)$, whose adjoint is $c^\dagger(t) = e^{i\overline{\omega} t} c^\dagger(0)$. Now, $c^\dagger(t)c(t) = e^{-i(\omega - \overline{\omega})\, t} c^\dagger(0) c(0)$. Of course, if $\omega \in \mathbb{R}$, then $c^\dagger(t)c(t) = c^\dagger(0)c(0)$, for all t. Suppose rather that $\omega = \omega_r + i\omega_i$, $\omega_r, \omega_i \in \mathbb{R}$. Then
>
> $$c^\dagger(t)c(t) = e^{-i(2i\omega_i)\, t} c^\dagger(0) c(0) = e^{2\omega_i t} c^\dagger(0) c(0),$$
>
> which is decaying only if $\omega_i < 0$. Otherwise it explodes. It is interesting to observe that in both cases, $\omega_i = 0$ or $\omega_i \neq 0$, $[h, c^\dagger c] = 0$. This might suggest that $c^\dagger c$ stay constant in time anyway, while the explicit solution of the Heisenberg equations of motion shows that it is not so. The reason is the following: When $h \neq h^\dagger$, as happens when $\omega_i \neq 0$, the use of the standard Heisenberg dynamics is not really necessarily the correct, or the most natural, choice. This problem was already discussed in Section 2.7.

Choosing the same initial conditions as in Figures 6.3–6.5, we get the following plots, seen in Figures 6.9–6.11.

The overall decay is evident, especially for the density of G_1. This is not surprising, since we have added the negative imaginary part exactly to the parameter measuring the inertia of G_1, which therefore reacts first. On the other hand, the decay of the density of G_2 seems rather slow. This is also expected, because of the large value of ω_4, which makes the inertia of G_2 rather large. For this reason not many drastic changes are expected for this compartment, at least in the short run – we need some time to observe serious differences in its density.

6.4.2 The Nonlinear Case

The numerical values of the parameters are exactly those of Section 6.3, except for ω_3, which is again replaced by $\omega_3 = 0.3 - 0.01\, i$. Our numerical computations show that, if we take $\Im(\omega_3) > 0$, even if very small, rather than damping we get the blow up of the densities. This is in agreement with what we have deduced before in this section (see our remark in Section 6.4.1). Figures 6.12–6.14 should be compared with Figures 6.6–6.8, which are deduced with exactly the same initial conditions and, except for $\Im(\omega_3)$, with the same values of the parameters of H.

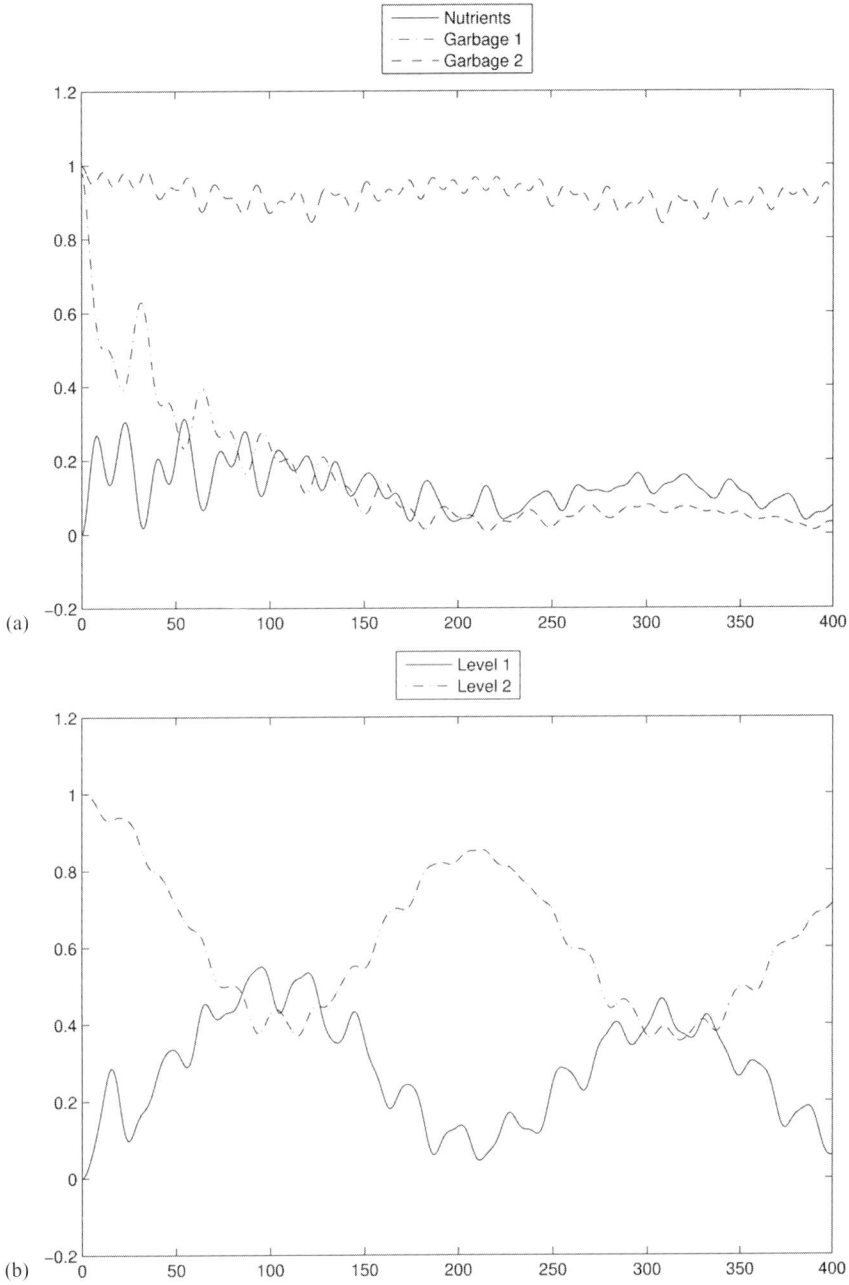

Figure 6.9 Densities of the nutrients, G_1 and G_2 (a) and of levels L_1 and L_2 (b). Initial conditions: nutrients and level L_1 empty, G_1, G_2 and level L_2 completely filled (Source: Bagarello and Oliveri [87]).

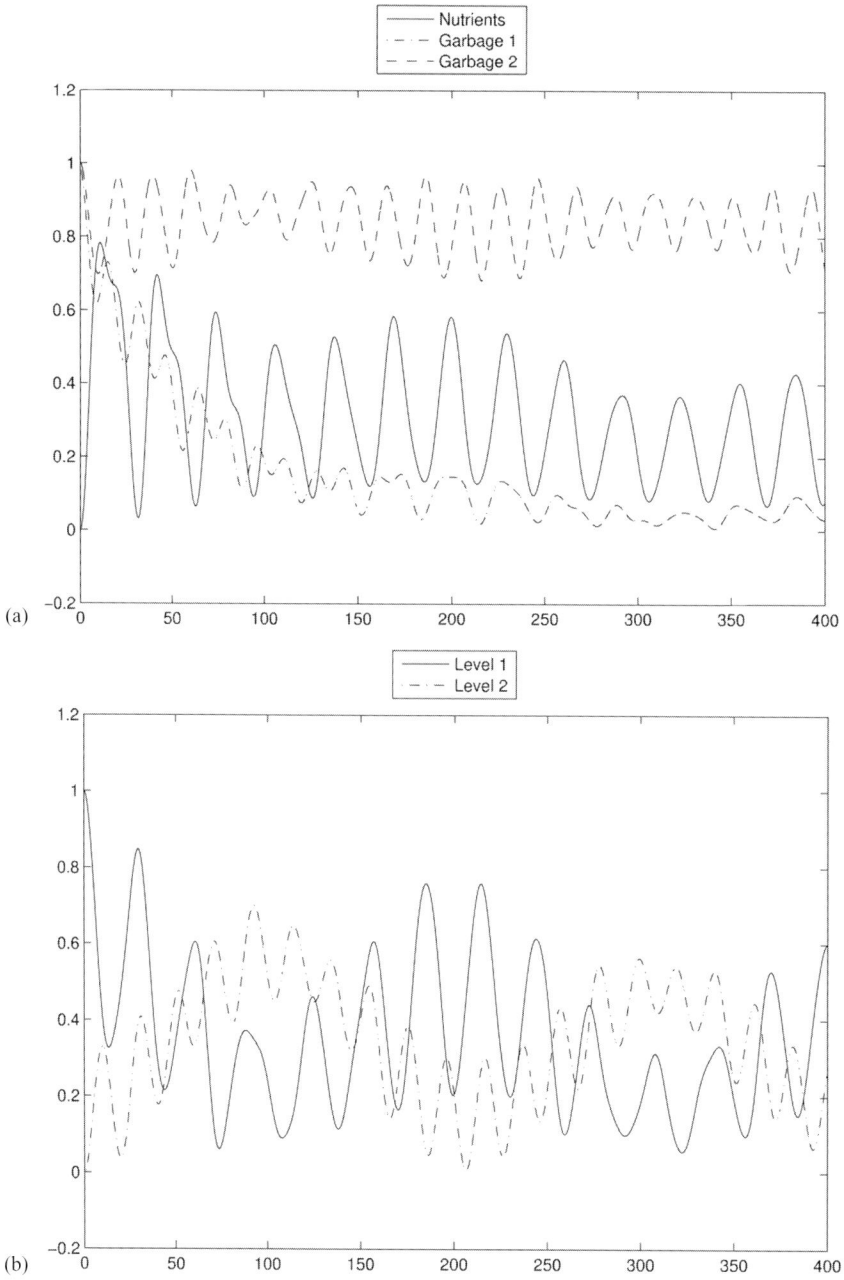

Figure 6.10 Densities of the nutrients, G_1 and G_2 (a) and of levels L_1 and L_2 (b). Initial conditions: nutrients and level L_2 empty, G_1, G_2 and level L_1 completely filled (Source: Bagarello and Oliveri [87]).

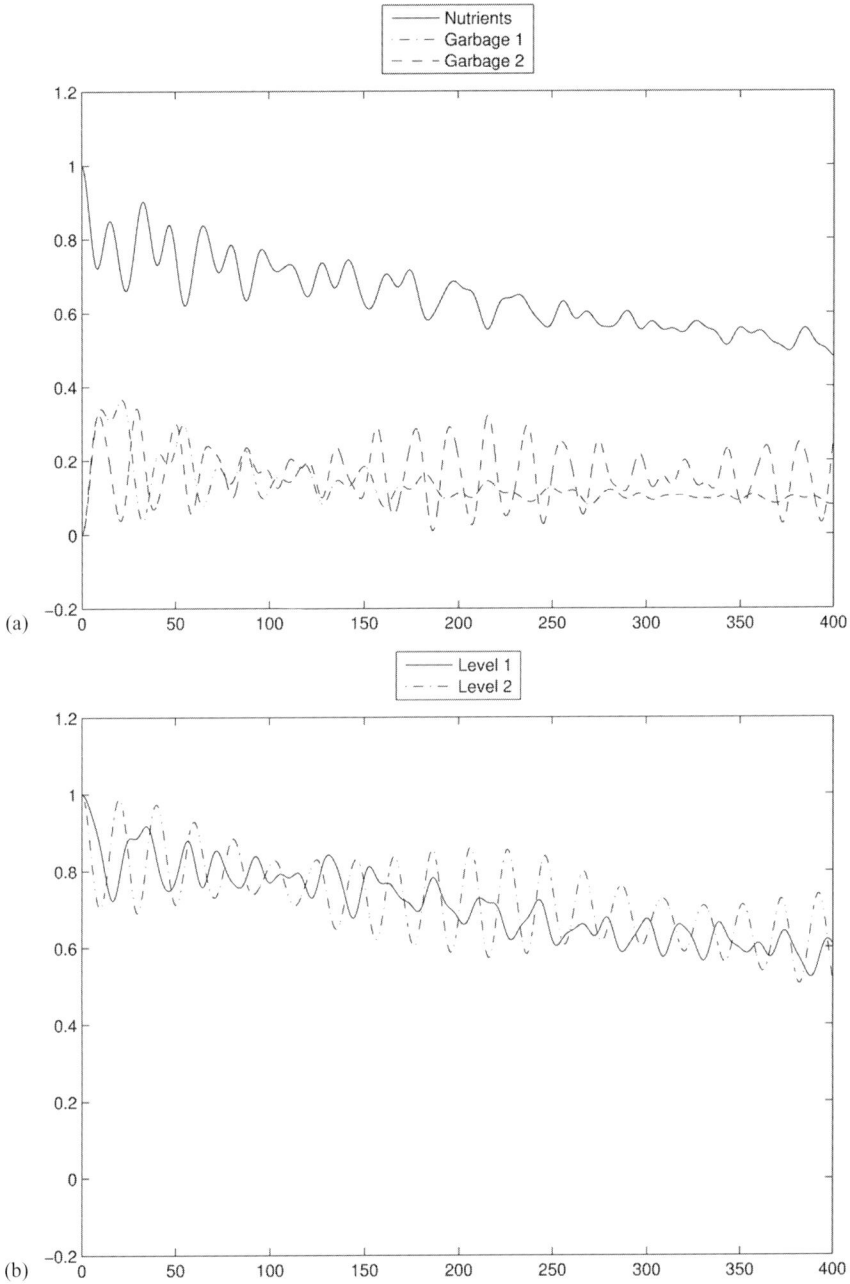

Figure 6.11 Densities of the nutrients, G_1 and G_2 (a) and of levels L_1 and L_2 (b). Initial conditions: nutrients, levels L_1 and L_2 completely filled, G_1 and G_2 empty (Source: Bagarello and Oliveri [87]).

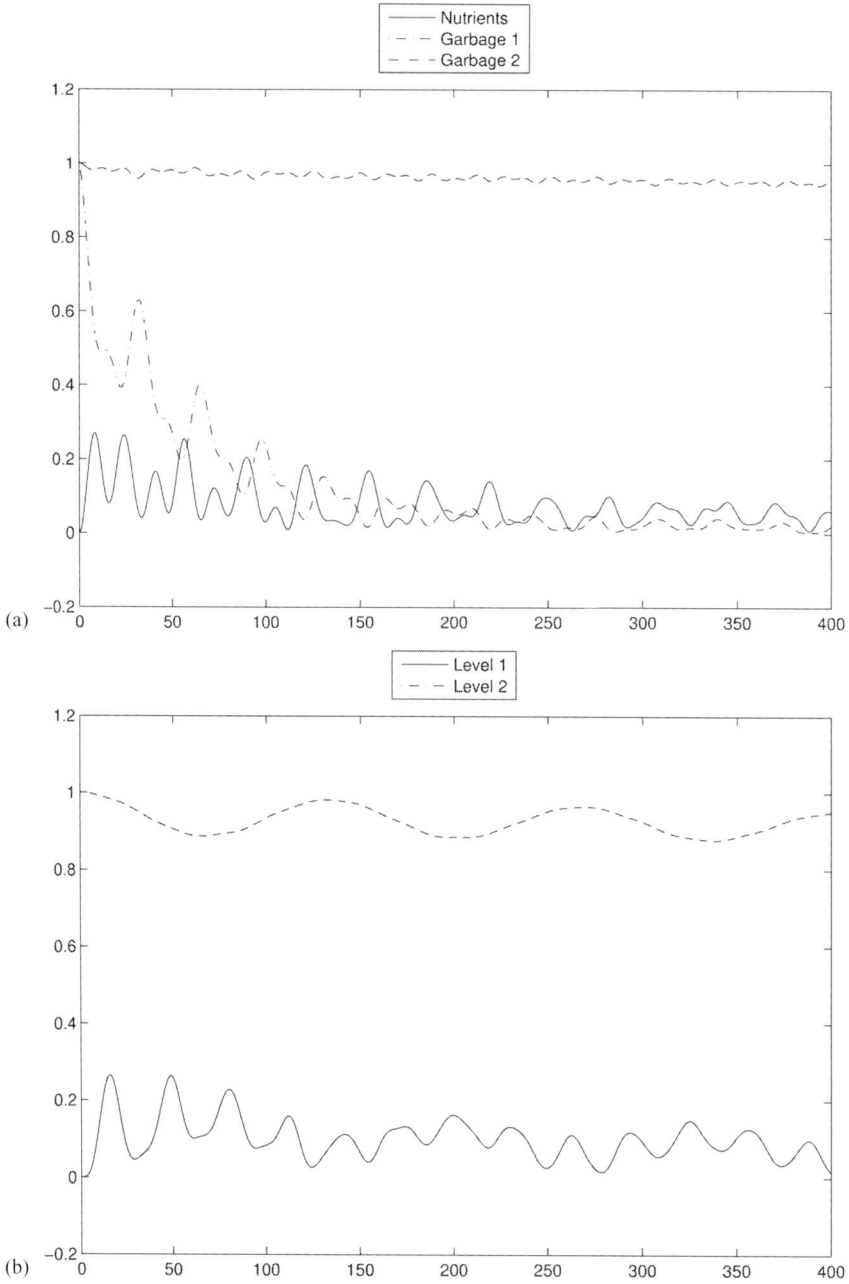

Figure 6.12 Densities of the nutrients, G_1 and G_2 (a) and of levels L_1 and L_2 (b). Initial conditions: nutrients and level L_1 empty, G_1, G_2 and level L_2 completely filled (Source: Bagarello and Oliveri [87]).

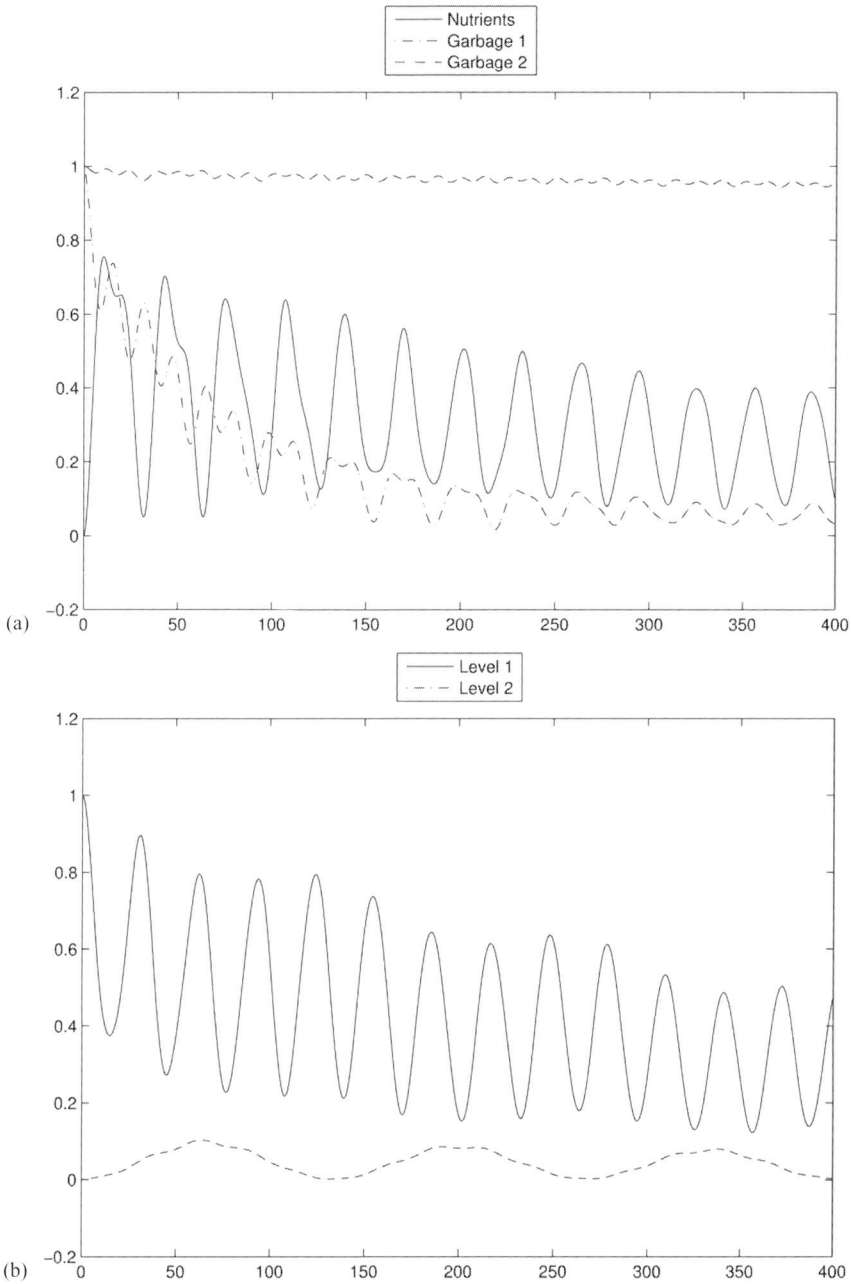

Figure 6.13 Densities of the nutrients, G_1 and G_2 (a) and of levels L_1 and L_2 (b). Initial conditions: nutrients and level L_2 empty, G_1, G_2 and level L_1 completely filled (Source: Bagarello and Oliveri [87]).

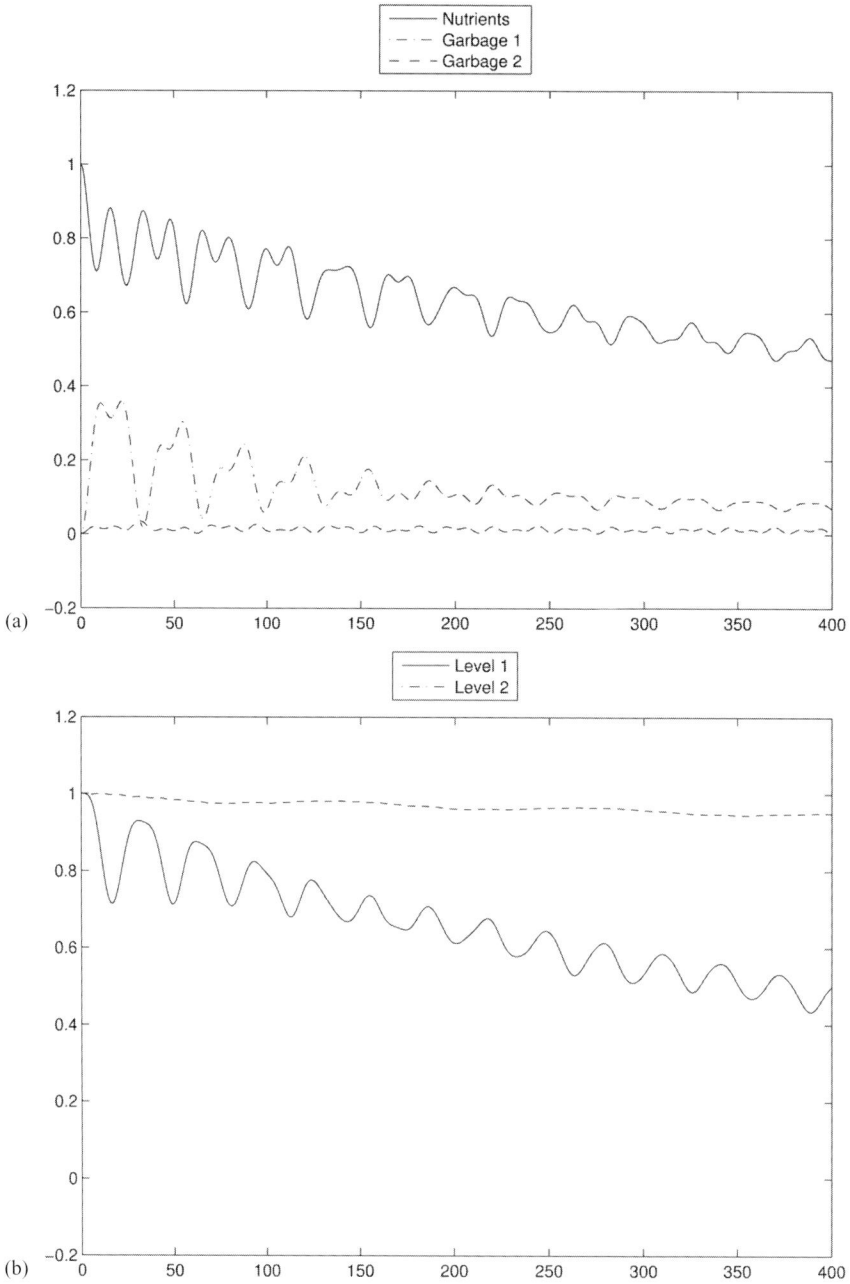

Figure 6.14 Densities of the nutrients, G_1 and G_2 (a) and of levels L_1 and L_2 (b). Initial conditions: nutrients and level L_2 empty, G_1, G_2 and level L_1 completely filled (Source: Bagarello and Oliveri [87]).

In all these figures the decay of the densities is again evident. This is even more evident when comparing Figures 6.12–6.14 with Figures 6.6–6.8. Once more, we observe that, due to the inertia of, say, G_2, its density decreases much slower than that of, say, G_1. The decay of level L_2 is also much slower than that of level L_1, again because of the different values of the inertia parameters.

The analysis considered in Chapter 4 suggests a possible way to improve the recycling efficiency of the ecological system, even in presence of some stress factor. The idea is simple, at least from a mathematical point of view: It is enough to add to ω_2 (or to some other ω_j, $j \neq 3$), a (very) small positive imaginary part. In this way, the densities of all the compartments can be restored, even when some stress factor acts on the system. Then, as in Chapter 4, we can conclude that small positive effects can efficiently balance higher stress factors and bring the system to some equilibrium.

6.5 Conclusions

This chapter has been devoted to show how ladder operators can be used to construct a self-adjoint Hamiltonian useful in the analysis of closed ecosystems. The same mechanism proposed in Chapter 4 to describe stress factors and positive effects also works well here, while nonlinearities of the kind introduced in (6.8) do not change the results much. This is easily understood since nonlinearities of any kind, for $N < \infty$, can only produce a self-adjoint finite-dimensional Hamiltonian matrix, which can only give rise to a periodic or quasi-periodic time evolution.

A possibly interesting extension of our approach would consist in removing from the Hamiltonians in (6.1) and in (6.5) those terms needed to make the Hamiltonians self-adjoint that have no realistic biological meaning. But this operation should be done with some extra care, of course. Chapter 7 is related exactly to a similar approach, in the analysis of tumor cell proliferation.

7

More on Biological Systems

7.1 Introduction

In some recent articles, a few authors used ladder operators to construct interesting models with a biological meaning. In [88, 89], for instance, these operators have been used to describe bacterial populations and their long-term survival by adopting a dynamical approach very similar to that discussed in these notes. Other authors have proposed two different models of stimulated cell division [90]: A first model in which the stimuli are expended in the division process and a second one in which the stimuli survive after this process. This last model is particularly interesting, since in a given regime it produces a dramatic growth of the population that the authors interpret as a cancerous growth phase. This result is deduced by using a Hamiltonian operator H that drives the dynamics, following our approach. In particular, their Hamiltonian is written in terms of ladder operators as follows:

$$H = \omega_n \hat{n} + \omega_m \hat{m} + \lambda \hat{m} \left(c^\dagger \hat{n} + \hat{n} c \right),$$

where $\hat{n} = c^\dagger c$, $\hat{m} = s^\dagger s$ and s and c obey CCR: $[c, c^\dagger] = [s, s^\dagger] = \mathbb{1}$, with all other commutators being zero. The parameters ω_n, ω_m and λ are real. Hence, $H = H^\dagger$. The biological meaning of these operators and of the Hamiltonian is the following: c, $c^\dagger$ and $\hat{n}$ are associated with the cell: c destroys a cell, $c^\dagger$ creates a cell and $\hat{n}$ counts the number of cells. The operators s, $s^\dagger$ and $\hat{m}$ are related to the stimulating agent, which forces the cell to divide. Since $[H, \hat{m}] = 0$, this stimulating agent does not disappear during the cell division and can be used again and again. Moreover, during the interaction, a term like $c^\dagger \hat{n}$ acts. The net effect of this term, which can be written as $c^{\dagger 2} c$, is that the number of cells, if there are any,[1] increases by one unit. Hence, cells may proliferate. The authors also propose a meaning to the adjoint term, $\hat{n} c$, in terms of the death of a cell.

---

[1] If this term acts on a vector with no cell, it is obvious that the effect of the operator in $c^{\dagger 2} c$ that acts first, c, is to destroy the vector.

194

The equations deduced out of H are not linear, and in fact the authors propose only a perturbative analysis of the solution and some interesting comments on possible phase transitions occurring in the model during the time evolution. We find this application very interesting and fascinating, and in the next section we will show how a possibly more complete model can be constructed, considering also the effects of some medical treatments.

7.2 Tumor Cell Proliferation

The basic mechanisms considered in our analysis are the following: (i) a healthy cell becomes sick (because of the presence of some degenerative factor), (ii) sick cells multiply, (iii) healthy cells also multiply but not at the same rate (because we believe that proliferation is faster for the sick cells than for the healthy cells) and (iv) a treatment for the disease is undertaken to try to reverse or stop the mutation.

An important aspect of the approach we will propose in this chapter is that we will use, from the beginning, a manifestly non-self-adjoint Hamiltonian H to drive the time evolution. The lack of self-adjointness will not be related to the appearance of some complex-valued parameters, as in Chapter 4 or in Chapter 6, for instance, but to the lack of some *conjugate terms* that would make little sense for our specific biological system S. The idea is the following: If a_1 and a_2 are any two operators relevant in the description of S, a Hermitian term contributing to H could be $a_1 a_2^\dagger + a_2 a_1^\dagger$. This, if a_j are ladder operators, represents a reversible exchange between, for example, agent 1 and agent 2. If we want to make this exchange irreversible, the natural choice is to replace this sum with a single term, $a_1 a_2^\dagger$: Agent 1 is giving something to agent 2 but not vice versa. In this way we clearly lose self-adjointness of the Hamiltonian, but we gain in its explicit interpretation. It is worth mentioning that, with a similar approach, the resulting operator H is not even PT-symmetric (see [24, 25, 91]): This is truly something different from what is usually considered in the literature.

The mathematical framework of the model we are going to introduce was constructed previously. We will use the ladder operators h, s and m already introduced in Section 2.3.4 to build up the Hamiltonian of the system. The related number operators are given in (2.29), and their common eigenstates are constructed as shown in (2.27).

The definition of the Hamiltonian is based on a series of ideas and on the mechanisms listed at the beginning of this section: First, when compared with [90], we will not consider any carcinogenic factor that causes the transition of a cell from a healthy to a sick state, since we will work under the assumption that this transition already occurs. Hence, this factor will not correspond to an explicit but rather to a hidden variable of the model. Hence, healthy cells become sick ($h \rightarrow s$). Second, these sick cells multiply by mitosis ($s \rightarrow s + s$). Healthy cells also multiply but possibly with a lower frequency ($h \rightarrow h + h$). The medical treatment destroys a sick cell and continues acting on other cells: $s + m \rightarrow m$. It is like a medicine that is given continuously to the patient.

These mechanisms can be easily implemented in terms of the operators h, s and m, their adjoints and their related number operators. A possible choice for the Hamiltonian, if we want this operator to be self-adjoint, is the following:

$$
\begin{cases}
\tilde{H} = \tilde{H}_0 + \tilde{H}_I + \tilde{H}_m \\
\tilde{H}_0 = \omega_h \hat{N}_h + \omega_s \hat{N}_s + \omega_m \hat{N}_m, \\
\tilde{H}_I = \mu_{hs}(h\hat{N}_h s^\dagger + s\hat{N}_h h^\dagger) + \mu_{hh}(h^\dagger \hat{N}_h + \hat{N}_h h) + \mu_{ss}(s^\dagger \hat{N}_s + \hat{N}_s s), \\
\tilde{H}_m = \mu_{sm}(s\hat{N}_s + \hat{N}_s s^\dagger)\hat{N}_m,
\end{cases}
\tag{7.1}
$$

where the parameters ω_α and $\mu_{\alpha\beta}$, $\alpha, \beta = h, s, m$, are assumed to be real and, for the moment, time-independent. Hence, $\tilde{H} = \tilde{H}^\dagger$. The meaning of $\tilde{H}_0$ is the following: first of all, it creates no interesting dynamics when all the $\mu_{\alpha\beta}$'s are zero since, in this case, $\hat{N}_h$, $\hat{N}_s$ and $\hat{N}_m$ commute with $\tilde{H}$, and then they become integrals of motion. This implies that, for instance, if we start with a situation where there are no tumor cells, then their number remains zero for all $t \geq 0$. Also, when some of the $\mu_{\alpha\beta}$'s are different from zero, the value of the ω_α describes a sort of inertia of the agent α: the larger this value, the smaller the variations of the mean value of the related number operator; see [1] for many concrete examples of this effect, which is also observed in several other applications scattered throughout this book.

The term $h\hat{N}_h s^\dagger$ in $\tilde{H}_I$ describes the mutation of healthy cells to sick ones. This is because one healthy cell is destroyed by h and one sick cell is created by $s^\dagger$. Of course this mutation is more likely when $\mathcal{S}$ is made of many healthy cells, and this is why the number operator $\hat{N}_h$ appears. It counts the number of healthy cells: if there are no such cells, the term $h\hat{N}_h s^\dagger$ simply does not contribute to the dynamics. Otherwise it does, with a strength that increases with the number of healthy cells. Its Hermitian conjugate, $s\hat{N}_h h^\dagger$, describes the opposite transition, from a cancer cell to a healthy one. This effect is like a spontaneous healing,[2] which is quite unexpected and not particularly meaningful from a biological point of view.

The terms $h^\dagger \hat{N}_h$ and $s^\dagger \hat{N}_s$ in $\tilde{H}_I$ describe duplications of the healthy and the sick cells, respectively. This duplication can occur only, of course, if at least one cell of that particular kind already exists, and the duplication is more frequent if these cells are many. This is modeled by the presence of $\hat{N}_h$ and $\hat{N}_s$ in the two terms. Of course, since we imagine that the duplication is faster for the sick cells rather

---

[2] In fact, no medicine is involved in the transition.

than the healthy ones, it is natural to take $\mu_{hh} < \mu_{ss}$. This is because of our general interpretation of coupling constants as a sort of generalized mobility (see our comments on $p_{\alpha,\beta}^{(a,b)}$ in Chapter 5, for instance). Once again, their Hermitian conjugates $\hat{N}_h h$ and $\hat{N}_s s$ look biologically unexpected. In fact, they would correspond respectively to the deaths of the healthy and sick cells. This could be understood as the effect of age, for instance, or, as mentioned in the previous paragraph, of some spontaneous healing. Nevertheless, it is quite strange to imagine that, for example, the sick cells are created and destroyed in S at the same rate.

Similar difficulties also arise when trying to interpret $\tilde{H}_m$: $s\hat{N}_s\hat{N}_m$ describes the death of a sick cell because of the action of the medicine, which, in turn, is not affected (i.e., it is not created or destroyed) while it is acting on the cell.[3] But H_m also contains the term $\hat{N}_s s^\dagger \hat{N}_m$: the number of sick cells increases, even if the medical treatment is acting on S. This is not what we expect, and certainly is not what we want.

7.2.1 A Possible Way Out

The conclusion of our preliminary analysis on $\tilde{H}$ shows that this Hamiltonian describes all the effects we are interested in considering for S and for other terms that are not actually biologically motivated. As we have stated previously (see also our results and comments in [1]), initial conditions can break down this apparent complete symmetry between natural and strange phenomena, forcing the system to go in a reasonable direction: an existing cell can be killed, but it cannot be created out of nothing, even if $\tilde{H}$ describes both processes! However, what we want to do here is to consider a different strategy, based on suitable choice of the Hamiltonian rather than on the initial conditions. In other words, we will now introduce a different Hamiltonian H, constructed as a sort of non-self-adjoint version of $\tilde{H}$, and we will deduce the dynamics of S by solving the Schrödinger equation for H. We will see that in this way, reasonable biological results can be deduced.

The Hamiltonian H that we will consider in the rest of the chapter is defined as follows:

$$\begin{cases} H = H_0 + H_I + H_m \\ H_0 = \omega_h \hat{N}_h + \omega_s \hat{N}_s + \omega_m \hat{N}_m, \\ H_I = \mu_{hs} h \hat{N}_h (s^\dagger + \hat{P}_{N_s}) + \mu_{hh} h^\dagger \hat{N}_h + \mu_{ss} s^\dagger \hat{N}_s, \\ H_m = \mu_{sm} s \hat{N}_s \hat{N}_m, \end{cases} \tag{7.2}$$

where we have introduced the following operator in H_I:

$$\hat{P}_{N_s} := \sqrt{N_s} \sum_{n_h, n_m} |\varphi_{n_h, N_s - 1, n_m}\rangle\langle\varphi_{n_h, N_s - 1, n_m}|. \tag{7.3}$$

---

[3] It is as if there were a huge source of medicine keeping the treatment always active.

Here we are using the Dirac notation. Each orthogonal projection operator $|\varphi_{n_h,N_s-1,n_m}\rangle\langle\varphi_{n_h,N_s-1,n_m}|$ in this sum acts on a generic vector $\Phi \in \mathcal{H}$ as follows:

$$(|\varphi_{n_h,N_s-1,n_m}\rangle\langle\varphi_{n_h,N_s-1,n_m}|)\Phi = \langle\varphi_{n_h,N_s-1,n_m},\Phi\rangle\,\varphi_{n_h,N_s-1,n_m}.$$

Except for the appearance of $\hat{P}_{N_s}$, H contains only those terms of $\tilde{H}$ that make sense in our previous analysis. But as is evident, $H \neq H^\dagger$. As for the operator $\hat{P}_{N_s}$, this has a nonzero effect only on those vectors whose expansion in terms of $\mathcal{E}$ contains some vector φ_{n_h,N_s-1,n_m}. Otherwise its action is trivial. In particular we see that, while $s^\dagger\varphi_{n_h,N_s-1,n_m} = 0$, $(s^\dagger + \hat{P}_{N_s})\varphi_{n_h,N_s-1,n_m} = \varphi_{n_h,N_s-1,n_m}$. This is useful to describe a sort of equilibrium for the time evolution of the system when the maximum number of sick cells is reached, which makes sense in an in vitro system. Of course, we don't really expect anything like this works in any living organism, where the first mechanism, $s^\dagger\varphi_{n_h,N_s-1,n_m} = 0$, seems more appropriate, since it can be seen as the death of the organism when the disease is too advanced. However, since the number of cells we can work with is not particularly large (see Section 7.3), our model cannot be used for living organisms, while it can still be relevant in the analysis of systems in vitro, which are those that we are restricted to. This motivates the appearance of $\hat{P}_{N_s}$ in H_I.

> *Remark:* This aspect is clearly a consequence of the fact that the Hilbert space $\mathcal{H}_s$ for the sick cells (see Section 2.3.4) is finite-dimensional and that $s^\dagger$ acts as an annihilation for the higher level of $\mathcal{H}_s$.

As is often the case when the Hamiltonian of the system is not self-adjoint, we prefer to work with the Schrödinger equation of motion rather than with its Heisenberg counterpart. Then we write

$$i\dot{\Psi}(t) = H\Psi(t), \tag{7.4}$$

whose formal solution is $\Psi(t) = e^{-iHt}\Psi(0)$, where $\Psi(0)$ is the initial state of the system. In our case, $\Psi(0)$ describes how many healthy and sick cells exist at $t = 0$, and if, when the evolution is starting, the medical treatment is active in the system. Hence, $\Psi(0)$ can be expressed in terms of the vectors in $\mathcal{E}$ in (2.27): $\Psi(0) = \sum_{n_h,n_s,n_m} c_{\vec{n}}(0)\varphi_{\vec{n}}$, where we have introduced the shorthand notation $\vec{n} = (n_h,n_s,n_m)$, with $n_h = 0,\ldots,N_h - 1$, $n_s = 0,\ldots,N_s - 1$ and $n_m = 0,\ldots,N_m - 1$.

In what follows we will always assume that $c_{\vec{n}}(0) = \delta_{\vec{n},\vec{n}^o}$, where $\vec{n}^o = \left(n_h^o, n_s^o, n_m^o\right)$. This means that, at $t = 0$, $\mathcal{S}$ is in a common eigenstate of its three number operators, so that we know exactly how many sick and healthy cells are in $\mathcal{S}$, and if the medical treatment is active or not. Here the other choices of $c_{\vec{n}}(0)$ are not particularly interesting for us. The Schrödinger equation, together with the orthonormality of the vectors in $\mathcal{E}$, can be written in terms of the time-evolved coefficients of the expansion of $\Psi(t)$, $c_{\vec{n}}(t)$, as

$$i\frac{dc_{\vec{n}}(t)}{dt} = \langle \varphi_{\vec{n}}, H\Psi(t) \rangle, \quad \forall \vec{n},$$

which in turns produces

$$\begin{aligned}
i\frac{dc_{n_h,n_s,n_m}(t)}{dt} &= \left(\omega_h n_h + \omega_s n_s + \omega_m n_m\right) c_{n_h,n_s,n_m} \\
&+ \Big([n_h < N_h - 1][n_s > 0]\mu_{hs}(n_h + 1)\sqrt{(n_h + 1)n_s}\, c_{n_h+1,n_s-1,n_m} \\
&+ [n_h < N_h - 1][n_s = N_s - 1]\mu_{hs}(n_h + 1)\sqrt{(n_h + 1)N_s}\, c_{n_h+1,N_s-1,n_m} \quad (7.5) \\
&+ \Big[n_h > 0\Big]\mu_{hh}(n_h - 1)\sqrt{n_h}\, c_{n_h-1,n_s,n_m} + \Big[n_s > 0\Big]\mu_{ss}(n_s - 1)\sqrt{n_s}\, c_{n_h,n_s-1,n_m}\Big) \\
&+ \Big(\Big[n_s < N_s - 1\Big]\mu_{sm}n_m(n_s + 1)\sqrt{n_s + 1}\, c_{n_h,n_s+1,n_m}\Big).
\end{aligned}$$

Here $[\Delta]$ is a logical operator returning 1 if the statement Δ is true and 0 if it is not. These equations must be solved using the initial conditions above, $c_{\vec{n}}(0) = \delta_{\vec{n},\vec{n}^o}$. Notice that, with this particular choice,

$$\|\Psi(0)\|^2 = \sum_{n_h,n_s,n_m} |c_{\vec{n}}(0)|^2 = 1.$$

However, there is no reason to expect that this normalization is maintained for $t > 0$, due to the fact that $H \neq H^\dagger$: it is quite likely that $\|\Psi(t)\|^2 = \sum_{n_h,n_s,n_m} |c_{\vec{n}}(t)|^2 \neq 1$, for $t > 0$.

The set in Equation (7.5) is a linear system of ordinary differential equations, at least if all the parameters of H are fixed. In this case, an analytic solution can be deduced, in principle. However, what is more interesting for us is to see what happens when some of the parameters of H depend on the density of the (sick and/or healthy) cells. This can be interesting and meaningful, since it can be used to describe, for instance, the fact that the medical treatment starts when the tumor has grown to a certain size and not before, possibly because its presence was not previously recognized. In this case system Equation (7.5) becomes, in general, non-linear, and its solution is not so simple and needs to be computed numerically. The results of our computations are given in the next section.

7.3 Numerical Results

The results discussed here are deduced fixing $N_m = 2$, $N_h = 50$ and $N_s = 150$. The choice $N_m = 2$ is because we are considering only two opposite situations for the medical treatment: medicine is given, or not. As for the choices of the numbers of healthy and sick cells, these are taken low (at least when compared with realistic in vitro systems), to keep the computational time under control. Moreover, the reason why we have taken N_h smaller than N_s is because the sick cells are rather aggressive, and then we assume they can grow in greater numbers (and faster) than the healthy ones.

In our simulations we have always fixed the following values of the parameters of H_0: $\omega_h = \omega_s = \omega_m = 10^{-2}$. These values are small with respect to those of the various parameters $\mu_{\alpha,\beta}$, since we do not want the inertia to cause the densities of the agents of S to stay (almost) constant in time. Further, we will quite often assume that $\mu_{hs} = 1$, and then we play with the other parameters. In particular, since tumor cells tend to remove vital space otherwise available to the healthy cells [92], we consider here a time-dependent production of the healthy cells, assuming that this production decreases in the presence of a greater number of tumor cells. More explicitly, we will take the strength parameter μ_{hh} in H_I directly related to the number of sick cells, $\langle \hat{N}_s(t) \rangle$, the time-dependent mean value of the number operator $\hat{N}_s$ on the (normalized version of) vector $\Psi(t)$:

$$\langle \hat{N}_s(t) \rangle = \left\langle \frac{\Psi(t)}{\|\Psi(t)\|}, \hat{N}_s \frac{\Psi(t)}{\|\Psi(t)\|} \right\rangle = \|s\Psi_n(t)\|^2. \tag{7.6}$$

The reason why we are using here the ratio $\Psi_n(t) = \frac{\Psi(t)}{\|\Psi(t)\|}$, rather than $\Psi(t)$ alone, is because, as already stressed, even if $\Psi(0)$ is normalized, $\Psi(t)$ does not need to be. However, $\|\Psi_n(t)\| = 1$ for all t, and this restores our usual interpretation of the mean value $\langle \hat{N}_s(t) \rangle$. In particular, using Equation (7.6) and condition $\sum_{n_h,n_s,n_m} |c_{\vec{n}}(0)|^2 = 1$, we can easily check that, at $t = 0$, $\langle \hat{N}_s(0) \rangle = 0, 1, 2, \ldots, N_s - 1$. Similar results can also be deduced for $\langle \hat{N}_h(t) \rangle$ and $\langle \hat{N}_m(t) \rangle$:

$$\langle \hat{N}_h(t) \rangle = \langle \Psi_n(t), \hat{N}_h \Psi_n(t) \rangle = \|h\Psi_n(t)\|^2, \quad \langle \hat{N}_m(t) \rangle = \langle \Psi_n(t), \hat{N}_m \Psi_n(t) \rangle = \|m\Psi_n(t)\|^2,$$

with $\langle \hat{N}_h(0) \rangle = 0, 1, 2, \ldots, N_h - 1$ and $\langle \hat{N}_m(0) \rangle = 0, 1$, when $t = 0$. The natural interpretation of these mean values is therefore the following: $\langle \hat{N}_s(t) \rangle$ and $\langle \hat{N}_h(t) \rangle$ are respectively the number of sick and healthy cells, while $\langle \hat{N}_m(t) \rangle = 0$ in absence of medical treatment and $\langle \hat{N}_m(t) \rangle = 1$ otherwise. As clearly shown, all these quantities, in general, are functions of t.

Remark: To read the results, it is useful to introduce the density-like operators

$$\hat{P}_h = \sum_{n_h,n_m} |\varphi_{n_h,0,n_m}\rangle\langle\varphi_{n_h,0,n_m}|, \qquad \hat{P}_s = \sum_{n_s,n_m} |\varphi_{0,n_s,n_m}\rangle\langle\varphi_{0,n_s,n_m}|, \tag{7.7}$$

which are sums of projection operators taking a generic vector $\Phi \in \mathcal{H}$ into, respectively, a healthy and a sick state:

$$\left(|\varphi_{n_h,0,n_m}\rangle\langle\varphi_{n_h,0,n_m}|\right) \Phi = \langle\varphi_{n_h,0,n_m}, \Phi\rangle \, \varphi_{n_h,0,n_m},$$
$$\left(|\varphi_{0,n_s,n_m}\rangle\langle\varphi_{0,n_s,n_m}|\right) \Phi = \langle\varphi_{0,n_s,n_m}, \Phi\rangle \, \varphi_{0,n_s,n_m}.$$

Hence $\hat{P}_h$ projects any state of the system in a subspace of $\mathcal{H}$ in which there are no tumor cells at all. For instance, if the state Φ_h of the system is a superposition of only healthy states, $\Phi_h = \sum_{n_h,n_m} c_{n_h,n_m} \varphi_{n_h,0,n_m}$, then $\hat{P}_h \Phi_h = \Phi_h$. However, if we consider a superposition of only sick states, $\Phi_s = \sum_{n_h,n_s>0,n_m} c_{n_h,n_s,n_m} \varphi_{n_h,n_s,n_m}$, then $\hat{P}_h \Phi_s = 0$. Moreover, simple computations also show that

$$0 \leq \langle \hat{P}_h(t) \rangle = \frac{\sum_{n_h,n_m} |c_{n_h,0,n_m}|^2}{\sum_{n_h,n_s,n_m} |c_{n_h,n_s,n_m}|^2} \leq 1, \qquad 0 \leq \langle \hat{P}_s(t) \rangle = \frac{\sum_{n_h,n_m} |c_{0,n_s,n_m}|^2}{\sum_{n_h,n_s,n_m} |c_{n_h,n_s,n_m}|^2} \leq 1,$$

for all t, which motivates why we can interpret $\langle \hat{P}_h(t) \rangle$ and $\langle \hat{P}_s(t) \rangle$ respectively as the probabilities of having only healthy, or only sick, cells in $\mathcal{S}$.

After this long digression, let us go back to μ_{hh} and to its relation with $\langle \hat{N}_s(t) \rangle$; we call $\tilde{\mu}_{hh}$ the value of the parameter μ_{hh} when there is no sick cell at all: $\langle \hat{N}_s \rangle = 0$. From now on, if not needed, we will no longer write explicitly the dependence on t of $\langle \hat{N}_s \rangle$, $\langle \hat{N}_h \rangle$ and $\langle \hat{N}_m \rangle$. This value should be the highest possible, since in this case healthy cells can easily duplicate. However, μ_{hh} is expected to decay for increasing $\langle \hat{N}_s \rangle$, since sick cells remove space and resources to the healthy ones. For this reason μ_{hh} is naturally treated as a function of $\langle \hat{N}_s \rangle$, $\mu_{hh}(\langle \hat{N}_s \rangle)$, and we put

$$\mu_{hh}\left(\langle \hat{N}_s \rangle\right) := \tilde{\mu}_{hh}\left(1 - \frac{\langle \hat{N}_s \rangle}{N_s - 1}\right).$$

Notice that $\mu_{hh}(\langle \hat{N}_s \rangle) = \tilde{\mu}_{hh}$ if $\langle \hat{N}_s \rangle = 0$, while $\mu_{hh}(\langle \hat{N}_s \rangle) = 0$ if $\langle \hat{N}_s \rangle = N_s - 1$, i.e., in the presence of the maximum number of sick cells. Notice also that since $\langle \hat{N}_s \rangle$ depends on time, $\mu_{hh}(\langle \hat{N}_s \rangle)$, and the Hamiltonian as a consequence, depends on time as well, even if we are not writing this dependence explicitly.

To check whether our model works well or not, we first check what happens in the absence of medical treatment. Hence we take $\mu_{sm} = 0$: this means that $H_m = 0$ (see [7.2]). Of course, what we expect is that, after some time, an original healthy system turns into a sick one. We consider three different scenarios, defined by different values of μ_{ss} and $\tilde{\mu}_{hh}$, while, as already stated, $\mu_{hs} = 1$ everywhere: the first scenario, $R1$, corresponds to the choice $\mu_{ss} = 0.5$ and $\tilde{\mu}_{hh} = 0.25$. In $R2$ we take $\mu_{ss} = 2$ and $\tilde{\mu}_{hh} = 1$, while in $R3$, $\mu_{ss} = 0.125$ and $\tilde{\mu}_{hh} = 0.0625$. We see that in all these cases, $\mu_{ss} = 2\tilde{\mu}_{hh}$. This is because sick cells duplicate faster than healthy ones. The difference between the various scenarios is in the relation of μ_{ss} and $\tilde{\mu}_{hh}$ with μ_{hs}, which is linked to the rapidity of the mutation process $h \to s$. For instance, in $R3$ proliferation of both tumor and healthy cells is a weak effect when compared to mutation, while in $R2$ the effect of tumor cells duplication is stronger than that of mutation, and so on. The initial conditions for all the simulations are $\langle \hat{N}_h \rangle = N_h - 1$ and $\langle \hat{N}_s \rangle = 0$, corresponding to $\Psi(0) = \varphi_{N_h-1,0,0}$. Hence, we have only healthy cells in the system, at $t = 0$. Also, we recall that no medicine is acting, so far.

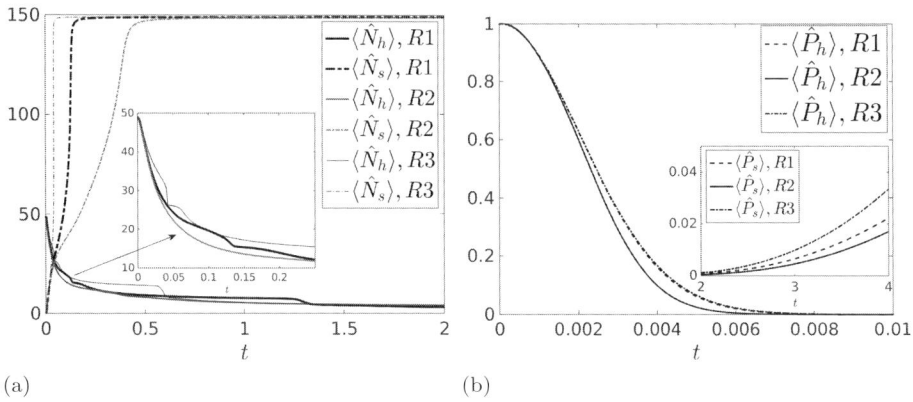

Figure 7.1 **(a)** Number $\langle \hat{N}_h \rangle, \langle \hat{N}_s \rangle$ of healthy and tumor cells configurations $R1$, $R2$ and $R3$. In the inset, the magnification of the early time evolutions of $\langle \hat{N}_h \rangle$. **(b)** Probability $\langle \hat{P}_h \rangle$ of having only healthy cells for the same scenarios. In the inset, the large time evolutions of the probability $\langle \hat{P}_s \rangle$ of having only tumor cells (Source: Bagarello and Gargano [26]).

The outcomes in terms of the number of healthy cells, $\langle \hat{N}_h \rangle$, and tumor cells, $\langle \hat{N}_s \rangle$, are shown in Figure 7.1(a), while the probabilities of having only healthy cells, $\langle \hat{P}_h \rangle$, or only tumor cells, $\langle \hat{P}_s \rangle$, are shown in Figure 7.1(b). Figure 7.1(a) shows that, independently of the scenario considered, the original healthy system evolves, as expected, into a fully sick state: we go from zero to 150 sick cells, while the number of healthy cells decreases significantly. We refer to [26] for a possible explanation of the discontinuities in the derivatives of $\langle \hat{N}_h \rangle$ in terms of the strength of the various terms in H. Here we want to observe that the plots in Figure 7.1(b) are in agreement with what we have just deduced: in fact, with our initial conditions, the probability of having a purely healthy system goes very rapidly to zero and, at the same time, the probability of finding only sick cells in S increases more and more, but slower.

Of course, these preliminary results are a good indication that the model is, at least, reasonable: the simulations show that, when the disease is not cured, the situation becomes worse and worse. What we want to understand next is whether the presence of H_m can help to heal the system and if so, how.

Our first attempt consists in taking μ_{sm} as time-independent. The initial conditions and the other parameters are those used in scenario $R1$. We show in Figure 7.2(a) and (b) the time evolutions of the mean values $\langle \hat{N}_h \rangle, \langle \hat{N}_s \rangle$ of healthy and tumor cells for different values of μ_{sm}. The results show that the action of a medical treatment reduces $\langle \hat{N}_s \rangle$ as the intensity of the treatment is increased; see Figure 7.2(b). This is exactly the result one expects for biological reasons. Conversely, the number $\langle \hat{N}_h \rangle$ is no more asymptotically going to zero, as in absence

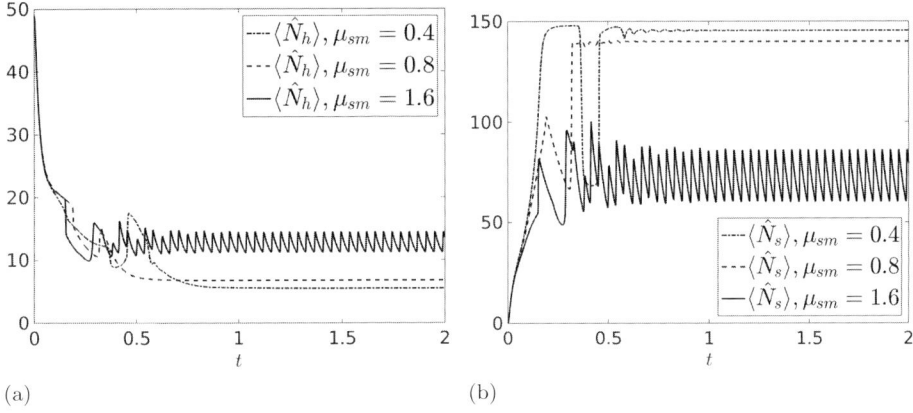

Figure 7.2 Number $\langle \hat{N}_h \rangle$ of healthy cells (a) and number $\langle \hat{N}_s \rangle$ of tumor cells (b), during the medical treatment scenario for various values of the parameter μ_{sm}. Initial conditions and other parameters are those used in configuration $R1$ (Source: Bagarello and Gargano [26]).

of medical treatment – see Figure 7.1(a) – and it stabilizes (or, better, oscillates) around a value that increases with μ_{sm}. This is related to the fact that the tumor cells do not saturate the system anymore, so that the production of healthy cells, given by the term $\mu_{hhh}h^\dagger \hat{N}_h$ in H_I, does not vanish anymore. S reaches a sort of equilibrium between healthy and sick cells: neither completely disappear from the system. Incidentally, this means that we have not achieved a complete healing, as we would like, but we see an improvement with respect to the $\mu_{sm} = 0$ situation. Of course, we could expect some further progress if we increased the value of μ_{sm}. However, for real systems, this could be impossible, since medicine against tumors usually has quite strong side effects. For this reason, what we want to do next is to understand what happens if we instead introduce some explicit time dependence in μ_{sm}, μ_{hs} and in μ_{ss}, and in H consequently.

Not surprisingly, there are many situations in which complete healing is not found. We refer to [26] for a detailed analysis of some of these cases. Here we focus on a particular setting in which complete healing is achieved.

First of all, we assume that the strength of the medical treatment has periodic peaks, as described by the following choice of $\mu_{sm}(t)$:

$$\mu_{sm}(t) = \tilde{\mu}_{sm} \sum_{k=1}^{M} \exp\left(-\left((t-k)/\sigma\right)^2\right), \quad \tilde{\mu}_{sm}, \sigma > 0.$$

Here $\tilde{\mu}_{sm}$ is the maximum value of $\mu_{sm}(t)$, M is related to the time interval when the medical treatment is active and σ to the width of the Gaussians appearing in $\mu_{sm}(t)$. Of course, peaks are found for $t = k$, $k = 1, 2, \ldots, M$. Further, we take

$$\mu_{hs}\left(\langle \hat{N}_m \rangle, \langle \hat{N}_s \rangle\right) = \tilde{\mu}_{hs}\exp\left(-\langle \hat{N}_m \rangle \frac{N_s - 1 - \langle \hat{N}_s \rangle}{\langle \hat{N}_s \rangle}\right),$$

and

$$\mu_{ss}\left(\langle \hat{N}_m \rangle, \langle \hat{N}_s \rangle\right) = \tilde{\mu}_{ss}\exp\left(-\langle \hat{N}_m \rangle \frac{N_s - 1 - \langle \hat{N}_s \rangle}{\langle \hat{N}_s \rangle}\right),$$

where $\tilde{\mu}_{hs}, \tilde{\mu}_{ss}$ are fixed constants, which we take as equal to one in the simulation considered here. It is clear that $\mu_{hs}(\langle \hat{N}_m \rangle, \langle \hat{N}_s \rangle)$ and $\mu_{ss}(\langle \hat{N}_m \rangle, \langle \hat{N}_s \rangle)$ depend on time (through $\langle \hat{N}_m \rangle$ and $\langle \hat{N}_s \rangle$) only when the medical treatment is acting ($\langle \hat{N}_m \rangle \neq 0$), and their amplitudes decrease when the number of tumor cells is low: $\mu_{hs}(\langle \hat{N}_m \rangle, \langle \hat{N}_s \rangle)$, $\mu_{ss}(\langle \hat{N}_m \rangle, \langle \hat{N}_s \rangle) \approx 0$ for $\langle \hat{N}_s \rangle \approx 0$. This means that mutation and proliferation of sick cells are small effects if $\langle \hat{N}_s \rangle$ is small. But when $\langle \hat{N}_s \rangle$ increases, these mechanisms become more and more relevant.

We present in Figure 7.3 the results of the numerical simulation obtained with initial condition $\Psi(0) = \varphi_{40,10,1}$, corresponding to $\langle \hat{N}_h \rangle = 40, \langle \hat{N}_s \rangle = 10$ and the other parameters as in $R1$. We have further fixed $\tilde{\mu}_{sm} = 2.5$, $\sigma = 0.25$ and $M = 4$ in $\mu_{sm}(t)$. The evolutions of $\langle \hat{N}_h \rangle$ and $\langle \hat{N}_s \rangle$ clearly show that the number of healthy cells increases whereas the number of tumor cells decreases, with a nonvanishing,

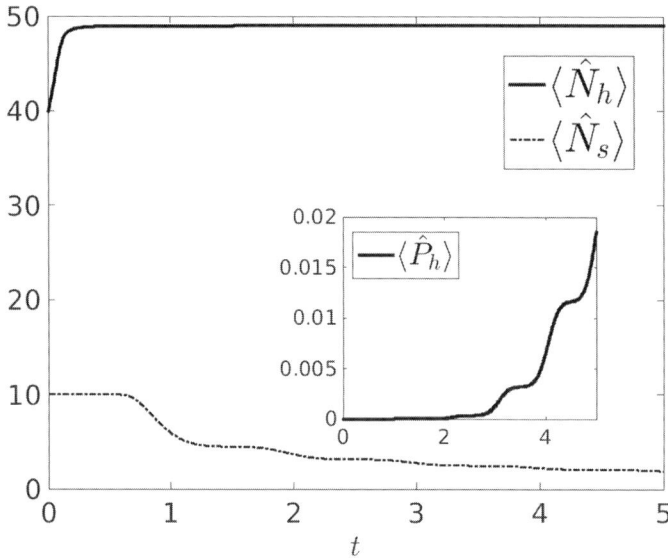

Figure 7.3 Numbers $\langle \hat{N}_h \rangle, \langle \hat{N}_s \rangle$ of healthy and tumor cells. In the inset, the probability $\langle \hat{P}_h \rangle$ of having only healthy cells in the system (Source: Bagarello and Gargano [26]).

and increasing, probability $\langle P_h \rangle$ that the system contains only healthy cells. This is what we were looking for: the disease appears fully under control. We conclude that, to achieve this result, the therapy must be time-dependent.

7.4 Conclusions

In this chapter we have shown how to use ladder operators in the description of the mutation of healthy into sick cells and of their proliferation, and we have found some cases that prevent the system from becoming completely sick and, more interesting, found a particular configuration that allows keeping the disease fully under control: the treatment should not be given uniformly – see $\mu_{sm}(t)$ – and the medicine must be chosen to keep low the effects of mutation and the proliferation of sick cells. This is possible mainly in the presence of an early diagnosis, where both these effects are still low and the treatment can act more efficiently to keep both effects low.

Other than for its (preliminary) biological meaning, this model is interesting because of the non-self-adjoint form of the Hamiltonian used to derive the time evolution of the system: the results we have deduced seem rather promising and suggest that a similar idea can also be useful and efficient in other contexts.

We should also mention that the use of parameters of H that depend on the number of cells, such as $\mu_{hh}(\langle \hat{N}_s \rangle)$, $\mu_{hs}(\langle \hat{N}_m \rangle, \langle \hat{N}_s \rangle)$ and $\mu_{ss}(\langle \hat{N}_m \rangle, \langle \hat{N}_s \rangle)$, makes the model highly nonlinear.[4] Therefore, and since $H \neq H^{\dagger}$, the time evolution of the observables of $\mathcal{S}$ and their mean values is not periodic or quasi-periodic, even if the Hilbert space $\mathcal{H}$ is finite-dimensional.

We end this chapter stressing once more that the model we have proposed is very basic, as it contains only a few essential mechanisms ruling the tumor growth. Many improvements are possible: some spatial dependency can be considered, the role of the starting time of the treatment could be considered in more details, more biologically motivated effects could be inserted in the Hamiltonian H and so on. But above all, a comparison of our results with biological findings is essential. These are only a few possible extensions for the particular application discussed in this chapter.

---

[4] In fact, the Hamiltonian in (7.2) is already not quadratic, even if the coupling constants are time-independent. But this new feature of $\mu_{\alpha\beta}$ makes the system "even more nonlinear".

8

Quantum Game of Life and Its (H, ρ)-Induced Dynamics

8.1 Introduction

We have already discussed in Chapters 2–5 that, in some particular applications, a pure Hamiltonian description of the dynamics is not the best choice, especially if we insist on requiring that the Hamiltonian H should be self-adjoint. This is because, in this way, we cannot naturally obtain an equilibrium for any system living in finite-dimensional Hilbert spaces. But this is also because there exist phenomena that cannot be described simply by considering more elaborated expressions for H, either because it is not clear that kind of term one should concretely consider in H, or because, even when the analytic expression of this term is clear, its adjoint should also be added to H, and this extra contribution could be somehow *unphysical*. This is, in fact, the main raison d'etre of Chapter 7, where the absence of these contributions was the key aspect of our analysis. In Section 2.8 we have discussed a different approach, showing how the Hamiltonian dynamics can be modified, by adding the effect of what we have called a *rule*, ρ. The effects of this (H, ρ)-induced dynamics have been already considered in Sections 3.4 and 5.3, in connection with politics and with escape strategies. Here we will discuss a different application of the rule to the so-called *game of life* (*GoL*) or, better to say, to a quantum version of it.

The GoL, originally proposed by J. Conway in 1970, can be thought of as a sort of dynamical system[1] S in which we are interested in the changes of the local densities of a given population $\mathcal{P}$ living in a lattice $\mathcal{R}$ [93, 94]. In the generic cell γ of $\mathcal{R}$, the density of the population changes according to what happens in the other cells surrounding γ itself (typically, the eight surrounding cells characterizing the so-called Moore neighborhood), $\mathcal{N}_\gamma$; in particular, the change in γ is driven by the sum of the densities of the populations in $\mathcal{N}_\gamma$. Each cell of $\mathcal{R}$ assumes only

---

[1] We use the word *sort* since there is no real continuous dynamics in the GoL, but sequences of discontinuous transitions from one state to another.

two possible values: 0 if the cell is in a dead state, 1 if the cell is alive. At each *generation*, any given cell undergoes possibly a transition according to specific rules based on its own state and on the states of the surrounding cells. Calling $s_\gamma[k]$ the value of cell γ at the k-th generation, then $s_\gamma[k] = 0, 1$ depending on whether cell γ is dead or alive. The status of the same cell at the next generation, $k + 1$, as usually assumed in the literature on the GoL, is fixed by the following equations:

$$s_\gamma[k + 1] = \begin{cases} 1 & \text{if } \left(s_\gamma[k] = 1 \wedge \left[\sum_{\alpha \in N_\gamma} s_\alpha[k] = 2 \vee \sum_{\alpha \in N_\gamma} s_\alpha[k] = 3 \right] \right) \vee \\ & \vee \left(s_\gamma[k] = 0 \wedge \sum_{\alpha \in N_\gamma} s_\alpha[k] = 3 \right) \\ 0 & \text{if } \left(s_\gamma[k] = 1 \wedge \left[\sum_{\alpha \in N_\gamma} s_\alpha[k] < 2 \vee \sum_{\alpha \in N_\gamma} s_\alpha[k] > 3 \right] \right) \vee \\ & \vee \left(s_\gamma[k] = 0 \wedge \sum_{\alpha \in N_\gamma} s_\alpha[k] \neq 3 \right) \end{cases} \tag{8.1}$$

The interpretation of these formulas is the following:

The state of the cell γ at generation $k + 1$ is *alive*, $s_\gamma[k + 1] = 1$, under one of the two following possibilities: (a1) the cell is already alive at generation k, $s_\gamma[k] = 1$, and the density of alive cells surrounding γ is not too small or too high: only two or three cells in N_γ should be alive. The idea is that, in this case, there is enough nutrient in γ to ensure it remains alive. Possibility (a2) is the following: the dead cell γ ($s_\gamma[k] = 0$) becomes alive, $s_\gamma[k + 1] = 1$, if it is surrounded by a reasonable number of alive cells (but not too many). This is a birth condition.

On the other hand, the state of the cell γ at generation $k + 1$ is *dead*, $s_\gamma[k + 1] = 0$, under one of the two following possibilities: (d1) the cell is alive at generation k, $s_\gamma[k] = 1$, but the density of the cells surrounding γ that are alive is either too small or too high. This possibility represents death due to under- or overpopulation. Possibility (d2) is the following: the status of γ does not change if the density of the cells in N_γ is different from 3, that is if N_γ has not a very special value of density.

Cells are usually updated synchronously, i.e., all cells in $\mathcal{R}$ undergo state transitions at the same time. However, in some articles, variations of the GoL have been proposed to implement also asynchronous evolutions (see, for instance [95]).

Remark: The rules proposed here are just one of the possible choices we could consider. Other possibilities can also be considered as well, of course, still keeping essentially unchanged the main aspects of the GoL. For instance, we could change the values of the density that fix what is meant by over- or underpopulated, or we could replace the Moore neighborhood with a different one, enlarging the neighborhood or making it narrower. This could be seen as an alternative way to make the over- or underpopulation effect more relevant. Of course, more serious differences could also be considered, changing significantly Equation (8.1). However, we will not discuss other variants of this particular GoL, since Equation (8.1) will be one of the essential ingredients of our quantum version of the GoL.

The main interest in the analysis of the GoL is to see how the system evolves or, more exactly, how the system changes when Equation (8.1) is applied several times, starting from some particular initial condition. In other words, one considers how alive and dead cells are distributed in $\mathcal{R}$ as a function of the generation index k. Incidentally, it should be underlined once more that, in the GoL as described here, there is no continuous time evolution really: the system is, at generation k, in a state $\mathcal{S}[k] = \{s_\alpha[k], \alpha = 1, 2, \dots, L^2\}$ and suddenly (i.e., when we apply Equation [8.1]), its state change into $\mathcal{S}[k + 1]$. For this reason in the literature the authors talk of generations rather than of time evolution.

What we will do here is to modify the game we are playing with by taking into account, together with the transition $\mathcal{S}[k] \to \mathcal{S}[k + 1]$, a continuous time evolution driven by a suitable Hamiltonian operator H, constructed, as usual, in terms of ladder operators, which modifies the state of the system according to the Heisenberg equations of motion. Then the idea is the following: at $t = 0$ the system $\mathcal{S}$ starts evolving according to H. Then, at $t = \tau$, for some fixed $\tau > 0$, a modified version of Equation (8.1) is applied to $\mathcal{S}$, producing a discontinuous change in some of the quantities defining the system. Then we use the new configuration we have obtained in this way as the new *initial condition* for $\mathcal{S}$, which evolves again driven by H, until the rule is not applied once more and so on. This is in the same line, of course, of what we have proposed in Section 2.8: a Hamiltonian evolution corrected by the periodic action of a set of rules acting over the system. In a few words, we are now going to consider an application of the (H, ρ)-induced dynamics.

Our quantum-like version of the GoL, hereafter *QGoL*, is based on the fact that, as everywhere in this book, the dynamical variables representing the whole system can be taken to be operator-valued, and the dynamics is deduced by introducing a suitable self-adjoint operator H that includes the effects of all possible interactions between the different parts of the physical system. In our QGoL, in order to describe cells of $\mathcal{R}$ that can be dead or alive, we use a function $n_{\gamma,k}(t)$, see for instance Equation (8.5). Rather than being a (double-indexed, time-independent) sequence of zeros and ones, as $\{s_\gamma[k]\}$, $n_{\gamma,k}(t)$ is, for each cell γ and generation k, a function of time taking values in the closed interval $[0, 1]$, and it can be interpreted as a density distribution of a single population, moving along $\mathcal{R}$: $n_{\gamma,k}(t)$

gives the density of the population at time t in the cell γ at the generation k. The dependence on two variables is a characteristic of the QGoL, connected to the fact that the changes of the systems are due both to the time evolution and to the action of the rule, responsible for the transition from generation k to generation $k+1$. For instance $n_{\gamma,k}(0) = 1$ if, at $t = 0$ and for generation k, cell γ is alive. Otherwise, if $n_{\gamma,k+1}(0) = 0$, cell γ is dead at $t = 0$ and at generation $k+1$. In this way the status of each cell is directly connected, for each t and for each k, with the value of the density $n_{\gamma,k}(t)$. This interpretation allows us to use, to deduce the time evolution of $n_{\gamma,k}(t)$, a Hamiltonian that should not differ much from the one used in Section 5.2, where our attention was focused on the time evolution of the densities of two interacting populations in a lattice. Then, if we just remove one population and we adjust some ingredient and the interpretation of the model, that Hamiltonian could be used also here. The details will be shown in the next section, where we will also describe our modified version of the rule in Equation (8.1). Therefore, at $t = 0$, the dynamics driven by H begins, and the (fictitious) population moves along $\mathcal{R}$, modifying continuously the density in the various cells: it may easily happen that, for some $t > 0$ and for some k, $n_{\gamma,k}(t)$ is neither zero nor one, while it stays surely between these values. It is like the cell is almost alive (if $n_{\gamma,k}(t) \simeq 1$) or almost dead (if $n_{\gamma,k}(t) \simeq 0$). The rule we will consider, see Section 8.2, will restore the strict *dead or alive* interpretation of the cells in $\mathcal{R}$: any time ρ is applied, the new densities of the cells are modified, in general, from intermediate values to new values that can only be zero or one. This means, incidentally, that the rule acts on the space of the states over the system, and it does not modify the parameters of the Hamiltonian. We will be more explicit on the definition of the rule, and on its effects, in the next section.

Before starting our analysis it may be worth noticing that what we are going to propose next is not the only possibility to construct a quantum version of the GoL. Many other possibilities exist, and we refer to [96–98], and references therein, for some of them. It is particularly interesting to notice that, in [97], the authors also use a Hamiltonian operator to define the time evolution of their QGoL and that, even more interesting, they use (bosonic) ladder operators to define the Hamiltonian.

8.2 The Quantum Game of Life

We are now ready to construct our QGoL, beginning with the description of the general framework. Then we will show how fermionic ladder operators can be used to construct the Hamiltonian of our game, and finally we will introduce our rule.

The system S is made by a population $\mathcal{P}$ distributed in the L^2 cells of a lattice $\mathcal{R}$. To each cell of the lattice is attached a fermionic variable, taking value 0 or 1 only,[2] and each possible configuration is given as a vector on the Hilbert space $\mathcal{H}$ constructed out of these fermionic operators as described below. The *observables* of the system S we are interested in are operators acting on $\mathcal{H}$, again expressed as suitable combination of the same ladder operators.

> *Remark:* As already observed, in comparison with the GoL, the population $\mathcal{P}$ here is really something different than just a set of dichotomous variables. In fact, in the GoL we just have cells with two possible status, dead or alive. These two situations correspond, for a given time t_0 and a certain generation k, to two values, zero and one, which can be seen as extreme values of some cell-dependent density function: a dead cell γ corresponds to a zero value of this function in γ, $n_{\gamma,k}(t_0) = 0$, while if γ is alive, then it corresponds to a value one of the same function, $n_{\gamma,k}(t_0) = 1$. As we will see, and as we have already observed before, this density function also takes values between zero and one, during the time evolution. This is because of the continuity (in time) of the Heisenberg dynamics.

At time zero, a cell may be dead or alive. As we have stated several times, these two states can be naturally represented by the numbers zero and one, respectively. This situation is well described by using a two-state vector φ_{n_α}, where α labels the cell, and $n_\alpha = 0, 1$. Therefore, if S is (partly) described by the vector φ_{0_β} this means that the cell β, at $t = 0$, is dead, while this cell is alive if its related vector is φ_{1_β}. Repeating once more the construction first discussed in Section 2.2, and several times afterwards, we can build up these vectors (two for each cell) by introducing a family of fermionic operators, one for each α, i.e., a family of operators $\{a_\alpha\}$ satisfying the following CAR:

$$\left\{a_\alpha, a_\alpha^\dagger\right\} = a_\alpha a_\alpha^\dagger + a_\alpha^\dagger a_\alpha = \mathbb{1}_\alpha, \qquad a_\alpha^2 = \left(a_\alpha^\dagger\right)^2 = 0.$$

Here $\mathbb{1}_\alpha$ is the identity operator on $\mathcal{H}_\alpha$, constructed as follows.

From $a_\alpha, a_\alpha^\dagger$ we can construct the operator $\hat{N}_\alpha = a_\alpha^\dagger a_\alpha$, which is the number operator for the cell α, and φ_{0_α}, which is the vacuum of a_α, i.e., the vector satisfying $a_\alpha \varphi_{0_\alpha} = 0$. Moreover, φ_{1_α} is simply $a_\alpha^\dagger \varphi_{0_\alpha}$, and $\hat{N}_\alpha \varphi_{n_\alpha} = n_\alpha \varphi_{n_\alpha}$, with $n_\alpha = 0, 1$. In this way, we have exactly the two vectors we were looking for, and the eigenvalues of the operator $\hat{N}_\alpha$ describe the status of the cell α (dead or alive). The Hilbert space $\mathcal{H}_\alpha$ *attached* to cell α is the two-dimensional set $\mathbb{C}^2$, endowed with its natural scalar product. Of course, $\{\varphi_{0_\alpha}, \varphi_{1_\alpha}\}$ is an o.n. basis for $\mathcal{H}_\alpha$, and $\mathbb{1}_\alpha$ is the

---

[2] Equivalently, we could of course use spin variables and work with Pauli matrices.

two-by-two identity matrix. Then, we define the vector of the full system S as the following tensor product:

$$\varphi_{\mathbf{n}} = \otimes_{\alpha=1}^{L^2} \varphi_{n_\alpha}, \qquad \mathbf{n} = (n_1, n_2, ..., n_{L^2}), \qquad (8.2)$$

which clearly describes the status of each cell in $\mathcal{R}$. This vector is an eigenstate of $\hat{N}_\alpha$, for all $\alpha \in \mathcal{R}$:

$$\hat{N}_\alpha \varphi_{\mathbf{n}} = n_\alpha \varphi_{\mathbf{n}}.$$

Incidentally we observe that $\hat{N}_\alpha$ should be identified with the following operator:

$$\tilde{N}_{[\alpha]} = \hat{N}_\alpha \otimes \left(\otimes_{\beta \neq \alpha} 1\!\!1_\beta \right),$$

which coincides with $\hat{N}_\alpha$ when restricted to $\mathcal{H}_\alpha$, and with the identity operator $1\!\!1_\beta$ on $\mathcal{H}_\beta$, $\beta \neq \alpha$. However, to keep the notation simple, we will simply use $\hat{N}_\alpha$ rather than $\tilde{N}_{[\alpha]}$. The Hilbert space $\mathcal{H}$ for S is then constructed by taking the linear span of all these vectors or, equivalently, the tensor product of all the $\mathcal{H}_\beta$'s. The scalar product in $\mathcal{H}$ is the natural one. In particular, in each cell it reduces to the scalar product in $\mathbb{C}^2$. The CAR in $\mathcal{R}$ extends those given above in a single cell:

$$\left\{ a_\alpha, a_\beta^\dagger \right\} = \delta_{\alpha,\beta} 1\!\!1, \qquad \forall \alpha, \beta, \qquad (8.3)$$

where $1\!\!1$ is the identity operator on $\mathcal{H}$ and a_α is now a $2^{L^2} \times 2^{L^2}$ matrix operator satisfying, in particular,

$$a_\alpha \varphi_{\mathbf{n}} = 0 \quad \text{if } n_\alpha = 0,$$

$$a_\alpha^\dagger \varphi_{\mathbf{n}} = 0 \quad \text{if } n_\alpha = 1.$$

Of course, $\hat{N}_\alpha = a_\alpha^\dagger a_\alpha$ is also a $2^{L^2} \times 2^{L^2}$ matrix.

As we have discussed before, we describe the change of status of each cell as if, during the time interval $[0, \tau[$, $\tau > 0$, part of the density distributed in $\mathcal{R}$ moves from one cell to another: it is like a diffusion phenomenon for S. With this is mind, a Hamiltonian that describes well the diffusion of a population in a closed region in terms of fermionic operators can be fixed here as follows:

$$H = \sum_{\alpha=1}^{L^2} a_\alpha^\dagger a_\alpha + \sum_{\alpha,\beta=1}^{L^2} p_{\alpha,\beta} \left(a_\alpha a_\beta^\dagger + a_\beta a_\alpha^\dagger \right), \qquad (8.4)$$

where $p_{\alpha,\beta}$ are parameters that we fix as follows: $p_{\alpha,\beta} = 1$ if $\alpha \neq \beta$ are neighboring cells,[3] and $p_{\alpha,\beta} = 0$ otherwise. With this choice, H is self-adjoint, i.e., $H = H^\dagger$. It is worth observing that H looks like a single population version of the Hamiltonian

---

[3] We consider for each cell the Moore neighborhood made, for internal cells, of the eight surrounding cells. Less cells obviously form the Moore neighborhood of a cell on the border.

introduced in Section 5.2, in our analysis of escape strategies. However, it is clear that the parameters $p_{\alpha,\beta}$ here have a much simpler analytic expression than that in (5.8), due to the different nature of the systems considered in Chapter 5 and here. Of course, this does not exclude that other choices of $p_{\alpha,\beta}$ in (8.4) could be considered instead of this. But, for the analysis we are interested in, this dichotomous alternative is sufficient.

The appearance of the Hamiltonian in the analysis of the GoL is what is new with respect to its classical treatment: in fact, before each new generation is constructed, we fix a *transient time* τ such that, during the time interval $[0, \tau]$, the population in $\mathcal{R}$ evolves as described by the Hamiltonian in Equation (8.4); hence, as time t increases, $t < \tau$, S is no longer in its initial state, $\varphi_{\mathbf{n}^1}$, but it is rather described by the vector $e^{-iHt}\varphi_{\mathbf{n}^1}$. This time-depending vector is, of course, a superposition of the vectors $\varphi_{\mathbf{n}}$ in Equation (8.2) with coefficients that, in general, depend on time: $e^{-iHt}\varphi_{\mathbf{n}^1} = \sum_{\mathbf{k}} c_{\mathbf{k}}^{(1)}(t)\varphi_{\mathbf{k}}$. Following our usual strategy, we relate the mean values of the cell-dependent number operators $\hat{N}_\alpha$ to the new densities of the population in each cell. More in detail, we put

$$\hat{N}_\alpha(t) := e^{iHt}\hat{N}_\alpha(0)e^{-iHt},$$

and then we take the mean values of this operator on the vector $\varphi_{\mathbf{n}^1}$ that describes the system at $t = 0$:

$$n_{\alpha,1}(t) = \left\langle \varphi_{\mathbf{n}^1}, \hat{N}_\alpha(t)\varphi_{\mathbf{n}^1} \right\rangle = \left\| a_\alpha e^{-iHt}\varphi_{\mathbf{n}^1} \right\|^2. \tag{8.5}$$

Because of the CAR in (8.3), the values $n_{\alpha,1}(t)$ are restricted to the range $[0, 1]$, for all α and all t. Hence, they can be endowed with a probabilistic meaning: for instance, if $n_{\alpha,1}(t) \simeq 0$ then the cell α, at the first generation, has a high probability to be in a dead state. Otherwise, if $n_{\alpha,1}(t) \simeq 1$, α is most probably alive. When t approaches τ, we apply the rule in (8.6) below synchronously to all the cells, i.e., to the set $\{n_{\alpha,1}(\tau)\}$ of the time evolution of the densities for the first generation computed at $t = \tau$, and this returns new values of the densities, $\{n_{\alpha,2}(0)\}$, in each cell of $\mathcal{R}$, the densities of the second generation at the initial time, which, by construction, can again only be either zero or one. In this way we deduce a new vector $\mathbf{n}^2$, which defines a new vector of $\mathcal{H}$, obtained through (8.2), $\varphi_{\mathbf{n}^2}$, which evolves once more for a time interval of length τ according to H, $e^{-iHt}\varphi_{\mathbf{n}^2}$, and then the rule acts again. This process is iterated for several generations, getting in this way a mixed evolution due to the joint action of the Hamiltonian and of the rule.

To conclude the description of the model, we still have to give the definition of the rule ρ. As we have anticipated, this is a deformed version of the one introduced in Equation (8.1), and it requires the introduction of a positive *tolerance parameter* σ, which measures how much we are moving away from the rules in Equation (8.1). Using a slightly different notation with respect to the one used in Section 8.1, mainly in order to stress more the differences between the GoL and its quantum version, we define

$$
\begin{cases}
\rho(n_{\alpha,k}(0) = 1) = 1 & \text{if} \quad \left(2 - \sigma \leq \sum_{\beta \in \mathcal{N}_\alpha} n_{\beta,k}(\tau) \leq 3 + \sigma\right), \\[3ex]
\rho(n_{\alpha,k}(0) = 1) = 0 & \text{if} \quad \left(\sum_{\beta \in \mathcal{N}_\alpha} n_{\beta,k}(\tau) < 2 - \sigma\right) \vee \left(\sum_{\beta \in \mathcal{N}_\alpha} n_{\beta,k}(\tau) > 3 + \sigma\right), \\[3ex]
\rho(n_{\alpha,k}(0) = 0) = 1 & \text{if} \quad \left(3 - \sigma \leq \sum_{\beta \in \mathcal{N}_\alpha} n_{\beta,k}(\tau) \leq 3 + \sigma\right), \\[3ex]
\rho(n_{\alpha,k}(0) = 0) = 0 & \text{if} \quad \left(\sum_{\beta \in \mathcal{N}_\alpha} n_{\beta,k}(\tau) < 3 - \sigma\right) \vee \left(\sum_{\beta \in \mathcal{N}_\alpha} n_{\beta,k}(\tau) > 3 + \sigma\right).
\end{cases}
\tag{8.6}
$$

Now, as in Section 8.1, the rule ρ produces the next initial state of the system: $n_{\alpha,k+1}(0) = \rho(n_{\alpha,k}(0))$, for all $k \geq 1$. The meaning of these formulas is the following: if the cell α is alive at the beginning of iteration k, i.e., if $n_{\alpha,k}(0) = 1$, then the cell starts the iteration $k + 1$ staying alive ($n_{\alpha,k+1}(0) = 1$), at least if, after the time evolution driven by H, at $t = \tau$, the total density in the Moore neighborhood of α is between $2 - \sigma$ and $3 + \sigma$. Of course, if $\sigma = 0$, we *approach* (8.1). However, we prefer to keep $\sigma > 0$ most of the time, to implement some extra uncertainty in our rule. This is because in this quantum version, we need to replace the sharp equalities in (8.1) with the inequalities above, since the time evolution of each $n_{\alpha,k}$ produces mean values which, in general, are neither zero nor one, but only belong to the interval $[0, 1]$. This shows how the role of σ is essential when we introduce H in the model, even if H and σ might appear not directly related, in our construction.

8.3 Some Results

Let us see now what our QGoL produces, in terms of time evolution of the dead-alive cells in $\mathcal{R}$, and of their distribution. We take $\mathcal{R}$ to be square, with L^2 cells. In all our simulations we have fixed $L = 33$, and the initial state of the system S is obtained by giving randomly a value zero or one to each cell. The size of the lattice is a good compromise between a reasonable computation time and a sufficiently large system. Moreover, we have tested that no serious differences arise when enlarging $\mathcal{R}$.

What we will discuss here is essentially what makes *quantum* the GoL in Section 8.1. In this perspective, we discuss in detail the role of both τ and σ.

These are really the main quantities to consider, since τ is the measure of the time interval in which the system evolves following the Heisenberg (or Schrödinger) evolution produced by H, while the value of σ describes essentially how much the rule ρ in Equation (8.6) differs from its classical version, (8.1).

8.3.1 The Parameters τ and σ

The parameter τ defines the time range of the (H, ρ)-induced dynamics of the system before the rule is applied: for $t < \tau$ the system evolves according to H, and ρ is applied only when $t = \tau$, and not before. This gives rise to the second generation of the system. Obviously, for $\tau = 0$ there is no Hamiltonian-driven dynamics at all. This means that, if we also take $\sigma = 0$, we have to recover the classical behavior of the GoL. To study how the parameters τ and σ modify this classical evolution, we restrict our analysis to evaluate, at the second iteration ($k = 2$), the following quantity:

$$\Delta(\tau, \sigma) = \frac{1}{L^2} \sum_{\alpha=1}^{L^2} |n_{\alpha,2}(0) - s_\alpha[2]|, \qquad (8.7)$$

where $n_{\alpha,2}(0)$ and $s_\alpha[2]$ are the states (or the densities at $t = 0$) in the cell α for the second generation for the QGoL and the GoL, respectively. We see that $\Delta(\tau, \sigma)$ is a global quantity that measures the difference between the two versions of the game. In fact, $\Delta(\tau, \sigma)$ looks like an averaged (on the size of $\mathcal{R}$) l_1-error norm between the states of the cells obtained using the quantum and the classical games of life.

It is clear that $\Delta(\tau, \sigma) = 0$ if and only if $n_{\alpha,2}(0) = s_\alpha[2]$ for all α, i.e., when the QGoL and the GoL produce the same results at the second generation: $G_2 = G_2^Q$, with an obvious simplifying notation. In principle, even if this is the case, the two third generations could be different: in general, we could still have $G_3 \neq G_3^Q$. However, this is not so if we know that $\Delta(\tau, \sigma) = 0$ *for all possible choices of the first generation*. In this case, in fact, if $G_2 = G_2^Q$, then $G_3 = G_3^Q$ as well, and so on. The reason is the following: first of all, we observe that $G_1 = G_1^Q$ always, since our analysis of the differences between GoL and QGoL, is strongly based on the assumption that the initial states of the two games, classical and quantum, coincide. Now, let us assume that $G_2 = G_2^Q$ for all possible choices of the first generation. Then, since the size of $\mathcal{R}$ is finite, and the Hilbert space is finite-dimensional, G_2 surely coincides with one of the possible configuration of the first generation. Then, in order to conclude that $G_3 = G_3^Q$ too, it is enough to notice that this third generation is nothing but the second generation obtained considering as the initial state G_2, that coincides with G_2^Q by assumption. Therefore they must also coincide. And the same holds true for the next generations, by iterating this argument.

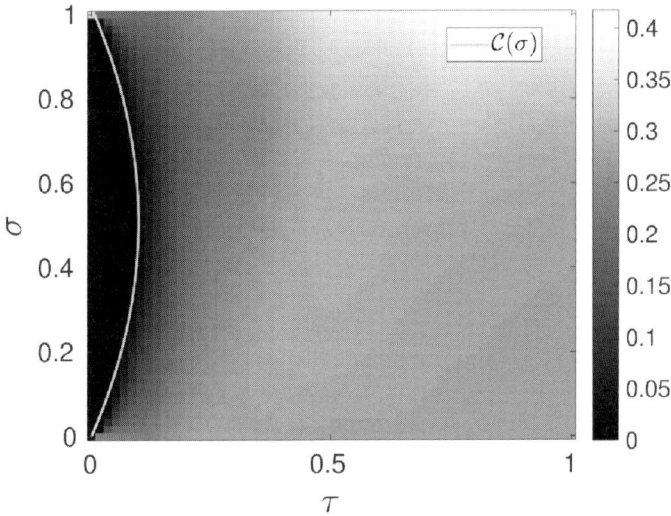

Figure 8.1 The function $\Delta(\tau,\sigma)$, and the gray range of its values (Source: Bagarello et al. [5]).

Remark: Since dim($\mathcal{H}$) $< \infty$, the number of possible configurations in the first generation for S could be large, but is finite.[4] Hence, in principle, we could check if $\Delta(\tau,\sigma) = 0$ or not, for any such configuration. Of course, this would not be possible for an infinitely extended lattice, or if we work with bosonic rather than with fermionic operators. But, likely enough, these situations are not relevant for us, here.

Instead of computing the second (and the other) generation for any possible initial condition, i.e., for any possible initial distributions of alive and dead cells, we have restricted our computations of $\Delta(\tau,\sigma)$ by considering 100 different random initial conditions for fixed τ and σ, and then averaging the related results. This is not sufficient to ensure that $G_3 = G_3^Q$ if $G_2 = G_2^Q$, but it is already a good indication that this can happen.

The function $\Delta(\tau,\sigma)$ is shown in Figure 8.1 for $0 \leq \tau,\sigma \leq 1$. The way in which τ affects Δ is clear: for most fixed values of σ, σ_0, $\Delta(\sigma_0,\tau)$ increases with τ. This is true for values of σ_0 close to zero ($\sigma_0 \lesssim 0.2$) or for σ_0 close to one ($\sigma_0 \gtrsim 0.6$). Otherwise, for $0.2 \lesssim \sigma_0 \lesssim 0.6$, we see that $\Delta(\sigma_0,\tau)$ increases first, and then decreases again, but it never goes back to zero. Therefore, in any case, the presence of τ produces a significant difference between the GoL and our QGoL. This difference is very small, at least for the second generation, only for τ close to zero, in a way that is almost independent of the value of σ: the darker region in Figure 8.1 shows, in fact, that $\Delta(\tau,\sigma)$ is very small if $\sigma \simeq 0.5$ and $\tau \simeq 0.05$. But, if $\sigma \simeq 0.1$, $\Delta(\tau,\sigma) \simeq 0$ only if τ is extremely small, much smaller than before. These results

---

[4] These would be very many, for our lattice $\mathcal{R}$ made by 33^2 cells, each of which can take values zero and one!

are in agreement with the fact that $\tau > 0$ corresponds to an effect that is absent in the classical situation, where no time evolution exists at all. However, they also suggest that there is the possibility of having not many differences for certain pairs of τ and σ, but not for many: these are exactly the values of τ and σ corresponding to the darker region in Figure 8.1, which is the region where the QGoL is closer to the GoL. This region is delimited by the quadratic curve $C(\sigma) = -0.337\sigma^2 + 0.384\sigma$, which approximates the contour level $\Delta(\tau = 0.1, \sigma = 0.5) = 0.02$.

A similar analysis can be repeated for the dependence of $\Delta(\tau, \sigma)$ on σ. For $\tau \lesssim 0.4$, $\Delta(\tau, \sigma)$ increases as σ approaches 0 or 1, while it decreases for intermediate values of σ. On the other hand, for $\tau \gtrsim 0.4$ the error increases with τ and with σ, taking its minimum value for $\sigma = 0$. However, in this region, even when $\sigma = 0$, the GoL and our QGoL are really different, already at the second generation, since $\Delta(\tau, \sigma)$ is never zero. This is essentially what one could expect. We also observe that the GoL is exactly recovered only when $\sigma = \tau = 0$: this is not surprising, since H does not act, and Equation (8.6) collapses into Equation (8.1).

8.4 Conclusions

The conclusion of our analysis is the following: the GoL can be enriched with more effects that we have called *quantum* since they are related to an Hamiltonian operator constructed in terms of ladder fermionic operators. This Hamiltonian produces a dynamics for the status of the cells of the lattice in some interval of time. Moreover, the introduction of the parameter σ produces an extra freedom in the computation of the next generation of the lattice, and this freedom is unavoidable exactly because of the presence of a continuous dynamics induced by the Hamiltonian. Therefore, even if at a first sight the role of σ appears simply as a slight modification of the original rule, as used in the GoL, it turns out that it is a quantum effect as well.

Many other possible extensions of the original Conway's game could be considered, both on a classical and at a quantum levels. For instance, the parameter σ could be taken to be a random quantity, which could be different in each of the 100 repetitions considered in our numerical simulations. Or we could consider a different form for the Hamiltonian H, taking care also of other effects, or still we could keep the same Hamiltonian but change the values of its parameters after each action of ρ, as we have discussed in Section 2.8.2. It would also be interesting to understand if the QGoL proposed here, or some modified version of it, could be applied in any real-life phenomenon.

We conclude this chapter suggesting the interested reader to read the article [5] for more results on this model of the QGoL, including a statistical analysis of the model and other aspects that were not discussed here.

9

Prehistoric Data Mining

9.1 Introduction

In most of the applications discussed so far in this book it may appear that, even if the general strategy proposed and applied could be considered interesting and new, what is not really evident is if it is also *efficient*. In particular, is our approach suitable to reproduce experimental data? And, even more, can a model constructed following our ideas be used to produce predictions? This is, in fact, what makes a model, besides interesting, a good model. According to this statement, the model proposed in [59] is good since it can reproduce some data observed in the Italian parliament. Also, the model proposed for desertification, considered in Chapter 4, is *potentially good*, since it is not so difficult to imagine possible experimental settings that can be used to fix the parameters of the model, and to use the detailed expression of the Hamiltonian deduced in this way to make predictions on desertification effects and to study concrete ways to avoid it. Something similar can be repeated, in principle, for most of the models proposed in this book. So far, however, comparison with real data has not been our main interest, quite often. We were more concerned on the possibility of using quantum ideas, and ladder operators in particular, to model some essential facts (not really the details!) of different macroscopic systems. However, we should also mention that some models already proposed in [1] are interesting from this point of view, since the results that we have deduced, in particular in the analysis of love affairs, describe quite well the main features of the real system. We will discuss more on comparison with real data also later, in Chapter 12.

The model we will describe in this chapter is, in our opinion, a particularly nice example on how mathematics can be useful in connection with something unusual and apparently far away from hard sciences. What will be discussed here is, in fact, a possible explanation of data connected to prehistoric demography in South America. And, surprisingly, the strategy that will be adopted, and that works quite well, is not particularly different from that we have proposed in Chapter 5, in a completely different context.

The analysis considered in this chapter originates from some experimental demographic data recently appearing in the literature, and concerning prehistoric South America [99]. These data contribute to the debate on the causes of conflict in human societies. For instance, some authors suggest that large-scale violence arose only with the spreading of agriculture and the building of complex societies. Other authors, including us, believe that migration due to resource competition also plays a crucial role in explaining the population dynamics, at least before the introduction of agriculture. The relation between population and resources indeed emerged as a key factor leading to migration and conflict once the carrying capacity of the environment, i.e., the maximum population size of the species that the environment can sustain, had been reached.

The analysis in [99] reconstructs the distribution of the human populations in South America in a time interval that goes from 14000 to 2000 years ago. Three phases are observed: the first phase shows a rapid expansion from 14000 to 9000 years ago, and can be understood as the continent colonization: people are reaching the continent and the population starts growing: the quantity of resources is large, compared to the populations. After this first period, a second phase of around 4000 years of a dynamical equilibrium around a carrying capacity, with no net growth, is observed: we are in a condition in which resources are enough for the people in the continent, but not for more. When the number of humans increases, the resources lack, and this causes the death of some percentage of humans. Then, resources are again enough to feed the population, which can increase (in density) a little bit, until resources are again not sufficient and so on. We stay close to the carrying capacity. Finally a new growth phase occurs due to the rising of agriculture that allows sustaining of a larger population [99, 100].

What we will discuss here is a possible explanation of the second phase, and in particular of the sharp oscillation with large amplitude that were observed in [99] around the carrying capacity. The idea behind our model is the following: South America can be modeled as a sort of two-dimensional lattice in which the different cells correspond to different geographic areas, possibly disconnected because of local ecological conditions. For this reason, the population is localized in the different cells and, in absence of particular reasons (in particular, if food is sufficient), they are not expected to move (or to move much) from one cell to another. In each cell we consider population and resources, and a sort of oscillating behavior is expected: when population grows, resources decrease and vice versa. So we expect to observe oscillations on the densities of the population in the different cells. However, when we average these oscillations on the whole lattice $\mathcal{R}$, there is no reason, a priori, for these oscillations to interact constructively, except if some further effect is properly considered. In other words, if we just consider the dynamics of the population and we average the *local densities*, i.e., the densities in the different cells, of the population on $\mathcal{R}$, we expect to find a smooth curve that does not correspond to what is described in [99]. And this will be, in fact, also the result of our analysis. However, if we add to the model the possibility of having some movement from one cell of $\mathcal{R}$ to another; that is, if we introduce migration in the model, then we are able to explain the finding in [99] in the intermediate time period above, from 9000 to 5000 years ago (or, as it is sometimes used in the literature on this topic, BP, which stands for *before present*). Hence, migration will be relevant to explain the data.

9.2 Introducing the Model

We model South America as a two-dimensional lattice $\mathcal{R}$ formed by L^2 different cells. Each cell includes two interacting populations $\mathcal{P}_a$ and $\mathcal{P}_b$, humans and natural resources. Since the period we are considering belongs to a pre agricultural situation, the natural resources represent the only sustenance for humans. Following our previous discussion, it is natural to imagine that, starting from some initial spatial distribution of $\mathcal{P}_a$ and $\mathcal{P}_b$, these evolve following a dynamics similar (at least, in part) to that of a predator-prey system. From an ecological point of view, this seems the most appropriate approach to represent a user-resource system with users depending on renewable biogenic resources [101].

We consider two opposite scenarios, one in which no migration is allowed, and a second one in which humans are allowed to migrate to neighboring cells. On the other hand, resources cannot diffuse across cells. Moreover, to make the model more realistic, migration of $\mathcal{P}_a$ from a cell α to a cell β is assumed to depend on a variable, K_α, which reflects the local (i.e., in cell α) ratio between the densities of resources and humans. More specifically, humans migrate at a rate that increases for decreasing K_α. This is because we can easily imagine that migration from α increases when local resources are low, while no migration is expected in presence of a large amount of resources in α. After humans migrate from α to β, resources in α tend to recover, since the quantity of users in α (the humans) decreases, while human density in β, after an initial obvious increment, can subsequently decline, possibly because of some overpopulation effect or because humans are moving somewhere else. This produces a reduction of the total human populations in α and β, which we can interpret as the effect of an increased mortality due to competition and/or conflict over local resources: the larger the density in α, the bigger the probability of some conflict between humans.

In order to deduce the dynamics of $\mathcal{P}_a$ we use fermionic operators and the general strategy introduced already in Chapter 2 and applied to different problems in population dynamics in Chapter 5 and in [1]. In fact, some of the ideas discussed in this chapter (predator-prey dynamics in a single cell and migration along $\mathcal{R}$) were already relevant for our analysis of escape strategies. To clarify our approach we discuss first what happens in a single cell, and then we propose a model for the whole lattice $\mathcal{R}$. In particular, we first introduce the operators, the vectors and the Hilbert spaces for the two situations, and then we propose a dynamics for them, analyzing the consequences of this proposal.

9.2.1 First Step: $\mathcal{R}$ Is a Single Cell

Let S be a system composed by two populations: the users $\mathcal{P}_a$ (i.e., the humans), and the natural resources $\mathcal{P}_b$. The system S can assume four basic states, while any other allowed state can be obtained as a combination of these: (i) $\varphi_{0,0}$, the *vacuum* of the system, corresponding to an (almost) complete absence of both humans and resources; (ii) $\varphi_{0,1}$ (resp. $\varphi_{1,0}$) corresponding to a low (resp. high) density of humans and abundance (resp. lack) of resources; (iii) $\varphi_{1,1}$ describes abundance of both human and resources.

We associate these basic states of the system with four independent mutually orthogonal vectors in a four-dimensional Hilbert space $\mathcal{H}_0$ endowed with the scalar product $\langle \cdot, \cdot \rangle$. This reflects essentially what already described in Sections 2.2 and 2.9, but with a *demographic* meaning in mind.

As in Section 2.9, a convenient way to construct the vectors $\varphi_{j,k}$ with $j, k = 0, 1$ makes use of two fermionic ladder operators, a and b, which together with their adjoints, $a^\dagger$ and $b^\dagger$, satisfy the CAR:

$$\{a, a^\dagger\} = \{b, b^\dagger\} = \mathbb{1}_4, \quad a^2 = b^2 = 0, \quad \{a^\sharp, b^\sharp\} = 0, \tag{9.1}$$

where $\{x, y\} = xy + yx$ is the usual anticommutator of x and y, $x^\sharp$ can be both x or $x^\dagger$, and $\mathbb{1}_4$ is the identity operator in the Hilbert space $\mathcal{H}_0 = \mathbb{C}^4$. We first introduce the vacuum $\varphi_{0,0}$ which satisfies $a \, \varphi_{0,0} = b \, \varphi_{0,0} = 0$, while the other vectors $\varphi_{j,k}$ can be constructed from $\varphi_{0,0}$ as usual:

$$\varphi_{1,0} := a^\dagger \varphi_{0,0}, \qquad \varphi_{0,1} := b^\dagger \varphi_{0,0}, \qquad \varphi_{1,1} := a^\dagger b^\dagger \varphi_{0,0}.$$

The set $\mathcal{F}_\varphi = \{\varphi_{j,k}, j, k = 0, 1\}$ is an o.n basis for $\mathcal{H}_0$, so that any vector Ψ of S can be expanded as $\Psi = \sum_{j,k=0}^{1} c_{j,k} \varphi_{j,k}$, with (complex-valued) coefficients $c_{j,k} = \langle \varphi_{j,k}, \Psi \rangle$. For the moment we are not considering any time evolution. Hence Ψ and the $c_{j,k}$'s are those computed at $t = 0$. Time dependence will be introduced later on. If we want to adopt the standard probabilistic interpretation of quantum mechanics to the present context, we can interpret Ψ as a state over S such that the probability to find $\mathcal{S}$ in the state $\varphi_{j,k}$ is given by $|c_{j,k}|^2$. Of course, since the probability of finding $\mathcal{S}$ in one of these states is one, then we must have the following constraint: $\sum_{j,k=0}^{1} |c_{j,k}|^2 = 1$.

Next we introduce the number operators $\hat{n}^{(a)} = a^\dagger a$ and $\hat{n}^{(b)} = b^\dagger b$ for the two populations $\mathcal{P}_a$ and $\mathcal{P}_b$. The vectors of $\mathcal{F}_\varphi$ are eigenstates of $\hat{n}^{(a)}$ and $\hat{n}^{(b)}$:

$$\begin{cases} \hat{n}^{(a)} \varphi_{n^{(a)},n^{(b)}} = n^{(a)} \varphi_{n^{(a)},n^{(b)}}, \\ \hat{n}^{(b)} \varphi_{n^{(a)},n^{(b)}} = n^{(b)} \varphi_{n^{(a)},n^{(b)}}, \end{cases} \tag{9.2}$$

with eigenvalues $n^{(a)}$ and $n^{(b)}$ that can be either 0 or 1. If Ψ is as above, using the fact that $n^{(a)} \varphi_{0,k} = 0$ and $n^{(b)} \varphi_{k,0} = 0$, $k = 0, 1$, we get

$$\hat{n}^{(a)} \Psi = \sum_{k=0}^{1} c_{1,k} \varphi_{1,k}, \qquad \hat{n}^{(b)} \Psi = \sum_{k=0}^{1} c_{k,1} \varphi_{k,1},$$

and the mean values of $\hat{n}^{(a)}$ and $\hat{n}^{(b)}$ on Ψ, $n_\Psi^{(a),(b)} = \langle \Psi, \hat{n}^{(a),(b)} \Psi \rangle$, can be rewritten as follows:

$$n_\Psi^{(a)} = \sum_{k=0}^{1} |c_{1,k}|^2, \qquad n_\Psi^{(b)} = \sum_{k=0}^{1} |c_{k,1}|^2. \tag{9.3}$$

We interpret $n_\Psi^{(a)}$ (resp. $n_\Psi^{(b)}$) as the density for the population $\mathcal{P}_a$ (resp. $\mathcal{P}_b$) when the system $\mathcal{S}$ is described by the state Ψ. Of course, because of the normalization requirement for Ψ, both $n_\Psi^{(a)}, n_\Psi^{(b)}$ are real numbers between 0 and 1. For instance, if at $t = 0$ the density of $\mathcal{P}_a$ is zero, then we must have $n_\Psi^{(a)} = 0$, and Equation (9.3) implies that $c_{1,0} = c_{1,1} = 0$. Hence Ψ can only be a combination of $\varphi_{0,0}$ and $\varphi_{0,1}$, $\Psi = c_{0,0} \varphi_{0,0} + c_{0,1} \varphi_{0,1}$. This corresponds, as it is obvious, to the (almost complete) absence of humans (zero density for $\mathcal{P}_a$): $\hat{n}^{(a)} \Psi = 0 \, \Psi$. Conversely, if $n_\Psi^{(a)} = 1$, then the same equation implies that $\sum_{k=0}^{1} |c_{1,k}|^2 = 1$, and therefore, due to the normalization condition for Ψ, $c_{0,0} = c_{0,1} = 0$. Therefore Ψ is necessarily a combination of $\varphi_{1,0}$ and $\varphi_{1,1}$, $\Psi = c_{1,0} \varphi_{1,0} + c_{1,1} \varphi_{1,1}$, and the density of the humans is the highest possible.

The dynamics of $\mathcal{P}_a$ and $\mathcal{P}_b$ we consider extends the one introduced in Section 2.9, which, as we have discussed several times, mimics some aspects of a predator-prey interaction: when the density of one population increases, the density of the other decreases and vice versa. However, contrarily to what we have done in Section 2.9, we will consider here the possibility of our Hamiltonian to depend explicitly on time, replacing its coefficients with suitable functions of time. For this reason, the Hamiltonian we consider is the following:

$$H_{SC}(t) = \omega^a(t) \, a^\dagger a + \omega^b(t) \, b^\dagger b + \lambda(t) \left(a^\dagger b + b^\dagger a \right), \tag{9.4}$$

which differ from that in (2.82) exactly because the real parameters of that Hamiltonian are replaced, here, by real-valued functions measuring the time-changing inertia of the populations, $\omega^a(t)$ and $\omega^b(t)$, and the time-dependent strength of the interaction, $\lambda(t)$. In this way we are considering the possibility that some of the characteristics of the populations can change continuously with time, which of course makes sense for the system we are describing.

> *Remark:* An alternative approach to introduce time-dependent parameters of the Hamiltonian would be to adopt a suitable rule ρ as in Section 2.8, and to replace the ordinary dynamics of S with an (H, ρ)-induced dynamics. However, this possibility will not be considered in this chapter, due to the very good agreement with the experimental data we obtain using the Hamiltonian in Equation (9.14) below.

The meaning of the various contributions in $H_{SC}(t)$ is completely analogous to that of those in (2.82): in absence of interaction ($\lambda(t) = 0$) the densities of $\mathcal{P}_a$ and $\mathcal{P}_b$ stay constant in the cell, while they change in opposite directions when $\lambda(t) \neq 0$: if the density of humans increases, the resources decrease, and vice versa. The amplitude of these (pseudo-)oscillations decreases for large values of $\omega^{a,b}(t)$, so that they can change with time according to the values of these functions. Of course, for the moment, one of the essential aspects of our model, i.e., the possibility of $\mathcal{P}_a$ to move along $\mathcal{R}$, is not relevant since there are no other cells where to move! And this is, in fact, the reason for the subscript SC for the Hamiltonian (9.4): it stands for *single cell*. For this reason, rather than deducing the dynamics for this simple case ($L = 1$) using $H_{SC}(t)$, we will next enrich $\mathcal{R}$ with more cells, and consider the dynamics for this larger system.

9.2.2 More Cells

Let us now consider a larger lattice, made of several cells, labeled by Greek letters. In particular, we use a square lattice consisting in L^2 cells. As suggested before, each cell corresponds to some geographical area of South America. In cell α the two populations $\mathcal{P}_a$ and $\mathcal{P}_b$ interact as described by (9.4). The main difference is that, now, $\mathcal{P}_a$ is also free to move around $\mathcal{R}$. Of course, due to its meaning within this model, $\mathcal{P}_b$ cannot move.

To construct the model we repeat what we have done in Chapter 5, and we introduce, in each cell $\alpha = 1, \ldots, L^2$, the fermionic operators a_α, $a_\alpha^\dagger$ and the related number operators $\hat{n}_\alpha^{(a)} = a_\alpha^\dagger a_\alpha$ for the humans $\mathcal{P}_a$, and b_α, $b_\alpha^\dagger$ and $\hat{n}_\alpha^{(b)} = b_\alpha^\dagger b_\alpha$ for the resources $\mathcal{P}_b$. As in Section 9.2.1, we suppose that these operators satisfy the standard CAR for several modes:

$$\left\{ a_\alpha, a_\beta^\dagger \right\} = \left\{ b_\alpha, b_\beta^\dagger \right\} = \delta_{\alpha\beta} \, \mathbb{1}, \qquad \forall \alpha, \beta \tag{9.5}$$

$$\left\{ a_\alpha^\sharp, a_\beta^\sharp \right\} = \left\{ b_\alpha^\sharp, b_\beta^\sharp \right\} = \left\{ a_\alpha^\sharp, b_\beta^\sharp \right\} = 0, \qquad \forall \alpha \neq \beta. \tag{9.6}$$

Moreover we assume that $a_\alpha^2 = b_\alpha^2 = 0$, for all α. In Equation (9.5) $\mathbb{1}$ is the identity operator on the Hilbert space $\mathcal{H}$ where the fermionic operators act, defined in analogy with $\mathcal{H}_0$ as the linear span of the vectors in (9.7). Notice that, clearly, $\mathcal{H}$ is finite-dimensional, with $\dim(\mathcal{H}) = 4^{L^2}$, which reduces to 4 if $L = 1$, as for the single cell Hilbert space $\mathcal{H}_0$ introduced in the previous section. As a consequence, the operators $a_\alpha^\sharp$, $b_\alpha^\sharp$ etc., are $4^{L^2} \times 4^{L^2}$ matrices.

We now introduce the vacuum of the system $\mathcal{S}$, that is a vector $\varphi_{\vec{0}_a, \vec{0}_b}$ which is annihilated by all the operators a_α, b_α:

$$a_\alpha \varphi_{\vec{0}_a, \vec{0}_b} = b_\alpha \varphi_{\vec{0}_a, \vec{0}_b} = 0, \quad \forall \alpha = 1, \dots, L^2.$$

Here $\vec{0}_a = \vec{0}_b = (0, 0, \dots, 0)$ are two L^2-dimensional vectors, with all zero entries.

Next we construct the states of the basis of $\mathcal{H}$ by acting with the operators $a_\alpha^\dagger, b_\alpha^\dagger$ over $\varphi_{\vec{0}_a, \vec{0}_b}$,

$$\varphi_{\vec{m}_a, \vec{m}_b} = \left(a_1^\dagger\right)^{m_a(1)} \cdots \left(a_{L^2}^\dagger\right)^{m_a(L^2)} \left(b_1^\dagger\right)^{m_b(1)} \cdots \left(b_{L^2}^\dagger\right)^{m_b(L^2)} \varphi_{\vec{0}_a, \vec{0}_b}, \tag{9.7}$$

where $\vec{m}_a = (m_a(1), \dots, m_a(L^2))$, $\vec{m}_b = (m_b(1), \dots, m_b(L^2))$ are all the possible L^2-dimensional vectors whose entries are only 0 or 1. This is because any other choice of $m_a(\alpha)$ or $m_b(\alpha)$ would destroy the state, because of the CAR. Notice that we can construct at most 2^{L^2} vectors $\vec{m}_a$ and 2^{L^2} vectors $\vec{m}_b$, so that we can build 4^{L^2} possible pairs $(\vec{m}_a, \vec{m}_b)$.

The set $\mathcal{F}_\varphi$ of all the vectors obtained in this way, $\mathcal{F}_\varphi = \{\varphi_{\vec{m}_a, \vec{m}_b}\}$, is an o.n. basis of $\mathcal{H}$. Moreover, introducing the local number operators, we have

$$\begin{cases} \hat{n}_\alpha^{(a)} \varphi_{\vec{m}_a, \vec{m}_b} = m_a(\alpha) \varphi_{\vec{m}_a, \vec{m}_b}, \\ \hat{n}_\alpha^{(b)} \varphi_{\vec{m}_a, \vec{m}_b} = m_b(\alpha) \varphi_{\vec{m}_a, \vec{m}_b}, \end{cases} \tag{9.8}$$

for all $\alpha = 1, \dots, L^2$. These eigenvalue equations clearly extend those in (9.2).

The vectors $\varphi_{\vec{m}_a, \vec{m}_b}$ can be interpreted similarly to the $\varphi_{j,k}$ in the previous section. The vacuum $\varphi_{\vec{0}_a, \vec{0}_b}$, for instance, describes a situation where only a few humans and very few resources are distributed all along the lattice. Analogously, the vector $\varphi_{\vec{k}_a, \vec{k}_b}$ with $\vec{k}_a = (1, 0, 0, \dots, 0)$ and $\vec{k}_b = (0, 0, \dots, 0, 1)$, describes a situation with a large amount of humans (but few resources) in the first cell and of resources (but few humans) in the last one, while the other cells are *almost empty*. We observe that, from Equation (9.8), the orthogonality of the set $\mathcal{F}_\varphi$ implies that

$$m_a(\alpha) = \left\langle \varphi_{\vec{m}_a, \vec{m}_b}, \hat{n}_\alpha^{(a)} \varphi_{\vec{m}_a, \vec{m}_b} \right\rangle, \qquad m_b(\alpha) = \left\langle \varphi_{\vec{m}_a, \vec{m}_b}, \hat{n}_\alpha^{(b)} \varphi_{\vec{m}_a, \vec{m}_b} \right\rangle. \tag{9.9}$$

A generic state $\Psi(t)$ of the system at time t can be written as a linear combination of the elements in $\mathcal{F}_\varphi$:

$$\Psi(t) = \sum_{\vec{m}_a, \vec{m}_b} c_{\vec{m}_a, \vec{m}_b}(t) \varphi_{\vec{m}_a, \vec{m}_b}, \tag{9.10}$$

where $c_{\vec{m}_a,\vec{m}_b}(t)$ are complex-valued functions of t such that $\sum_{\vec{m}_a,\vec{m}_b}$ $|c_{\vec{m}_a,\vec{m}_b}(t)|^2 = 1$, in order to preserve the probabilistic interpretation of the $\Psi(t)$. The time-dependent coefficients $c_{\vec{m}_a,\vec{m}_b}(t)$ could be deduced from $\Psi(t)$ as follows:

$$c_{\vec{m}_a,\vec{m}_b}(t) = \left\langle \varphi_{\vec{m}_a,\vec{m}_b}, \Psi(t) \right\rangle.$$

Of course, if one of the $c_{\vec{m}_a,\vec{m}_b}(0)$ is equal to one, then all the other coefficients $c_{\vec{m}_a,\vec{m}_b}(0)$ in Equation (9.10) must be zero, and the sum above for $\Psi(t)$, at $t = 0$, will collapse to a single contribution. In other words, in this case the system S is, at $t = 0$, in an eigenstate of all the (local) density operators $\hat{n}_\alpha^{(a)}$ and $\hat{n}_\alpha^{(b)}$. Of course, there is no reason why this situation should be true also for $t > 0$. Moreover, in the general situation in which more than just one $c_{\vec{m}_a,\vec{m}_b}(0)$ are nonzero, the vector $\Psi(0)$ is no longer an eigenstate of all these operators, and the expansion in (9.10) does not reduce to a single term, even for $t = 0$. And this is even more evident for $t > 0$. In this case, extending what we have done in Equation (9.4), we need to compute the following mean values:

$$n_{\Psi,\alpha}^{(a)}(t) = \left\langle \Psi(t), \hat{n}_\alpha^{(a)}\Psi(t) \right\rangle = \|a_\alpha \Psi(t)\|^2, \tag{9.11}$$

$$n_{\Psi,\alpha}^{(b)}(t) = \left\langle \Psi(t), \hat{n}_\alpha^{(b)}\Psi(t) \right\rangle = \|b_\alpha \Psi(t)\|^2, \tag{9.12}$$

which are phenomenologically interpreted as densities of humans and resources in the cell α at time t, for a system described, at $t = 0$, by $\Psi(0)$. Of course, to compute $n_{\Psi,\alpha}^{(a)}(t)$ and $n_{\Psi,\alpha}^{(b)}(t)$, we need to deduce first $\Psi(t)$, which can be found once we fix $\Psi(0)$ and the dynamics of the system. This will be done in the next section, by introducing a Hamiltonian that extends that in (9.4) and introduces the possibility that humans migrate along $\mathcal{R}$.

Remark: It might be worth noticing that $n_{\Psi,\alpha}^{(a)}(t)$ and $n_{\Psi,\alpha}^{(b)}(t)$ are related to the mean values computed in Equation (9.9). In fact, we can easily see that

$$n_{\Psi,\alpha}^{(a)}(0) = n_a(\alpha), \qquad n_{\Psi,\alpha}^{(b)}(0) = n_b(\alpha),$$

if $\Psi(0) = \varphi_{\vec{n}_a,\vec{n}_b}$. Otherwise, if $\Psi(0)$ is some more complicated linear combination of vectors $\varphi_{\vec{n}_a,\vec{n}_b}$, they do not necessarily coincide.

9.2.3 Evolution of the System

As we have seen, to determine the densities in Equations (9.11) and (9.12), we need first to compute the time evolution of the state of the system $\Psi(t)$, expanded in terms of $\mathcal{F}_\varphi$ as in Equation (9.10). Hence, the knowledge of $\Psi(t)$ is guaranteed once the $c_{\vec{m}_a,\vec{m}_b}(t)$ are found.

These time-depending functions $c_{\vec{m}_a,\vec{m}_b}(t)$ can be deduced knowing their initial values by solving the Schrödinger equation

$$i\frac{\partial \Psi(t)}{\partial t} = H(t)\,\Psi(t), \tag{9.13}$$

where $H(t)$ is the Hamiltonian operator describing the dynamics of humans and resources for the whole system. We assume that $H(t)$ is a sum of contributions like those in $H_{SC}(t)$ in Equation (9.4), one for each cell, plus a migration contribution $H_M(t)$. More in details, the Hamiltonian of the system is the following:

$$H(t) = \sum_{\alpha \in \mathcal{R}} H_\alpha(t) + H_M(t), \tag{9.14}$$

where

$$H_\alpha(t) = \omega_\alpha^a(t)a_\alpha^\dagger a_\alpha + \omega_\alpha^b(t)b_\alpha^\dagger b_\alpha + \lambda_\alpha(t)\left(a_\alpha^\dagger b_\alpha + b_\alpha^\dagger a_\alpha\right), \tag{9.15}$$

$$H_M(t) = \sum_{\alpha \neq \beta} p_{\alpha,\beta}\gamma_{\alpha,\beta}(t)a_\alpha a_\beta^\dagger. \tag{9.16}$$

Here the functions $\omega_\alpha^a(t)$, $\omega_\alpha^b(t)$ and $\lambda_\alpha(t)$ have the same meaning as the analogous ones for $H_{SC}(t)$, while $p_{\alpha,\beta} \neq 0$ if α and β are neighboring cells, and $p_{\alpha,\beta} = 0$ if they are not. As for $\gamma_{\alpha,\beta}(t)$, this is a measure of the mobility of the humans. This is in agreement with what we have deduced in Chapter 5, formula (5.8), where ρ_a was introduced exactly to measure the speed of the movement of the first population inside the room, while trying to reach the exit. The fact that $\gamma_{\alpha,\beta}(t)$ here depends on time is reasonable, since the humans can move fast in some particular conditions, and be slower, or much slower, in other conditions (for instance, when there is no real reason to leave their original cell and move along $\mathcal{R}$).

The reason why $H_M(t)$ describes migration is because of the presence of the term $a_\alpha a_\beta^\dagger$, which describes movement of humans from cell α to cell β: in fact $a_\beta^\dagger$ increases the density of humans in β, while a_α decreases their density in α. Of course, such a migration is possible only for nonzero values of $p_{\alpha,\beta}$: no migration is possible (in a single step) if α and β are not neighboring. We observe that no analogous term for resources is considered in $H_M(t)$. This is because we have assumed, from the very beginning, that humans can move, but resources are geographically localized. Of course, such an assumption will not be reasonable if we wanted to model, as possible resources, sheeps, cows or any group of animals that follows the humans during their migration. But nothing like this was possible during the prehistoric age we are considering here.

To complete the description of $H(t)$ we still have to define the analytic expression for the functions $\omega_\alpha^{a,b}(t)$, $\lambda_\alpha(t)$ and $\gamma_{\alpha,\beta}(t)$. As we have already observed, we assume that humans try to leave a given cell where they are originally localized if the amount of resources in that particular cell is not sufficient to feed all of them. This is one criterium we use to fix these functions. We also expect that, the less nutrients in α, the faster the migration of humans. To turn this general idea into formulae, we introduce a ratio between resource and human densities. This quantity depends on Ψ and on the cell α, and it is a function of time:

$$K_{\Psi,\alpha}(t) = \frac{n_{\Psi,\alpha}^{(b)}(t)}{n_{\Psi,\alpha}^{(a)}(t)}. \tag{9.17}$$

Humans tend to leave cell α if (or when) $K_{\Psi,\alpha}(t)$ is low enough, and the speed of this migration is expected to increase when $K_{\Psi,\alpha}(t)$ decreases.

Another mechanism that we expect should occur in a system like ours, and that we want to implement in our Hamiltonian, is the following: natural resources tend to proliferate in a cell when this cell has not many humans, i.e., in case of low human density. In other words, resources refill themselves if $K_{\Psi,\alpha}(t)$ is large. Otherwise, i.e., if there are too many humans for few resources, they tend to exhaust. This is particularly clear in absence of migration, of course, where humans and resources are forced to stay in a single cell.

These assumptions suggest us to choose the following functions in $H(t)$:

$$\omega_{\alpha}^{a}(t) = \sigma_{\alpha} \left(\exp^{-\frac{1}{(K_{\Psi,\alpha}(t)\tau_\alpha)^2}} K_{\Psi,\alpha}(t) \right)^{1/2}, \tag{9.18}$$

$$\omega_{\alpha}^{b}(t) = \sigma_{\alpha} \left(\exp^{-\left(\frac{K_{\Psi,\alpha}(t)}{\tau_\alpha}\right)^2} K_{\Psi,\alpha}(t)^{-1} \right)^{1/2}, \tag{9.19}$$

$$\lambda_{\alpha}(t) = \omega_{\alpha}^{a}(t) + \omega_{\alpha}^{b}(t) + \mu_{\alpha}, \tag{9.20}$$

$$\gamma_{\alpha,\beta}(t) = \omega_{\alpha}^{b}(t) + \omega_{\beta}^{b}(t), \tag{9.21}$$

where $\sigma_\alpha, \tau_\alpha$, and μ_α are real parameters.[1] From (9.18) to (9.21), it follows that the higher the value of $K_{\Psi,\alpha}(t)$, the higher (resp. the smaller) the value of $\omega_{\alpha}^{a}(t)$ (resp. $\omega_{\alpha}^{b}(t)$). This reflects the fact that, if in cell α there is a lot of nutrients compared with the density of the humans, the humans have no particular reason to move away. And in fact, in this case, the inertia function for humans in α, $\omega_{\alpha}^{a}(t)$, is high. On the opposite side, if in cell α we have a lot of humans compared with the amount of nutrients ($K_{\Psi,\alpha}(t)$ small), $\omega_{\alpha}^{a}(t)$ is small. Then, the inertia of $\mathcal{P}_a$ is small, and for this reason humans can easily move from cell α to a different cell. Of course, since the Hamiltonian (9.14) does not include any migration term for $\mathcal{P}_b$, the value

---

[1] Since the right-hand sides of Equations (9.18)–(9.21) depend on Ψ, the left-hand sides also depend on Ψ. Hence, for instance, rather than $\omega_{\alpha}^{a}(t)$, we should write $\omega_{\Psi,\alpha}^{a}(t)$. But we prefer to avoid this notation, since (see [9.16]), $\omega_{\alpha}^{a}(t)$ could also be fixed a priori, with no reference whatsoever to Ψ. We will come back on this particular aspect of the Hamiltonian later on.

(high or small) of $\omega_\alpha^b(t)$ does not affect any movement of the resources along $\mathcal{R}$. It can modify only the amplitude of the oscillations of the density of $\mathcal{P}_b$ inside each cell, modeling the ability of the resources not to be affected much by their use by humans (the density of resources does not oscillate much).

Formula (9.20) shows that the strength of the interaction between $\mathcal{P}_a$ and $\mathcal{P}_b$ in each cell is assumed to be proportional to the sum of the inertia of the two species, plus a constant contribution μ_α: the interaction between $\mathcal{P}_a$ and $\mathcal{P}_b$ does not really depend on which species have a larger inertia, in cell α, but only on their sum $\omega_\alpha^a(t) + \omega_\alpha^b(t)$. Moreover, the presence of $\mu_\alpha > 0$ in Equation (9.20) guarantees a sort of minimal predator-prey dynamics even when both $\omega_\alpha^a(t)$ and $\omega_\alpha^b(t)$ are close to zero, which may happen in some particular situation. This is what can be deduced (see formulas [9.18] and [9.19]), for small values of σ_α and intermediate values of $K_{\Psi,\alpha}(t)$.

Formula (9.21) relates $\omega_\alpha^b(t)$ and $\omega_\beta^b(t)$ to the mobility of $\mathcal{P}_a$ from cell α to cell β. This might look a bit strange, since mobility refers to $\mathcal{P}_a$ and not to $\mathcal{P}_b$, while the right-hand side of (9.21) refers to $\mathcal{P}_b$ and not to $\mathcal{P}_a$. However, this is not really so, since Equations (9.18) and (9.19) produce a link between $\omega_\beta^b(t)$ and $\omega_\beta^a(t)$. Therefore, Equation (9.21) could be rewritten, in principle, in terms of $\omega_\alpha^a(t)$ and $\omega_\beta^a(t)$, but we prefer to adopt this simpler expression for $\gamma_{\alpha,\beta}(t)$. Incidentally we observe that $\gamma_{\alpha,\beta}(t)$ is symmetric with respect to the exchange $\alpha \leftrightarrow \beta$. This is related, of course, to the fact that our $H(t)$ is self-adjoint.

Finally, the values of $\sigma_\alpha, \tau_\alpha$ can be used as an extra freedom for our analysis: for instance, by decreasing τ_α, Equations (9.18) and (9.19) show that only very high or very low values of $K_{\Psi,\alpha}(t)$ can produce values of $\omega_\alpha^a(t)$ or $\omega_\alpha^b(t)$ that are significantly different from zero. Obviously, σ_α magnifies or reduces these values.

> *Remark:* It may be worth noticing that the model, despite the analytic expressions in (9.14)–(9.16), which suggest a quadratic expression for $H(t)$, is highly nonlinear. The reason is that the time-dependent functions needed in the definition of $H(t)$ are constructed by means of quantities, see the expression of $K_{\Psi,\alpha}(t)$ in Equation (9.17) in particular, which are directly related to the number operators for $\mathcal{P}_a$ and $\mathcal{P}_b$, and to their mean values. Therefore, if we consider for instance the simplest free contribution for $\mathcal{P}_b$ in $H(t)$, $\omega_\alpha^b(t) b_\alpha^\dagger b_\alpha$, even if this looks quadratic in the ladder operators $b_\alpha^\sharp$, it is not, since $\omega_\alpha^b(t)$ depends (also) on $n_{\Psi,\alpha}^{(b)}(t)$ and $n_{\Psi,\alpha}^{(a)}(t)$. A similar idea was already investigated in Chapter 7, in connection with tumor cells proliferation.

Going back to the Schrödinger equation for $\Psi(t)$, the orthogonality conditions for the vectors $\varphi_{\vec{m}_a,\vec{m}_b}$ produce the following set of ordinary differential equations for the coefficients $c_{\vec{m}_a,\vec{m}_b}(t)$ in the expansion (9.10):

$$i\frac{\partial c_{\vec{m}_a,\vec{m}_b}(t)}{\partial t} = \langle \varphi_{\vec{m}_a,\vec{m}_b}, H(t)\Psi(t)\rangle. \tag{9.22}$$

Once these equations are solved (numerically), we are able to construct the time-dependent densities in each cell α as in Equations (9.11) and (9.12), and we get

$$n_{\Psi,\alpha}^{(a)}(t) = \sum_{\vec{m}_a,\vec{m}_b} \left| c_{\vec{m}_a,\vec{m}_b}(t) \right|^2 m_\alpha^{(a)},$$

$$n_{\Psi,\alpha}^{(b)}(t) = \sum_{\vec{m}_a,\vec{m}_b} \left| c_{\vec{m}_a,\vec{m}_b}(t) \right|^2 m_\alpha^{(b)}. \tag{9.23}$$

Notice that in these sums we need to have $m_\alpha^{(a)} = m_\alpha^{(b)} = 1$ to have a nontrivial contribution to the resulting sum. Otherwise we do not have such a contribution in the right-hand sides of (9.23). Notice also that both $n_{\Psi,\alpha}^{(a)}(t)$ and $n_{\Psi,\alpha}^{(b)}(t)$ are bounded by one for all t. This is due to the normalization condition on $\Psi(t)$, and to the fact that, in general $m_\alpha^{(c)} \leq 1$, $c = a,b$. Moreover, since $H(t) = H(t)^\dagger$, it is possible to show that, with obvious notation,

$$N_\Psi(t) = n_\Psi^{(a)}(t) + n_\Psi^{(b)}(t) = \sum_\alpha \left(n_{\Psi,\alpha}^{(a)}(t) + n_{\Psi,\alpha}^{(b)}(t) \right) = n_\Psi^{(a)}(0) + n_\Psi^{(b)}(0) = N_\Psi(0), \tag{9.24}$$

even in this case, where $H(t)$ depends explicitly on time. In fact, let us observe first that $N_\Psi(t)$ can be rewritten, adopting the Schrödinger representation, as

$$N_\Psi(t) = \left\langle \Psi(t), \hat{N}(0)\Psi(t) \right\rangle,$$

where $\hat{N}(0) = \sum_\alpha (a_\alpha^\dagger a_\alpha + b_\alpha^\dagger b_\alpha)$ is the initial time global number (or density) operator for $\mathcal{P}_a$ and $\mathcal{P}_b$ (which, obviously, does not depend on Ψ). Then, using (9.13) and the properties of the scalar product (linearity in its second variable and anti-linearity in the first), we get

$$\dot{N}_\Psi(t) = i \left\langle \Psi(t), [H(t), \hat{N}(0)]\Psi(t) \right\rangle.$$

This formula is also a consequence of the self-adjointness of $H(t)$: $H(t) = H^\dagger(t)$. Now, it is easy to check that $[H(t), \hat{N}(0)] = 0$, and therefore $\dot{N}_\Psi(t) = 0$, as we had to prove. Formula (9.24) has a direct consequence: if the global population density $n_\Psi^{(a)}(t)$ increases, the global resources density $n_\Psi^{(b)}(t)$ decreases, and vice versa. This reflects, at a global level, the predator-prey mechanism assumed in each cell.

> *Remark:* Of course this result is strongly related to the particular nature of the time dependence we are considering for $H(t)$. In fact, this is all contained in the parameters of $H(t)$, which have been replaced by very special functions of time. No further possibility is considered here.

We introduce now the mean $K = \frac{n_\Psi^{(a)}(0) + n_\Psi^{(b)}(0)}{2}$, and we use K as a sort of carrying capacity for the system, since, recalling that $N_\Psi(t)$ stays constant, (9.24), the various terms in the Hamiltonian work to reduce $n_\Psi^{(a)}(t)$ (and to increase $n_\Psi^{(b)}(t)$) when

$n_\psi^{(a)}(t) > K$, and to act in the opposite way when $n_\psi^{(a)}(t) < K$. Notice that we are using K, rather than $K_{\Psi,\alpha}$, since, as we will show in few lines, the starting point of our numerical computations will be exactly to fix directly the explicit value of K, rather than deducing it.

The values of the parameters used in our numerical simulations to reproduce the empirical data in [99] are $\sigma_\alpha = 12.5$, $\tau_\alpha = 0.35$, $\mu_\alpha = 0.25$ and $p_{\alpha,\beta} = 1$, $\forall \alpha, \beta$. We further fix $K = 0.5$. The coefficients in (9.10) are chosen in such a way that the initial human density $n_{\Psi,\alpha}^{(a)}(0)$, in each cell, randomly differs at most by 10% from K/L^2. We further require the *global condition* $n_\psi^{(a)}(0) = K$. The idea behind this choice is to reproduce, in this way, the particular prehistoric situation when the carrying capacity was reached in the continent (i.e., at a global level). This configuration is what fixes the initial time of our analysis. In fact, as discussed in Section 9.1, we are not interested in explaining all the data in [99], but only those that refer to the period when the carrying capacity was reached, but agriculture was not yet discovered. Lastly, the densities of the resources are fixed as $n_{\Psi,\alpha}^{(b)}(0) = (2K/L^2 - n_{\Psi,\alpha}^{(a)}(0))$.

9.3 Results

We are now ready to deduce and discuss the results of our assumptions, and of the model constructed out of them. In particular we will separate our analysis, considering first what happens in absence of migration, and then we will show the effects of migration. In mathematical terms, we will first fix to zero the value of all the $p_{\alpha,\beta}$ in $H_M(t)$, so that $H(t) = \sum_\alpha H_\alpha(t)$, and then we will make a different choice, allowing movements for humans between neighboring cells. Not surprisingly, these different choices will present different numerical difficulties: in the first case each cell is disconnected from any other cell, and therefore the full lattice $\mathcal{R}$ will be a replica (with possibly different initial conditions and different values of the parameters) of just a one-cell lattice. For this reason we will be able to work with rather large lattices. On the opposite side, in presence of migration, different cells are in fact different, since they *see their neighboring cells*, and we are forced to work with relatively small lattices to keep under control the computational time.

9.3.1 No-Migration Scenario

In absence of migration our results are not compatible with what is described in Figure 9.3(b) in [99], where some big spikes and valleys are observed in the temporal region after colonization and before the discovery of agriculture, as we have already discussed in Section 9.1. What we observe here is something completely different, and these results do not really depend on the size of the lattice.

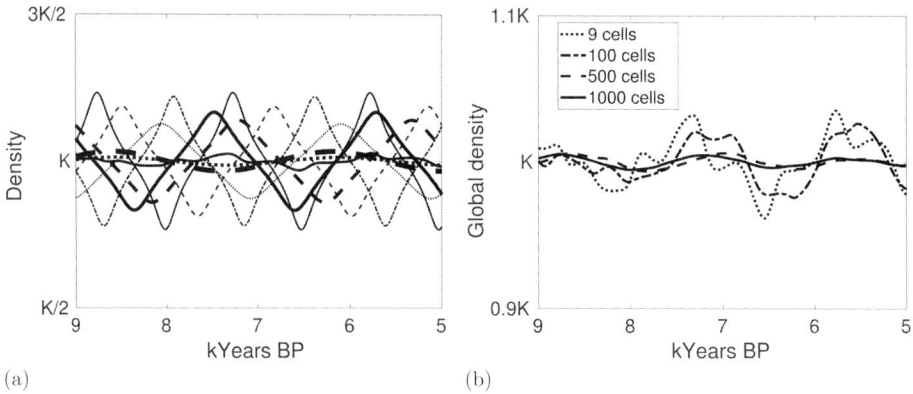

Figure 9.1 Model dynamics in absence of migration. (a) Global human population density (dashed thicker curve), and nine times the single-cell human population density (other curves) in a nine-cell system. (b) Global human population
densities in systems composed by different number of cells. The time interval
goes from 9000 to 5000 years BP (Source: Bagarello et al. [4]).

In fact, we get the same conclusions for a number of cells going from 9 to 1000: in
each cell we observe oscillations in the human density, and these oscillations are
simply smeared out when we consider the global population density $n_\psi^{(a)}(t)$, i.e.,
the sum of the densities of the humans in the whole lattice (the dashed thicker line
in Figure 9.1[a]). This is because, in this no-migration scenario, oscillations tend to
compensate each other. Figure 9.1(a) clearly shows this effect: almost all the dotted and dashed curves describe the global human densities as deduced, assuming
that the nine cells[2] of $\mathcal{R}$ are identical, with identical parameters and identical initial conditions of the densities of $\mathcal{P}_a$ and $\mathcal{P}_b$. Hence almost all the curves we plot
are just nine times the analogous density deduced for the single cell. On the other
hand, the dashed thicker line describes again the global human density, but this is
computed assuming slightly different initial conditions for the populations in the
various cells. These are fixed, as explained before, by adding a small amount of ran-
domness to the initial densities in each cell, around the carrying capacity. Hence, in
this case, cells are really different. It is evident from the figure that, in this last and
more realistic situation, the result is rather different from the one in [99]. Moreover,
Figure 9.1(b) shows that this smearing effect increases with the size of the lattice:
the larger this size, the smaller the oscillations. As a result, the global density
curve became dramatically different from the one for South America between

---

[2] Such a small lattice is useful when we consider the scenario with migration, to avoid numerical difficulties.
In the no-migration scenario, on the other hand, this choice is not so crucial and in fact much larger lattices
are also considered see Figure 9.1(b). The dotted curve in Figure 9.1(b) is particularly relevant, since it is the
one which should be compared with Figure 9.2, deduced by working again with a nine-cell lattice, but in
presence of migration.

9000 and 5500 years BP in [99]. This suggests that some coordination mechanism, able to keep in phase the local oscillations, should be relevant to reproduce the experimental data.

9.3.2 Migration Scenario

In the second scenario, migration is possible: humans can move from a cell to a neighboring one. The interesting aspect of our model is that the parameters of the Hamiltonian can be chosen to fit well the empirical data presented in [99]. As already discussed, due to computational constraints, simulations of the second scenario are shown here only for an L^2 lattice with $L = 3$. However, we should mention that slightly larger lattices can also be considered, and they do not produce any significant difference in the conclusions. From the point of view of the interpretation of the model, what we are doing by considering only a few cells is to divide South America in to a few macro-areas, considered more or less homogeneous.

Even in presence of migration, when considered at the scale of the single cell, we observe an oscillating behavior of the human density as in the previous scenario, albeit sharper. However, unlike the previous case, oscillations do not compensate each other at the level of the whole lattice, and the global population density curve exhibits peaks and drops that appear consistent with the ones empirically estimated for South America. These peaks and drops are evident in Figure 9.2, where we have fixed $p_{\alpha,\beta} = 1$ if α and β are neighboring cells, and $p_{\alpha,\beta} = 0$ otherwise.

It should be understood that what is plotted in Figure 9.2 and in [99] cannot be really directly compared: just to cite one major and obvious difference between the two approaches, our results are deduced with a dimensionless model, and what we

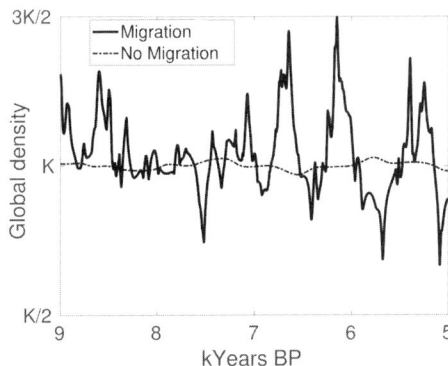

Figure 9.2 Model dynamics in a nine-cell system with and without migration (Source: Bagarello et al. [4]).

plot is what we have called the density of humans. On the other hand, what is plotted in [99] is the result of the reconstruction of the density of population as deduced from some archeological discoveries. However, and quite interestingly, we can still compare the qualitative behaviors of the two curves, and these look quite similar. In particular, we can compare the amplitude of the oscillations in the empirical curve and in the one produced by our model, with respect to the carrying capacity, under the migration scenario: in our model the global population density oscillations belong to the time interval $[0.66\,K, 1.50\,K]$, while in Figure 3(b) in [99] they fall in the $[0.70\,K, 1.57\,K]$ range: they are quite similar. In addition, the number of the most relevant peaks and valleys for the two curves essentially coincide. So, what we get is indeed very close to what is observed in [99], for the time interval considered in our analysis. And we also want to stress once more that the presence of migration is crucial to have such a similarity. Otherwise, see Section 9.3.1, our conclusions would be quite different from the empirical results in [99]. This is also clearly described by the curve in black in Figure 9.2, which is completely different from the other curve, deduced in absence of migration.

A sensitivity analysis can be performed for our model, using a wide range of parameters configurations. We have seen that, in absence of migration, under no condition among those we have considered it is possible to produce the same (or similar) peaks as those observed in [99] and shown also in Figure 9.2. Therefore, in our model migration emerges as a crucial mechanism able both to exacerbate the local oscillations and to keep them synchronized at the global level. This can be observed in Figure 9.3, which shows the dependence of the global human population

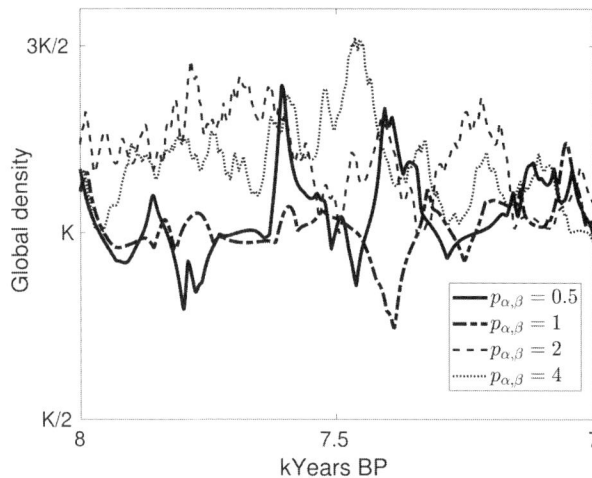

Figure 9.3 Global human population density for different values of the migration parameters $p_{\alpha,\beta}$ (Source: Bagarello et al. [4]).

density in $\mathcal{R}$ for different values of the migration parameters $p_{\alpha,\beta}$. We observe that all the nonzero choices of $p_{\alpha,\beta}$ considered here give rise to peaks and valleys, as in [99]. However, the choice considered in Figure 9.2, where the $p_{\alpha,\beta}$ are simply zero or one, is the one fitting better the empirical data. Incidentally, it can be useful to observe that the time intervals in Figures 9.2 and 9.3 are different, and the curve for $p_{\alpha,\beta} = 1$ in Figure 9.3 coincides with the continuous one in Figure 9.2, but only in the common time interval.

9.4 Conclusions

We have shown that the model discussed in this chapter reproduces well the empirical data, but it must be fine-tuned first, to work at its best. In particular, the analytical form of $H(t)$ and of each of its ingredients is essential: choices that are different from those in Equations (9.18)–(9.21) could not work as well as the ones we have adopted, even if they can still produce interesting results.

We have also seen that migration of $\mathcal{P}_a$ is essential to explain the experimental data, since otherwise the results look very different from those in [99].

From a mathematical point of view, maybe the most interesting aspect of our model is the presence of *hidden nonlinearities* in $H(t)$, which can be handled numerically by solving the Schrödinger equation of motion for the system. To keep the computational time under control, we have been forced to consider small lattices in presence of migration, while no such limitation exists when there is no migration, since all the cells in $\mathcal{R}$ are copies of a single one, and all the cells are disjoint. We have also checked that our conclusions are stable under (small) changes in the size of $\mathcal{R}$, in particular when we increase it.

We refer to [4] for more details, both from the mathematical and from the anthropological sides.

10

A Simple Model of Information in Stock Markets

10.1 Introduction

Quite often, when dealing with macroscopic systems involving humans (or sufficiently evolved organisms), it is not really possible to ignore the role of the information in a procedure of decision-making. We have already met and discussed this
aspect in Chapter 3, and in other applications in [1]. This chapter is devoted to an
analysis of information in connection with simplified stock markets described in
terms of bosonic ladder operators, of the same kind we have considered in [1].

In particular, to avoid useless computations, we will consider an extremely simplified stock market made of just two traders τ_1 and τ_2, and we will try to describe
what happens *before* the trading begins, i.e., in the phase in which the information
begins to circulate in the market and is used by the traders to decide their strategies.
We will show that, within our framework and under our assumptions, in absence of
information the *portfolio operators* of the traders[1] commute with the Hamiltonian
and, for this reason, they stay constant. This suggests that the interesting dynamics
of the system is really due to the information. It is important to stress that, as it will
appear clear in the following discussion, this will be a non-perturbative result.

We will also discuss what happens *after* trading has begun. In this case, however, since the model is nonlinear, and the ladder operators involved in our analysis
satisfy the CCR rather than the CAR,[2] we will be able to deduce only an approximated solution. Nevertheless, even if just approximated, this solution allows us to
get some interesting conclusions concerning the model.

---

[1] Portfolio of trader τ_j is defined here as the sum of the cash and of the value of the shares τ_j possesses, see [1].
[2] This makes the dimensionality of the Hilbert space infinite, and the mathematical treatment of the system is, of course, by far more complicated.

10.2 Before the Trading Begins

What we are interested in this section is the *preparation of the traders*: τ_1 and τ_2 are reached by some information, and they start wondering what they should do when the market opens. Then there is no trading between the two, and for this reason the price of the share (in our oversimplified market, other than having just two traders, we also have only one kind of shares!) is not so important. Here it is fixed to be one. The main ingredients of the system are the following: the *cash*, the *shares*, the *bad-quality information* (BQI) and the *set of external rumors and facts* (ERFs) used by each trader to create the information. Of course, the use of BQI rather than *good-quality information* is just a matter of convenience. In fact, in our setting, it is easier to write the Hamiltonian using BQI rather than its good-quality counterpart. But, in principle, both choices could be equally adopted.

Cash, shares, BQI and ERFs form, all together, the system $\mathcal{S}$. They are represented in terms of different ladder operators that are assumed to obey the CCR. More in detail, c_j, s_j and i_j are the lowering operators related to the cash, number of shares and BQI respectively. The index $j=1,2$ refers to the trader. The family of operators $r_j(k)$, $j = 1, 2$ and $k \in \mathbb{R}$ are lowering operators of the ERFs. Of course, they also depend on a real index, since we are assuming here that ERFs are created by several factors (virtually infinite!) that are modeled by considering a continuous external environment interacting with τ_1 and τ_2, and used by these to increment or lower their BQI, and then to decide their strategies. These operators satisfy the following rules:

$$[s_j, s_k^\dagger] = [c_j, c_k^\dagger] = [i_j, i_k^\dagger] = \mathbb{1}\,\delta_{j,k}, \quad [r_j(k), r_l^\dagger(q)] = \mathbb{1}\,\delta_{j,l}\,\delta(k-q), \quad (10.1)$$

$j, k = 1, 2$, $k, q \in \mathbb{R}$, all the other commutators being zero. Here $\mathbb{1}$ is the identity operator on the Hilbert space $\mathcal{H}$, defined later. Then, introducing the number operators $\hat{K}_j = c_j^\dagger c_j$, $\hat{S}_j = s_j^\dagger s_j$, $\hat{I}_j = i_j^\dagger i_j$ and $\hat{R}_j(k) = r_j^\dagger(k) r_j(k)$, we can write the Hamiltonian of our open stock market as follows:

$$\begin{cases} H = H_0 + H_{int}, \\ H_0 = \sum_{j=1}^{2} (\omega_j^{(s)} \hat{S}_j + \omega_j^{(c)} \hat{K}_j + \Omega_j \hat{I}_j + \int_{\mathbb{R}} \Omega_j^{(r)}(k) \hat{R}_j(k)\, dk), \\ H_{int} = \sum_{j=1}^{2} \left[\lambda_{inf} \left(i_j \left(s_j^\dagger + c_j^\dagger \right) + i_j^\dagger (s_j + c_j) \right) + \gamma_j \int_{\mathbb{R}} (i_j^\dagger r_j(k) + i_j r_j^\dagger(k))\, dk \right], \end{cases}$$

$$(10.2)$$

where all the parameters, $\omega_j^{(s)}$, $\omega_j^{(c)}$, Ω_j, λ_{inf} and γ_j, are real, as well as the functions $\Omega_j^{(r)}(k)$, $j = 1, 2$. Then, $H = H^\dagger$.

The Hilbert space of the full system $\mathcal{S}$, $\mathcal{H}$, is the tensor product of two smaller Hilbert spaces, $\mathcal{H}_{sm}$ and $\mathcal{H}_{res}$, related respectively to the discrete operators $c_j^\sharp$, $s_j^\sharp$ and $i_j^\sharp$ ($\mathcal{H}_{sm}$), and to the operators $r_j^\sharp(k)$ ($\mathcal{H}_{res}$). $\mathcal{H}_{sm}$ is built in analogy to what discussed in Section 2.1, taking the closure of the linear span of all the common eigenstates of the number operators $\hat{K}_j$, $\hat{S}_j$ and $\hat{I}_j$, $j = 1, 2$. They are constructed as usual: we first introduce the vacuum $\varphi_\mathbf{0}$ of the lowering operators[3] c_j, s_j and i_j, $j = 1, 2$, $c_j \varphi_\mathbf{0} = s_j \varphi_\mathbf{0} = i_j \varphi_\mathbf{0} = 0$, then we act on this with (arbitrary) powers of the raising operators $c_j^\dagger$, $s_j^\dagger$ and $i_j^\dagger$ and we finally normalize the result as in (2.2). In this way we get a vector, $\varphi_{S_1,K_1,I_1,S_2,K_2,I_2}$, which we simply indicate as φ_G, with the shorthand notation $G = \{S_1, K_1, I_1, S_2, K_2, I_2\}$. These numbers are exactly the eigenvalues of the operators $\hat{S}_j$, $\hat{K}_j$ and $\hat{I}_j$.

The construction of $\mathcal{H}_{res}$ is more complicated, and not really needed here in all its details. In fact, all we need are the properties of some states defined for the operators on $\mathcal{H}_{res}$, rather than $\mathcal{H}_{res}$ itself. These states will be introduced later on in this section, analogously to what we have done first in Section 2.7 and then in Sections 3.2 and 3.3, for instance. We refer the interested reader to [102] for the explicit construction of $\mathcal{H}_{res}$.[4]

Going back to H, this is the essential ingredient to describe the time evolution of the portfolio $\hat{\Pi}_j$ of the trader $j - th$. This is an operator defined, see [1], in a very simple minded way: $\hat{\Pi}_j = \hat{K}_j + \hat{S}_j$, $j = 1, 2$. In other words, the portfolio is just the global richness of the trader. The aim of our analysis is to deduce $\hat{\Pi}_j(t)$ and its mean value on some suitable state which describes the system at $t = 0$. We will see that this is a difficult task already for our Hamiltonian, which does not appear particularly complicated,[5] and for this reason we will look only for approximated solutions.

The Hamiltonian contains a free *canonical* part H_0, commuting with all the number operators introduced before: $[H_0, \hat{K}_j] = [H_0, \hat{S}_j] = [H_0, \hat{I}_j] = [H_0, \hat{R}_j(k)] = 0$, for all $j = 1, 2$ and for all $k \in \mathbb{R}$. This implies that, if $\lambda_{inf} = \gamma_j = 0$, so that $H = H_0$, nothing particularly interesting happens for these particular observables of the market. The essential part of the Hamiltonian is H_{int}. This contains two different terms. The first term, proportional to λ_{inf}, describes the following: when the BQI increases (because of $i_j^\dagger$), the value of the portfolio decreases (because of $s_j + c_j$) and vice versa. This is because the BQI confuses the trader and surely does not help him to make the right choices. The second term in H_{int}, proportional to γ_j, describes how the ERFs are related to the BQI. Considering, for instance, the contribution $i_j r_j^\dagger(k)$ in H_{int}, this shows that the BQI decreases (so that the trader is *less, but better informed*) when a larger amount of news, rumors, etc. reaches the trader. If the amount of ERFs decreases, the BQI increases, and as we have seen this has direct consequences on the portfolio of the trader. We see that we are considering,

---

[3] Here $\mathbf{0}$ stands for the six-dimensional zero vector $(0, 0, 0, 0, 0, 0)$.

[4] An alternative strategy to construct $\mathcal{H}_{res}$ is to use the GNS construction, see Section 2.6, using the state ω_{res} below as its main ingredient. Again, this goes beyond the scope of this book and will not be discussed further.

[5] Notice that, in particular, H is quadratic.

in fact, two different forms of information. The first one, the ERFs, is the sum of all possible informational inputs arising from everything and everyone surrounding τ_j. For this reason, we use a reservoir (which has an infinite number of degrees of freedom) to model ERFs. Our second kind of information, the BQI, is the result of the processing of the ERFs by τ_j. And, because of this, it is not attached to any environment, but only to τ_j himself.

> *Remark:* We have already discussed in Sections 2.7 and 3.2 that the continuous dependence on k of the operators of the reservoir $r_j^\sharp(k)$ could be replaced by a, maybe more appropriate, discrete one, by making use of the Poisson summation formula along the computations. However, as before, we prefer to work with the continuous version of the reservoir.

Notice also that no interaction between τ_1 and τ_2 is considered in Equation (10.2), in agreement with our working assumptions: we are not interested, for the moment, in describing any trading between the traders, but only the pre-trading phase.

With our choice of H it is easy to check that the operators $\hat{M}_j = \hat{S}_j + \hat{K}_j + \hat{I}_j + \hat{R}_j = \hat{\Pi}_j + \hat{I}_j + \hat{R}_j$, $j = 1, 2$ are preserved during the time evolution. Here $\hat{R}_j = \int_{\mathbb{R}} r_j^\dagger(k) r_j(k)\, dk$. This is because $[H, \hat{M}_j] = 0, j = 1, 2$. Hence what stays constant in time is not the total amount of cash or the global number of shares, as in the closed stock markets analyzed in [1], but it is rather the sum of the portfolio, the BQI and the global ERFs of each single trader. This implies that our model admits two integrals of motion.

> *Remark:* Of course, all these quantities are assumed here to be dimensionless, as always in this book. This is the simplest way to construct *global operators* as sums of operators related to different observables of the system (as the cash and the shares, for instance).

The Heisenberg differential equations of motion for the lowering operators are the following:

$$
\begin{cases}
\frac{d}{dt} s_j(t) = -i\omega_j^s s_j(t) - i\lambda_{inf}\, i_j(t), \\
\frac{d}{dt} c_j(t) = -i\omega_j^c c_j(t) - i\lambda_{inf}\, i_j(t), \\
\frac{d}{dt} i_j(t) = -i\Omega_j i_j(t) - i\lambda_{inf}(s_j(t) + c_j(t)) - i\gamma_j \int_{\mathbb{R}} r_j(k,t)\, dk \\
\frac{d}{dt} r_j(k,t) = -i\Omega_j^{(r)}(k)\, r_j(k,t) - i\gamma_j\, i_j(t).
\end{cases}
\tag{10.3}
$$

To solve the system, we first rewrite the last equation in its integral form:

$$
r_j(k,t) = r_j(k) e^{-i\Omega_j^{(r)}(k)t} - i\gamma_j \int_0^t i_j(t_1) e^{-i\Omega_j^{(r)}(k)(t-t_1)}\, dt_1,
\tag{10.4}
$$

and we replace this in the differential equation for $i_j(t)$. Assuming that $\Omega_j^{(r)}(k)$ is linear in k, $\Omega_j^{(r)}(k) = \Omega_j^{(r)} k$, see Section 2.7 for instance, after some algebra we deduce that

$$\frac{d}{dt}i_j(t) = -\left(i\Omega_j + \frac{\pi\gamma_j^2}{\Omega_j^{(r)}}\right)i_j(t) - i\gamma_j \int_{\mathbb{R}} r_j(k)e^{-i\Omega_j^{(r)}kt}\,dk - i\lambda_{inf}(s_j(t) + c_j(t)).$$

$$(10.5)$$

This differential equation is coupled to the first two equations in (10.3). This makes finding an exact solution of the system rather hard, but not impossible. However, to make the situation simpler, from now on we will assume that the last contribution in Equation (10.5) can be neglected, when compared to the other ones. In other words, we are taking λ_{inf} to be very small. Of course, to make this approximation rigorous, we should check how big the error we do is by neglecting this term concretely. This aspect of our approach will not be discussed here, since it is outside our purposes. However, we should stress that the approximation proposed here is slightly better than simply considering $\lambda_{inf} = 0$ in H in (10.2), since we will still keep track of this term in the first two equations in Equation (10.3). Solving now Equation (10.5) in its simplified expression, and replacing the solution $i_j(t)$ in the first formula in Equation (10.3), we find

$$s_j(t) = e^{-i\omega_j^s t}\left(s_j(0) - i\lambda_{inf}\alpha_j(t)\,i_j(0) - \lambda_{inf}\gamma_j \int_{\mathbb{R}} r_j(k)\,\eta_{2,j}(k,t)\,dk\right), \qquad (10.6)$$

where we have defined

$$\alpha_j(t) = \frac{e^{(i\omega_j^s - \Gamma_j)t} - 1}{i\omega_j^s - \Gamma_j}, \qquad \eta_{2,j}(k,t) = \int_0^t \eta_{1,j}(k,t_1)e^{(i\omega_j^s - \Gamma_j)t_1}\,dt_1,$$

with

$$\Gamma_j = i\Omega_j + \frac{\pi\gamma_j^2}{\Omega_j^{(r)}}, \qquad \eta_{1,j}(k,t) = \frac{e^{(\Gamma_j - i\Omega_j^{(r)}k)t} - 1}{\Gamma_j - i\Omega_j^{(r)}k}.$$

Of course, from Equation (10.3), we understand that a completely analogous solution can be found for $c_j(t)$. The only difference is that ω_j^s should be replaced everywhere by ω_j^c.

The states of the system are constructed as always when we deal with open systems: for each operator of the form $X_{sm} \otimes Y_{res}$, where X_{sm} is an operator of the stock market and Y_{res} an operator of the reservoir, we have

$$\langle X_{sm} \otimes Y_{res}\rangle = \langle\varphi_G, X_{sm}\varphi_G\rangle\,\omega_{res}(Y_{res}),$$

where φ_G has been introduced before, and $\omega_{res}(.)$ is a state satisfying

$$\omega_{res}(\mathbb{1}_{res}) = 1, \quad \omega_{res}(r_j(k)) = \omega_{res}(r_j^\dagger(k)) = 0,$$

$$\omega_{res}(r_j^\dagger(k)r_l(q)) = N_j^{(r)}(k)\,\delta_{j,l}\delta(k - q),$$

for a suitable function $N_j^{(r)}(k)$, see Section 2.7. Also, $\omega_{res}(r_j(k)r_l(q)) = 0$, for all j and l, and for all real k and q. Then $S_j(t) = \langle s_j^\dagger(t)s_j(t)\rangle$ assumes the following expression:

$$S_j(t) = S_j(0) + \lambda_{inf}^2 I_j(0)|\alpha_j(t)|^2 + \lambda_{inf}^2 \gamma_j^2 \int_{\mathbb{R}} N_j^{(r)}(k)|\eta_{2,j}(k,t)|^2 \, dk, \qquad (10.7)$$

where $I_j(0) = \langle i_j^\dagger(0)i_j(0)\rangle = I_j$ and $S_j(0) = S_j$ are fixed by the quantum numbers of φ_G. The expression for $k_j(t) = \langle c_j^\dagger(t)c_j(t)\rangle$ is completely analogous to the one in (10.7), with ω_j^s replaced by ω_j^c, and the portfolio of τ_j, $\pi_j(t) =< \hat{\Pi}_j(t) >$ is simply the sum of $S_j(t)$ and $k_j(t)$. What we are interested in here is the variation of $\pi_j(t)$ over long time scales:

$$\delta\pi_j := \lim_{t,\infty} \pi_j(t) - \pi_j(0).$$

Equation (10.7) shows that, if γ_j is small enough, the integral contribution is not expected to contribute much to $\delta\pi_j$. When this is true, we find that

$$\delta\pi_j \simeq \lambda_{inf}^2 I_j \left(\Omega_j^{(r)}\right)^2 \left(\frac{1}{\pi^2\gamma_j^4 + (\omega_j^s - \Omega_j)^2(\Omega_j^{(r)})^2} + \frac{1}{\pi^2\gamma_j^4 + (\omega_j^c - \Omega_j)^2(\Omega_j^{(r)})^2} \right).$$
$$(10.8)$$

Remark: Formula (10.8) clearly shows the effects of our approximations, and the limit of our model. We cannot really interpret $\delta\pi_j$ as a real increment of the portfolio of τ_j. Otherwise, both τ_1 and τ_2 would only become richer and richer! In fact, $\delta\pi_j \geq 0$ for $j = 1, 2$, independently of the choice of the parameters. This is quite strange, of course, but not really impossible in our model, since $\hat{\Pi}_1 + \hat{\Pi}_2$ is not an integral of motion. However, rather than considering the details of $\delta\pi_j$, what we do next is just to compare $\delta\pi_1$ and $\delta\pi_2$, in order to understand who is the most efficient trader, in our scheme, and why.

What is interesting for us is to consider, at $t = 0$, is the two traders as equivalent: $\omega_1^c = \omega_2^c =: \omega^c$, $\omega_1^s = \omega_2^s =: \omega^s$, $\Omega_1^{(r)} = \Omega_2^{(r)}$ and the initial conditions are $S_1 = S_2$, $K_1 = K_2$ and $I_1 = I_2$. The reason for this choice is that, in this way, we are really focusing on the role of the information on τ_1 and τ_2, which only differ since Ω_1 is taken larger than Ω_2: $\Omega_1 > \Omega_2$.[6] For this reason the traders are identical under many aspects: they differ only because of the inertia of the BQI and, see the analysis which follows, because of the values of the interaction parameters γ_1 and γ_2. To avoid further complications, we limit our considerations to the case in which

---

[6] The case $\Omega_1 < \Omega_2$ can be easily deduced by exchanging the role of Ω_1 and Ω_2. The case $\Omega_1 = \Omega_2$ is not particularly interesting for us, since otherwise τ_1 and τ_2 would be almost indistinguishable.

Ω_1 and Ω_2 are sufficiently larger than ω^c and ω^s. This implies that $|\omega^c - \Omega_2| < |\omega^c - \Omega_1|$ and $|\omega^s - \Omega_2| < |\omega^s - \Omega_1|$.

With this in mind, we will now consider three different cases: (i) $\gamma_1 = \gamma_2$; (ii) $\gamma_1 > \gamma_2$; (iii) $\gamma_1 < \gamma_2$. In this way, we are admitting possible different interaction strengths for the two traders between the environment and the BQI term in *H*.

Let us consider the first situation (i): $\gamma_1 = \gamma_2$. This means that the interaction between BQI and ERFs has the same strength for the two traders. From (10.8), we see that $\delta\pi_1 < \delta\pi_2$. This can be understood as follows: the only difference between τ_1 and τ_2 is in the value of Ω_1 and Ω_2, which describe the inertia of the BQI for the two traders. Since $\Omega_1 > \Omega_2$, and since $I_1 = I_2$, the second trader can lower his value of the BQI, while this is more difficult for τ_1, since his inertia is larger. Hence τ_1 remains badly informed, and he is not able to increment his portfolio as much as τ_2 does.

Of course, if we have both $\gamma_1 > \gamma_2$ and $\Omega_1 > \Omega_2$, case (ii), (10.8) shows that, again and a fortiori, $\delta\pi_1 < \delta\pi_2$. Other than the same *inertia effect* we have just deduced in case (i) for the BQI, the inequality on the interaction parameters γ_1 and γ_2 suggests that the interaction between the BQI and the ERFs is quite delicate: even if this interaction is stronger for τ_1, the presence of this larger inertia for the BQI is sufficient to destroy the possible effect of the ERFs. The presence of the reservoir is like a *second-order effect*. This is probably connected to the various approximations we have introduced along the way.

Case (iii) is different, since the inequalities $\gamma_1 < \gamma_2$ and $\Omega_1 > \Omega_2$ act in an opposite ways in (10.8). What we can say is that, for fixed Ω_1 and Ω_2, there exists some value of (γ_1, γ_2) such that, instead of having $\delta\pi_1 < \delta\pi_2$, we will have exactly the opposite inequality, $\delta\pi_1 > \delta\pi_2$. In other words, in case (iii) we can have both inequalities, depending on the relations between the pairs (Ω_1, Ω_2) and (γ_1, γ_2).

> *Remarks:* (1) It is not difficult to extend the model to more traders. The only difficulty would be to read the results, but writing the Hamiltonian and deducing the equations of motions are easy. However, this is not really relevant for us here, and will not be done. Much more interesting would be to remove some (or all) of the approximations we have introduced in our analysis. This is very hard, but not impossible.
>
> (2) Formulas in Equation (10.3) describe also the dynamics of the direct and reflected information going around the market, that is, of the ERFs and of the BQI. In fact, these could be deduced by looking for the time dependence of $i_j(t)$ and $r_j(k,t)$ and taking the mean values of their related number operators. Here, however, information is not interesting by itself, but because of the consequences produced by the procedure of decision-making of the traders, in particular with regard to the values of their portfolios.

10.3 After the Trading Has Begun

In the previous section we have shown how the Hamiltonian (10.2) produces, out of two originally (i.e., at $t = 0$) almost equivalent traders, two traders who, when the stock market opens, are no longer equivalent. They have used the information they got (and elaborated) to improve their financial condition. In this section we want to discuss what happens when τ_1 and τ_2 are also trading. This will imply the addition of a further term in the Hamiltonian of the market, and this will make an already hard situation even harder. However, we will be able to deduce an interesting non-perturbative result, proving the relevance of the information within the context of our stock market: more explicitly, we will show that, when the information Hamiltonian is zero, the portfolio operators of the traders commute with the Hamiltonian, even in the presence of a trading term. This suggests that the relevant dynamics is really caused by the information: in absence of information, the traders cannot choose any clever strategy, and they simply don't trade, to avoid any risk.

The trading term in the Hamiltonian is written under the assumption that the price of the share is fixed, and it is equal to one. Then we define

$$H_{tr} = \lambda \left(s_1 c_1^\dagger s_2^\dagger c_2 + s_1^\dagger c_1 s_2 c_2^\dagger \right), \tag{10.9}$$

for some real λ. Then

$$H_{full} = H + H_{tr}, \tag{10.10}$$

where H is the Hamiltonian introduced in Equation (10.2). The meaning of H_{tr} is the following: the term $s_1 c_1^\dagger s_2^\dagger c_2$ describes the fact that τ_1 is selling a share to τ_2. For this reason, the number of the shares in his portfolio decreases by one unit (and this is due to the presence of s_1) while his cash increases by one unit (because of $c_1^\dagger$), in agreement with our assumption on the price of the share. After the interaction, τ_2 has one more share (because of $s_2^\dagger$), but one less unit of cash (as a consequence of the action of c_2). Of course, H_{tr} also contains the adjoint contribution, which describes the opposite situation: τ_2 sells a share to τ_1 and gets money because of this operation. This suggests calling $s_j c_j^\dagger$ and $s_j^\dagger c_j$ *the selling and buying operators*, respectively. The interaction parameter λ measures the strength of the interaction between the traders. In particular, if $\lambda = 0$, τ_1 and τ_2 just do not interact, and we go back to the analysis in Section 10.2.

The Hamiltonian H_{full} commutes with the operators $\hat{M}_j = \hat{\Pi}_j + \hat{I}_j + \hat{R}_j$ introduced before, but not with the total cash and number of share operators $\hat{K} = \sum_{j=1}^{2} \hat{K}_j$ and $\hat{S} = \sum_{j=1}^{2} \hat{S}_j$: $[H_{full}, \hat{K}] \neq 0$, $[H_{full}, \hat{S}] \neq 0$. Hence, we are allowing here for bankruptcy or for the possibility that money comes in and out of the market. The market is not closed, and not only for the presence of the informational environment. Moreover, shares can also be added to or withdrawn from the market.

As before, we concentrate on the time evolution of the portfolios of the traders and on their mean values. For that we need to compute the differential equations of motion driving the system, and to solve them (when possible). The system that replaces the one in (10.3) is

$$
\begin{cases}
\frac{d}{dt} s_k(t) = -i\omega_k^s s_k(t) - i\lambda_{inf}\, i_k(t) - i\lambda c_k(t) s_{\overline{k}}^\dagger(t) c_{\overline{k}}^\dagger(t), \\[2mm]
\frac{d}{dt} c_k(t) = -i\omega_k^c c_k(t) - i\lambda_{inf}\, i_k(t) - i\lambda s_k(t) c_{\overline{k}}^\dagger(t) s_{\overline{k}}^\dagger(t), \\[2mm]
\frac{d}{dt} i_k(t) = -i\Omega_k i_k(t) - i\lambda_{inf}(s_k(t) + c_k(t)) - i\gamma_k \int_{\mathbb{R}} r_k(q,t)\, dq \\[2mm]
\frac{d}{dt} r_k(q,t) = -i\Omega_k^{(r)}(q)\, r_k(q,t) - i\gamma_k\, i_k(t).
\end{cases}
\tag{10.11}
$$

Here we use the following notation: $\overline{k} = 1$ if $k = 2$, while $\overline{k} = 2$ if $k = 1$. With respect to the equations deduced in Equation (10.3), we see that two nonlinear contributions appear in the first two equations above, as a consequence of H_{tr}. This is because H_{tr} is quartic in the ladder operators. Not surprisingly, we are not able to solve the system exactly.[7] Still, we will deduce an approximated solution for the system, and we will comment on that.

However, before finding this approximated solution, something interesting (and non-perturbative) can be deduced in absence of information, i.e., when we fix $\lambda_{inf} = 0$ in (10.10). In this case the full Hamiltonian can be written as

$$
H_{full} = H_0 + H_{tr} + \sum_{j=1}^{2} \gamma_j \int_{\mathbb{R}} (i_j^\dagger r_j(k) + i_j r_j^\dagger(k))\, dk.
\tag{10.12}
$$

Then, it is easy to check that $[H_{full}, \hat{\Pi}_j] = 0$, $j = 1, 2$. Hence, $\hat{\Pi}_j(t) = \Pi_j(0)$ for all $t \in \mathbb{R}$: even if the cash and the shares of the two traders may change in time, their portfolios do not. This result looks reasonable since, as already noticed, our traders do not receive information from outside the market, and then they are not in a good position to decide what to do (*Sell or buy? This is the problem! And I have no way to decide...*). This conclusion is confirmed further by noticing that, when $\lambda_{inf} \neq 0$, the portfolio operators do not commute anymore with H_{full}:

$$
\left[H_{full}, \hat{\Pi}_j \right] = \lambda_{inf} \left(i_j^\dagger (s_j + c_j) - i_j (s_j^\dagger + c_j^\dagger) \right),
$$

$j = 1, 2$. This makes the dynamics of $\hat{\Pi}_j(t)$, in this case, nontrivial. Hence, it is exactly the presence of λ_{inf} that makes the model interesting, while the interaction between τ_1 and τ_2, see H_{tr}, has no effect in this computation.

> *Remark:* The Hamiltonian in Equation (10.12) contains both ERFs and BQI. Its analytic expression shows that, even if we take $\lambda_{inf} = 0$, the BQI still appears in H through the contribution proportional to γ_j, see (10.2). The reason why this is not

---

[7] This is obvious: we already had serious difficulties in finding an exact solution when $\lambda = 0$. So why should we be able to do that when $\lambda \neq 0$?

enough to affect the dynamics of the portfolios of the traders is because each $\hat{\Pi}_j$ can be modified, in our model, only because of the effect of the term $i_j(s_j^\dagger + c_j^\dagger)$, and of its Hermitian conjugate in H_{int} (see Equation [10.2]). When the term responsible of this change is missing, which is the case when $\lambda_{inf} = 0$, this effect simply disappears, and $\hat{\Pi}_j$ must stay constant.

Of course, this feature of the model is strongly related to the fact that, in our present assumptions, the price of the share does not change in time. In fact, when this working assumption is removed, then the portfolio of, say, τ_1, should be defined in a more appropriate way as $\hat{\Pi}_1(t) := \hat{K}_1(t) + \hat{P}(t)\hat{S}_1(t)$, $\hat{P}(t)$ being the value of the share at time t. In general, this operator does not need to commute anymore with H, even when $\lambda_{inf} = 0$, and for this reason it would no longer be constant.

Let us now go back to system in Equation (10.11), and let's try to solve it, assuming that $\lambda_{inf} \neq 0$, not to make the dynamics trivial. Since the last two formulas in Equation (10.11) coincide with those in Equation (10.3), we repeat here what we did first in Section 10.2. In particular, we get for $i_j(t)$ the same formula as in Equation (10.5), and as before we assume λ_{inf} is small enough to replace Equation (10.5) with the following approximated equation:

$$\frac{d}{dt}i_j(t) = -\left(i\Omega_j + \frac{\pi\gamma_j^2}{\Omega_j^{(r)}}\right)i_j(t) - i\gamma_j \int_{\mathbb{R}} r_j(k)e^{-i\Omega_j^{(r)}kt}\,dk, \tag{10.13}$$

whose solution is

$$i_j(t) = e^{-\left(i\Omega_j + \frac{\pi\gamma_j^2}{\Omega_j^{(r)}}\right)t}\left(i_j(0) - i\gamma_j \int_{\mathbb{R}} r_j(q)\rho_j(q,t)\,dq\right), \tag{10.14}$$

where

$$\rho_j(q,t) = \int_0^t e^{\left[i(\Omega_j - \Omega_j^{(r)}q) + \frac{\pi\gamma_j^2}{\Omega_j^{(r)}}\right]t_1}\,dt_1 = \frac{e^{\left[i(\Omega_j - \Omega_j^{(r)}q) + \frac{\pi\gamma_j^2}{\Omega_j^{(r)}}\right]t} - 1}{i\left(\Omega_j - \Omega_j^{(r)}q\right) + \frac{\pi\gamma_j^2}{\Omega_j^{(r)}}}.$$

The differential equations for $s_j(t)$ and $c_j(t)$ can now be deduced by replacing $r_j(k,t)$ in (10.4) and $i_j(t)$ in Equation (10.14) back in system (10.11), and they look now as follows:

$$\begin{cases} \dot{s}_1(t) = -i\omega_1^s s_1(t) - i\lambda c_1(t)s_2(t)c_2^\dagger(t) - i\lambda_{inf}\, i_1(t), \\ \dot{s}_2(t) = -i\omega_2^s s_2(t) - i\lambda c_2(t)s_1(t)c_1^\dagger(t) - i\lambda_{inf}\, i_2(t), \\ \dot{c}_1(t) = -i\omega_1^c c_1(t) - i\lambda s_1(t)c_2(t)s_2^\dagger(t) - i\lambda_{inf}\, i_1(t), \\ \dot{c}_2(t) = -i\omega_2^c c_2(t) - i\lambda s_2(t)c_1(t)s_1^\dagger(t) - i\lambda_{inf}\, i_2(t). \end{cases} \tag{10.15}$$

We see that, even if we have neglected the term proportional to λ_{inf} when going from Equation (10.5) to Equation (10.13), this is not the same as taking $\lambda_{inf} = 0$ in H_{full} from the very beginning, as we have discussed before, since this would make the model not particularly interesting from a dynamical point of view. The fact that λ_{inf} still plays an important role is evident, since the equations in Equation (10.15) depend explicitly on λ_{inf}. We also observe that the system in Equation (10.15) is still nonlinear, despite our approximations. Similar nonlinearities were considered in Section 3.5, but in a much simpler fermionic setting, for which it was still possible to deduce an exact solution of the equations of motion. This is not the case here, since the Hilbert space for the traders is infinite-dimensional.

> *Remark:*   So far, the order of the various operators appearing in the right-hand side of these equations is not important since they all commute between them, at equal time: $[c_1(t), s_1^\dagger(t)] = [s_1(t), s_2(t)] = 0$, for all $t \in \mathbb{R}$, and so on. This is a consequence of the analogous commutation rule at $t = 0$, and of the fact that the time evolution is unitarily implemented by H: $X(t) = e^{iHt}X(0)e^{-iHt}$, for each dynamical variable X. This commutativity will be lost from the approximated solutions of Equation (10.15), as we will see.

10.3.1 The Perturbative Solution of the Equations

Since we are not able to solve Equation (10.15) analytically, we need to use some perturbation scheme. The one we adopt works as follows. First of all, we define new variables $\sigma_j(t) := s_j(t)e^{i\omega_j^s t}$ and $\theta_j(t) := c_j(t)e^{i\omega_j^c t}$, $j = 1, 2$. To simplify the treatment a little bit, we also assume that $\lambda = \lambda_{inf}$. From an economical point of view, this simply means we are assuming that the interaction and the information terms in H have the same strength. Then Equations (10.15) become

$$\begin{cases} \dot{\sigma}_1(t) = -i\lambda \left(\sigma_2(t)\theta_1(t)\theta_2^\dagger(t)e^{i\hat{\omega}t} + i_1(t)e^{i\omega_1^s t} \right), \\ \dot{\sigma}_2(t) = -i\lambda \left(\sigma_1(t)\theta_1^\dagger(t)\theta_2(t)e^{-i\hat{\omega}t} + i_2(t)e^{i\omega_2^s t} \right), \\ \dot{\theta}_1(t) = -i\lambda \left(\sigma_1(t)\sigma_2^\dagger(t)\theta_2(t)e^{-i\hat{\omega}t} + i_1(t)e^{i\omega_1^c t} \right), \\ \dot{\theta}_2(t) = -i\lambda \left(\sigma_1^\dagger(t)\sigma_2(t)\theta_1(t)e^{i\hat{\omega}t} + i_2(t)e^{i\omega_2^c t} \right), \end{cases} \qquad (10.16)$$

where $\hat{\omega} = \omega_1^s - \omega_2^s - \omega_1^c + \omega_2^c$. Now we solve (10.16) iteratively, assuming that λ is small. The zeroth approximation in λ is quite simple, since is deduced by taking $\lambda = 0$: hence $\dot{\sigma}_j^{(0)}(t) = \dot{\theta}_j^{(0)}(t) = 0$, for $j = 1, 2$. Therefore, with obvious notation, $\sigma_j^{(0)}(t) = \sigma_j^{(0)}(0) = s_j$ and $\theta_j^{(0)}(t) = \theta_j^{(0)}(0) = c_j$, $j = 1, 2$. Next we insert these results in the right-hand side of system (10.16) to deduce the first order approximation for $\sigma_j(t)$ and $\theta_j(t)$. By introducing the new operators

$$I_j^s(t) := \int_0^t i_j(t_1)e^{i\omega_j^s t_1}\,dt_1, \qquad I_j^c(t) := \int_0^t i_j(t_1)e^{i\omega_j^c t_1}\,dt_1,$$

which can be computed from (10.14), considering the initial conditions $\sigma_j^{(1)}(0) = s_j$, $\theta_j^{(1)}(0) = c_j$, and assuming that $\hat{\omega} \neq 0$, we get

$$
\begin{cases}
\sigma_1^{(1)}(t) = s_1 - \frac{\lambda}{\hat{\omega}}\left(e^{i\hat{\omega}t} - 1\right) s_2 c_1 c_2^{\dagger} - i\lambda I_1^s(t), \\
\sigma_2^{(1)}(t) = s_2 + \frac{\lambda}{\hat{\omega}}\left(e^{-i\hat{\omega}t} - 1\right) s_1 c_1^{\dagger} c_2 - i\lambda I_2^s(t), \\
\theta_1^{(1)}(t) = c_1 + \frac{\lambda}{\hat{\omega}}\left(e^{-i\hat{\omega}t} - 1\right) s_1 s_2^{\dagger} c_2 - i\lambda I_1^c(t), \\
\theta_2^{(1)}(t) = c_2 - \frac{\lambda}{\hat{\omega}}\left(e^{i\hat{\omega}t} - 1\right) s_1^{\dagger} s_2 c_1 - i\lambda I_2^c(t).
\end{cases}
\tag{10.17}
$$

These operators give a better approximation of the exact solution of Equation (10.16), than their zeroth order versions. Unfortunately, we cannot be satisfied with this result. In fact, to deduce the (classical) functions $n_j(t)$ and $k_j(t)$, describing the number of shares and the cash of τ_j at time t, we have to compute the mean values[8]

$$
n_j(t) := \left\langle \varphi_{\mathcal{G}_0}, s_j^{\dagger}(t) s_j(t) \varphi_{\mathcal{G}_0} \right\rangle
$$

$$
= \left\langle \varphi_{\mathcal{G}_0}, \sigma_j^{\dagger}(t) \sigma_j(t) \varphi_{\mathcal{G}_0} \right\rangle \simeq \left\langle \varphi_{\mathcal{G}_0}, (\sigma_j^{(1)}(t))^{\dagger} \sigma_j^{(1)}(t) \varphi_{\mathcal{G}_0} \right\rangle,
$$

and

$$
k_j(t) := \left\langle \varphi_{\mathcal{G}_0}, c_j^{\dagger}(t) c_j(t) \varphi_{\mathcal{G}_0} \right\rangle \simeq \left\langle \varphi_{\mathcal{G}_0}, (\theta_j^{(1)}(t))^{\dagger} \theta_j^{(1)}(t) \varphi_{\mathcal{G}_0} \right\rangle.
$$

Here the vector $\varphi_{\mathcal{G}_0}$ is

$$
\varphi_{\mathcal{G}_0} = \frac{1}{\sqrt{n_1! n_2! k_1! k_2! I_1! I_2!}} \left(s_1^{\dagger}\right)^{n_1} \left(s_2^{\dagger}\right)^{n_2} \left(c_1^{\dagger}\right)^{k_1} \left(c_2^{\dagger}\right)^{k_2} \left(i_1^{\dagger}\right)^{I_1} \left(i_2^{\dagger}\right)^{I_2} \varphi_{\mathbf{0}},
$$

and $\varphi_{\mathbf{0}}$ is the vacuum of s_j, c_j and i_j: $s_j \varphi_{\mathbf{0}} = c_j \varphi_{\mathbf{0}} = i_j \varphi_{\mathbf{0}} = 0$, $j = 1, 2$, as in Section 10.2. The explicit choice of the numbers n_1, n_2, k_1, k_2, I_1 and I_2 depends, of course, on the original (i.e., at $t = 0$) status of the two traders: for example, n_1 is the number of share that τ_1 has at $t = 0$, k_1 are the units of cash in his portfolio, at this same time, while I_1 is what measures the amount of his BQI. Easy computations show that, at this order in λ, $n_j(t) = n_j(0) = n_j$ and $k_j(t) = k_j(0) = k_j$, so that each portfolio stays constant in time: $i_j(t) = n_j(t) + k_j(t) = \pi_j(0)$, $j = 1, 2$. The conclusion is therefore that, if we want to get some nontrivial dynamics, we need to go, at least, at the second order in the perturbation expansion.

This second-order approximation is deduced in the same way: we replace the first-order solution we have just found in the right-hand side of system Equation (10.16), and then we simply integrate on time, requiring now that $\sigma_j^{(2)}(0) = s_j$ and $\theta_j^{(2)}(0) = c_j$. However, we need to stress that, as already stated, because of this approximation we get problems of ordering of the operators. In fact, as we have already discussed, while $\sigma_2(t)\theta_1(t)\theta_2^{\dagger}(t) = \theta_1(t)\sigma_2(t)\theta_2^{\dagger}(t) = \sigma_2(t)\theta_1(t)\theta_2^{\dagger}(t)$, these

---

[8] In the formulas for $n_j(t)$ and $k_j(t)$ here we are focusing on the role of the system of the traders. In fact, we should also add the contribution ω_{res}, which is relevant due to the presence of $I_j^s(t)$ and $I_j^c(t)$. We prefer this simpler notation. Incidentally we observe that we are calling here $n_j(t)$ what, in (10.7) was called $S_j(t)$. This is to stress the difference between the two cases: no trading versus trading.

equalities are false when we replace the operators with their first- or second-order approximations. This can be checked by a direct computation, already at the first order. Using (10.17) we see, in fact, that, for instance

$$\left[\sigma_1^{(1)}(t), \sigma_2^{(1)}(t)\right] = -2\left(\frac{\lambda}{\hat{\omega}}\right)^2 (1 - \cos(\hat{\omega}\,t))\, s_1 s_2 \left(-c_1 c_1^\dagger + c_2 c_2^\dagger\right),$$

which is nonzero. For this reason we adopt here the following *normal ordering rule*: every time we have products of operators, we order them considering first s_1 or $s_1^\dagger$, then s_2 or $s_2^\dagger$, c_1 or $c_1^\dagger$ and, finally, c_2 or $c_2^\dagger$. In particular the equations in Equation (10.17) are already written in this normal-ordered form. Needless to say, this is an arbitrary choice and needs not to be, in principle, the best option. However, the use of normal ordering procedures is rather common in quantum mechanics for systems with infinite degrees of freedom, and for this reason we have adopted a similar procedure here.

After some lengthy but straightforward computations we get the following results:

$$\begin{cases} \sigma_1^{(2)}(t) = s_1 - i\lambda \left(-i\eta_1(t)X_1 + I_1^s(t)\right) - i\lambda^2 \left(Q_1(t) + \overline{\eta_2(t)}\, Y_1\right) \\ \sigma_2^{(2)}(t) = s_2 - i\lambda \left(i\overline{\eta_1(t)}\, X_2 + I_2^s(t)\right) - i\lambda^2 \left(Q_2(t) + \eta_2(t)Y_2\right) \\ \theta_1^{(2)}(t) = c_1 - i\lambda \left(i\overline{\eta_1(t)}\, X_3 + I_1^c(t)\right) - i\lambda^2 \left(Q_3(t) + \eta_2(t)Y_3\right) \\ \theta_2^{(2)}(t) = c_2 - i\lambda \left(-i\eta_1(t)X_4 + I_2^c(t)\right) - i\lambda^2 \left(Q_4(t) + \overline{\eta_2(t)}\, Y_4\right), \end{cases} \qquad (10.18)$$

where we have introduced the following quantities, needed to simplify the notation:

$$\eta_1(t) := \frac{e^{i\hat{\omega}t} - 1}{\hat{\omega}}, \qquad \eta_2(t) = \int_0^t \eta_1(t_1)e^{-i\hat{\omega}t_1}dt_1 = \frac{1}{\hat{\omega}}\left(t - i\overline{\eta_1(t)}\right),$$

$$X_1 := s_2 c_1 c_2^\dagger, \quad X_2 := s_1 c_1^\dagger c_2, \quad X_3 := s_1 s_2^\dagger c_2, \quad X_4 := s_1^\dagger s_2 c_1,$$

$$Y_1 := s_1(c_1^\dagger c_1 c_2 c_2^\dagger + s_2 s_2^\dagger c_2 c_2^\dagger - c_1 c_1^\dagger s_2 s_2^\dagger), \quad Y_2 := s_2(-s_1 s_1^\dagger c_1^\dagger c_1 + s_1 s_1^\dagger c_2^\dagger c_2 - c_1 c_1^\dagger c_2^\dagger c_2),$$

$$Y_3 := c_1(s_1 s_2^\dagger c_2^\dagger c_2 - s_2 s_2^\dagger c_2^\dagger c_2 - s_1 s_1^\dagger s_2^\dagger s_2), \quad Y_4 := c_2(s_1^\dagger s_1 s_2 s_2^\dagger + s_1^\dagger s_1 c_1^\dagger c_1 - s_2^\dagger s_2 c_1^\dagger c_1),$$

as well as the following time-dependent operators:

$$G_1(t) := -i\left(-s_2 c_1 I_2^c(t)^\dagger + s_2 c_2^\dagger I_1^c(t) + c_1 c_2^\dagger I_2^s(t)\right),$$

$$G_2(t) := -i\left(s_1 c_1^\dagger I_2^c(t) - s_1 c_2 I_1^c(t)^\dagger + c_1^\dagger c_2 I_1^s(t)\right),$$

$$G_3(t) := -i\left(s_1 s_2^\dagger I_2^c(t) + s_2^\dagger c_2 I_1^s(t) - s_1 c_2 I_2^s(t)^\dagger\right),$$

$$G_4(t) := -i\left(s_1^\dagger s_2 I_1^c(t) + s_1^\dagger c_1 I_2^s(t) - s_2 c_1 I_1^s(t)^\dagger\right),$$

and

$$Q_j(t) := \begin{cases} \int_0^t G_j(t_1)e^{i\hat{\omega}t_1}dt_1, & j = 1, 4, \\ \int_0^t G_j(t_1)e^{-i\hat{\omega}t_1}dt_1, & j = 2, 3. \end{cases}$$

We can now compute the mean values of $\sigma_j^{(2)\dagger}(t)\sigma_j^{(2)}(t)$ and $\theta_j^{(2)\dagger}(t)\theta_j^{(2)}(t)$ on the state $\langle \varphi_{\mathcal{G}_0}, \cdot \varphi_{\mathcal{G}_0} \rangle$. Another approximation is adopted at this stage: Equation (10.14) shows that the contribution of the reservoir, $i\gamma_j \int_{\mathbb{R}} r_j(q)\rho_j(q,t)\,dq$, is $O(\gamma_j)$ with respect to the other contribution, $i_j(0)$. For this reason, assuming that γ_j is small enough, we approximate $i_j(t)$ with $e^{-\left(i\Omega_j + \frac{\pi\gamma_j^2}{\Omega_j^{(r)}}\right)t} i_j(0)$.

Then, up to the second order in λ, we get

$n_1(t)$

$$\simeq n_1 + \frac{2\lambda^2}{\hat{\omega}^2}(1-\cos(\hat{\omega}t))[n_1(k_1n_2-k_1k_2-n_2k_2-k_2)+n_2k_1(1+k_2)] + \lambda^2 I_1 |\eta_3^s(t)|^2,$$

$n_2(t)$

$$\simeq n_2 + \frac{2\lambda^2}{\hat{\omega}^2}(1-\cos(\hat{\omega}t))[n_2(n_1k_2-k_1k_2-k_1n_1-k_1)+n_1k_2(1+k_1)] + \lambda^2 I_2 |\eta_4^s(t)|^2,$$

$k_1(t)$

$$\simeq k_1 + \frac{2\lambda^2}{\hat{\omega}^2}(1-\cos(\hat{\omega}t))[k_1(n_1k_2-n_1n_2-n_2k_2-n_2)+n_1k_2(1+n_2)] + \lambda^2 I_1 |\eta_3^c(t)|^2,$$

$k_2(t)$

$$\simeq k_2 + \frac{2\lambda^2}{\hat{\omega}^2}(1-\cos(\hat{\omega}t))[k_2(k_1n_2-n_1n_2-n_1k_1-n_1)+k_1n_2(1+n_1)] + \lambda^2 I_2 |\eta_4^c(t)|^2.$$

These results confirm, in particular, that the first nontrivial contributions to $n_j(t)$ and $k_j(t)$ in our perturbation scheme are quadratic in λ. Here we have defined

$$\eta_k^s(t) = \int_0^t e^{\left[i(\omega_{k-2}^s - \Omega_{k-2}) - \frac{\pi\gamma_{k-2}^2}{\Omega_{k-2}^{(r)}}\right]t_1}\,dt_1 = \frac{e^{\left[i(\omega_{k-2}^s - \Omega_{k-2}) - \frac{\pi\gamma_{k-2}^2}{\Omega_{k-2}^{(r)}}\right]t} - 1}{i(\omega_{k-2}^s - \Omega_{k-2}) - \frac{\pi\gamma_{k-2}^2}{\Omega_{k-2}^{(r)}}},$$

for $k = 3,4$. The other function $\eta_k^c(t)$ is identical to $\eta_k^s(t)$, with the only difference that ω_{k-2}^s is replaced by ω_{k-2}^c. If we now compute the variation of the portfolios, $\delta\pi_j(t) := \pi_j(t) - \pi_j(0)$, we find that

$$\delta\pi_1(t) = \lambda^2 I_1 \left(|\eta_3^s(t)|^2 + |\eta_3^c(t)|^2\right), \quad \delta\pi_2(t) = \lambda^2 I_2 \left(|\eta_4^s(t)|^2 + |\eta_4^c(t)|^2\right). \quad (10.19)$$

The limit of these expressions is evident: first both $\delta\pi_1(t)$ and $\delta\pi_2(t)$ are positive, for each t, which is quite unlikely, for the same reasons discussed in Section 10.2, in absence of trading, where a similar result was also deduced. Secondly, we see that the only *quantum numbers* that appear in the formulas are I_1 and I_2, which are related to the BQI of the two traders. But the perturbative result suggests that, the higher this value, the higher the related value of $\delta\pi_l$. This is not what we can expect, since the BQI should not be able to help one trader to raise the value of his portfolio more than the other, which is reached by a lower level of BQI. For these reasons formulas in Equation (10.19) can be considered only as the starting point for a deeper analysis, in which, for instance, we should consider higher powers in λ or propose some different approximation scheme.

However, already at this level it is interesting to notice that, because of the analytical expressions for $\eta_3^s(t)$ etc.,

$$|\eta_3^s(t)|^2 = \frac{e^{-\frac{2\pi\gamma_1^2}{\Omega_1^{(r)}}t} - 2e^{-\frac{\pi\gamma_1^2}{\Omega_1^{(r)}}t}\cos(\omega_1^s - \Omega_1)t + 1}{(\omega_1^s - \Omega_1)^2 + \frac{\pi^2\gamma_1^4}{\Omega_1^{(r)2}}},$$

both $\delta\pi_1(t)$ and $\delta\pi_2(t)$ admit a nontrivial asymptotic value, $\delta\pi_j(\infty) = \lim_{t,\infty} \delta\pi_j(t)$. In fact, using the above formula for $|\eta_3^s(t)|^2$ and the analogous formulas for $|\eta_3^c(t)|^2$, $|\eta_4^s(t)|^2$ and $|\eta_4^c(t)|^2$, we get

$$\delta\pi_1(\infty) = \lambda^2 I_1 \left(\frac{1}{(\omega_1^s - \Omega_1)^2 + \frac{\pi^2\gamma_1^4}{\Omega_1^{(r)2}}} + \frac{1}{(\omega_1^c - \Omega_1)^2 + \frac{\pi^2\gamma_1^4}{\Omega_1^{(r)2}}} \right), \tag{10.20}$$

and

$$\delta\pi_2(\infty) = \lambda^2 I_2 \left(\frac{1}{(\omega_2^s - \Omega_2)^2 + \frac{\pi^2\gamma_2^4}{\Omega_2^{(r)2}}} + \frac{1}{(\omega_2^c - \Omega_2)^2 + \frac{\pi^2\gamma_2^4}{\Omega_2^{(r)2}}} \right), \tag{10.21}$$

which essentially coincide with formulas in Equation (10.8). The first evident conclusion is that $\delta\pi_1(\infty) + \delta\pi_2(\infty) \neq 0$. This is possible, in principle, since the total amount of cash and the total number of shares are not required to be constant in time, in our model. Therefore, there is no reason to expect that the gain for τ_1 become the loss for τ_2, or vice versa. However, as we have already observed in Section 10.2, the fact that both $\delta\pi_1(\infty)$ and $\delta\pi_2(\infty)$ are positive is still more evidence of the fact that our model, or more probably the approximations introduced to deal with it, needs to be further refined.

10.4 Conclusions

This chapter was devoted to the analysis of a simple (but not too simple!) stock market in the presence of information of two different kinds: a *bad information*, which is directly related to the variables i_j, $i_j^\dagger$ and $\hat{I}_j$ and, therefore, is a character-istic of trader τ_j, and a *good one*, which is related to the environment, i.e., to the variables $r_j(q)$, $r_j^\dagger(q)$ and $\hat{R}_j(q)$. The differentiation of information into good and bad, or between *private* and *public*, can be found back in early work in finance. For instance, the so-called "Kyle measure" [103] was proposed to give an indication of how the level of private information compares to the level of so-called noise trading (which could be caused by bad information).

Another possibility to include information in stock markets by adopting quantum mechanical ideas has been explored in terms of Bohmian quantum mechanics, in which the information is associated with a pilot wave function satisfying a certain Schrödinger equation. In this way one gets a *mental force*, which is added to the other *hard forces* acting on the market. Together, these forces produce a Newton-like equation of motion driving the dynamics of the system. We refer to [104] for the details of this approach.

Our analysis was divided into two parts: we first considered what happens in the absence of trading, while in Section 10.3 we also considered the effect of this trading in the dynamics of the system. In both cases, to deduce an analytical form of the portfolios of the traders, we introduced several approximations which, un-fortunately, make the results of our analysis not particularly realistic. However, we still believe that the results of this chapter have a certain interest and can be further refined, especially in its linear version. We also believe that our non-perturbative result given in Section 10.3, on the relevance of information in the time evolution of the portfolios, is an interesting aspect of the model, which can be taken as the start-ing point for a new, and more detailed, analysis of financial systems in presence of information.

We refer to [105, 106] for more results and comments on this model and for other, slightly different systems, in which the information is treated using some alternative (but related) strategy.

11

Decision-Making Driven by the Environment

11.1 Introduction

Very often, in daily life and in extremely different conditions, we understand how difficult making some decision can be. Therefore, it is not really surprising to see that describing, or modeling, a procedure of decision-making can be rather hard. However, the literature is quite rich in different contributions and proposals on this topic, and this is because applications are relevant in several fields of social life, in a broad sense. Decision-making is relevant in game theory, in economics, in population dynamics and in biology as well as in many other fields of science. We refer to [107–113] as a partial list of contributions, all based on quantum ideas. More applications can be found, for instance, in [114–116], where, in particular, the authors introduce some suitable *utility function*, which has to be maximized during the procedure of decision-making.

We adopt here a different approach, following the same strategy already proposed in Chapters 3 and 10. Our idea is that a decision is reached after interaction with some environment, which can be *physical*, as in Chapter 3, or *informational*, as in Chapter 10, or *mental*, as in this chapter.

This same idea is shared by other authors who model the interaction with the environment, and the related dynamics, by means of some suitable master equation depending on very few parameters, [117], or by some Hamiltonian-like finite-dimensional matrix that produces a time evolution *restricted* to some finite time interval, defined by some suitable *decision time* [67], which can be understood as that particular instant of time in which, independently of other aspects of the system, the agent has to make its decision. These two possibilities are useful, since they are easy enough to produce explicit solutions, but, in our opinion, are in fact too easy: it is quite unlucky that the input of a huge environment can be modeled by just two or three parameters, and the explicit choice of the decision time, as we have already discussed in this book, appears to be essential in the deduction of concrete results: different choices can produce really different conclusions. For these reasons, we strongly prefer to use environments as the basic source of our modeling approach for creating decisions, and we want these environments to be described in some sufficiently detailed way, that is, by some Hamiltonian that describes, at least, the essential ingredients of the environments themselves. However, we further require this description to be sufficiently simple to produce analytical (or numerical) results, (possibly) in agreement with what one could expect, or with the experimental data, if any. These two requirements have already driven our choice of the Hamiltonians of the reservoirs in our previous applications, as in Chapter 3, where the reservoirs were used to model electors, and the Hamiltonian was taken to be quadratic in the ladder operators.

In this chapter we will consider the role of the reservoir more at a general level within a generic procedure of decision-making.

11.2 The Environment and Its Hamiltonian

As already stated, in the model presented in this section, agents are not aiming to maximize their utility (which is not even defined, here), but they instead adapt their behavior to information gained from other agents and from the environment. To describe this process of adaptivity, we adopt the strategy proposed in Section 2.7. In particular, the *environment-adaptivity* of the agents will be encoded in the dynamics of *quantum decision operators*, dynamics driven, as usual, by a suitable Hamiltonian describing the interactions between the various agents of the system $\mathcal{S}$, and their interactions with the environment, whose meaning will be discussed later. This is achieved by considering the agents as a part of an open quantum system.

In this section we will study a very general model in decision-making presented in the form of a game played by two players $\mathcal{G}_1$ and $\mathcal{G}_2$ interacting with two environments, $\mathcal{R}_1$ and $\mathcal{R}_2$. These players can be agents of a market, e.g., corporations or traders of a stock market, political parties or yet ingredients of some biological or social system. A decision is taken only after the interactions $\mathcal{G}_j \leftrightarrow \mathcal{R}_j$ and $\mathcal{G}_1 \leftrightarrow \mathcal{G}_2$ occur. The reflections of the agents generated by these interactions modify the agents' mental states, driving them toward some decision.

Quite often, in the applications discussed so far, see in particular Sections 3.2 and 3.3, at the beginning of the process of decision-making, $t = 0$, each agent is in a *sharp* or (with a slight abuse of language) *pure state*. This means that each agent knows, at $t = 0$, exactly which is his own choice. This is expressed by the fact that the vector of the parties is a common eigenstate of the number operators $\hat{P}_j(t)$ at $t = 0$, $j = 1, 2, 3$. However, while interacting with other agents or with the environments, his initial choice could be modified, and the agent have to reconsider again what to do.

Of course, the fact that the agent's decision is not affected by any uncertainty at $t = 0$ is a simplifying assumption that often does not match reality. For instance, when we consider traders in a financial market, they might have no clear idea of what to do when the market opens, especially in the absence of information, see Chapter 10. Or, a political party can be inclined to form some alliance, but not completely sure that this would be the best choice, see Chapter 3.

For this reason we present here a model in which some uncertainty can be present already at the beginning of the story. This is different from what we have done in many applications along the years and also in this book, where the agents are often assumed to be, at $t = 0$, in some eigenstate of a suitable complete set of observables of the system. Here, rather than this, we will consider the possibility that, initially, an agent is also being in a state of uncertainty. From a mathematical point of view, this just means that the agent is represented as a superposition of quantum(-like) pure states corresponding to the sharp choices. Such a superposition encodes a very deep uncertainty that cannot be modeled in the framework of classical probability [20, 22]. We will show later that the original uncertainty of decisions induces a kind of interference between possible decisions of agents. In fact, we will see that, in comparison with what found in [15] and [12], see also Chapter 3, the dynamical behavior of what we call the decision functions changes drastically. In particular, a sort of overall noise appears because of the presence of some interference effects in the system connected to the initial uncertainty.

The two players of our game, $\mathcal{G}_1$ and $\mathcal{G}_2$, can operate, at $t = 0$, only two choices, *zero* and *one*. Hence, we have four different initial possibilities, which we associate with four different, and mutually orthogonal, vectors in a four-dimensional Hilbert space $\mathcal{H}_\mathcal{G}$. These vectors are $\varphi_{0,0}$, $\varphi_{1,0}$, $\varphi_{0,1}$ and $\varphi_{1,1}$. The first vector, $\varphi_{0,0}$, describes the fact that, at $t = 0$, the two players have both chosen zero. Sometimes, in the literature on the two-prisoners dilemma [108, 117], rather than $\varphi_{0,0}$ we find the following equivalent notation: $0_1 0_2$. Of course, such a choice can change during the time evolution of the system. Analogously, $\varphi_{0,1}$ describes the fact that, at $t = 0$, the first player has chosen zero, while the second has chosen one. This corresponds to $0_1 1_2$, in the notation of [108, 117] and so on. We construct these states by using two fermionic ladder operators, i.e., two operators b_1 and b_2, satisfying the following CAR:

$$\left\{b_k, b_l^\dagger\right\} = \delta_{k,l}\, 1\!\!1, \qquad \{b_k, b_l\} = 0, \qquad (11.1)$$

where $k, l = 0, 1$. Here $1\!\!1$ is the identity operator in $\mathcal{H}_\mathcal{G}$ and, as usual, $\{x, y\} = xy + yx$ is the anticommutator between x and y. Then we take $\varphi_{0,0}$ as the vacuum of b_1 and b_2: $b_1\varphi_{0,0} = b_2\varphi_{0,0} = 0$, and we build up the other vectors out of it:

$$\varphi_{1,0} = b_1^\dagger\varphi_{0,0}, \qquad \varphi_{0,1} = b_2^\dagger\varphi_{0,0}, \qquad \varphi_{1,1} = b_1^\dagger b_2^\dagger\varphi_{0,0}.$$

The set $\mathcal{F}_\varphi = \{\varphi_{k,l}, k, l = 0, 1\}$ is an o.n. basis for $\mathcal{H}_G$: $\langle \varphi_{k,l}, \varphi_{n,m} \rangle = \delta_{k,n} \, \delta_{l,m}$. Then, any general mental state of the system $\mathcal{S}_G$ (i.e., of the two players), for $t = 0$, is a linear combination

$$\Psi_0 = \sum_{k,l=0}^{1} \alpha_{k,l} \varphi_{k,l}, \tag{11.2}$$

where the coefficients $\alpha_{k,l}$ can be deduced from Ψ_0 as in $\alpha_{k,l} = \langle \varphi_{k,l}, \Psi_0 \rangle$ and we assume that $\sum_{k,l=0}^{1} |\alpha_{k,l}|^2 = 1$, in order to normalize the total probability described by the wave function. Indeed, for instance, we interpret $|\alpha_{0,0}|^2$ as the probability that $\mathcal{S}_G$ is, at $t = 0$, in a state $\varphi_{0,0}$, i.e., that both $\mathcal{G}_1$ and $\mathcal{G}_2$ have chosen zero. Analogous interpretations can be given to the square modules of the other coefficients.

> *Remark:* It could be interesting to notice that $\varphi_{0,0}$ is not *a special vector*, when compared with the others in $\mathcal{F}_\varphi$. This is important, since otherwise the state $0_1 0_2$ of the system would be peculiar, while there is no reason for that, in fact. In fact, we could construct $\mathcal{F}_\varphi$ by adopting a different but equivalent order: we could first define $\varphi_{1,1} \in \mathcal{H}_G$ as that nonzero vector that is annihilated by both $b_1^\dagger$ and $b_2^\dagger$, $b_1^\dagger \varphi_{1,1} = b_2^\dagger \varphi_{1,1} = 0$, and then we can construct the other vectors of $\mathcal{F}_\varphi$ by acting on $\varphi_{1,1}$ with b_1 and b_2. In this case, $\varphi_{1,1}$ could be called, with some fantasy, the *anti-vacuum*, to highlight the fact that this vector is still annihilated by some ladder operators ($b_1^\dagger$ and $b_2^\dagger$) but for different reasons. In fact, while $b_1 \varphi_{0,0} = b_2 \varphi_{0,0} = 0$ because we are trying to destroy a particle, where there is no particle at all; $b_1^\dagger \varphi_{1,1} = b_2^\dagger \varphi_{1,1} = 0$ since we are trying to put together two identical fermions. But this is forbidden by the Pauli exclusion principle. Incidentally we observe that no anti-vacuum could be introduced in the same way for bosons, since a bosonic raising operator never annihilates any eigenvector of the related number operator.

An explicit representation of these vectors and operators is the following: $\varphi_{k,l} = \varphi_k^{(1)} \otimes \varphi_l^{(2)}$, with $\varphi_0 = \begin{pmatrix} 1 \\ 0 \end{pmatrix}$ and $\varphi_1 = \begin{pmatrix} 0 \\ 1 \end{pmatrix}$. Then, for instance,

$$\varphi_{1,0} = \varphi_1^{(1)} \otimes \varphi_0^{(2)} = \begin{pmatrix} 0 \\ 1 \end{pmatrix} \otimes \begin{pmatrix} 1 \\ 0 \end{pmatrix} = \begin{pmatrix} 0 \\ 0 \\ 1 \\ 0 \end{pmatrix},$$

$$\varphi_{1,1} = \varphi_1^{(1)} \otimes \varphi_1^{(2)} = \begin{pmatrix} 0 \\ 1 \end{pmatrix} \otimes \begin{pmatrix} 0 \\ 1 \end{pmatrix} = \begin{pmatrix} 0 \\ 0 \\ 0 \\ 1 \end{pmatrix},$$

and so on. The matrix form of the operators b_j and $b_j^\dagger$ are also quite simple. We find

$$b_1 = \begin{pmatrix} 0 & 1 \\ 0 & 0 \end{pmatrix} \otimes \begin{pmatrix} 1 & 0 \\ 0 & 1 \end{pmatrix} = \begin{pmatrix} 0 & 0 & 1 & 0 \\ 0 & 0 & 0 & 1 \\ 0 & 0 & 0 & 0 \\ 0 & 0 & 0 & 0 \end{pmatrix},$$

$$b_2 = \begin{pmatrix} 1 & 0 \\ 0 & 1 \end{pmatrix} \otimes \begin{pmatrix} 0 & 1 \\ 0 & 0 \end{pmatrix} = \begin{pmatrix} 0 & 1 & 0 & 0 \\ 0 & 0 & 0 & 0 \\ 0 & 0 & 0 & 1 \\ 0 & 0 & 0 & 0 \end{pmatrix},$$

and so on.

Let now $\hat{n}_j = b_j^\dagger b_j$ be the number operator of $\mathcal{G}_j$: the CAR above implies that $\hat{n}_1 \varphi_{k,l} = k\varphi_{k,l}$ and $\hat{n}_2 \varphi_{k,l} = l\varphi_{k,l}$, $k,l = 0,1$. Then, because of what we discussed before, the eigenvalues of these operators correspond to the choice operated by the two players at $t = 0$. For instance, $\varphi_{1,0}$ corresponds to the choice $1_1 0_2$, since "one" is the eigenvalue of $\hat{n}_1$ and "zero" is the eigenvalue of $\hat{n}_2$. It is natural, therefore, to call $\hat{n}_1$ and $\hat{n}_2$ the *strategy operators* (at $t = 0$), since they describe, at $t = 0$, the strategies of the two players. Moreover, for similar reasons, since b_j and $b_j^\dagger$ modify the attitude of $\mathcal{G}_j$, they can be called the *reflection operators*. Of course, in the presence of more options (not just "zero" and "one"), the CAR could not be the optimal choice. In this case, maybe, the truncated version of the CCR described in Section 2.3.2 could be more useful. Of course, if our problem has an infinite number of possible answers, then we should rather use the CCR.

Our main effort now consists in *giving a dynamics* to the strategy operators $\hat{n}_j$, following our standard scheme, see Chapter 2. As usual, we first need to introduce a Hamiltonian H for the system. H is what we need to deduce the dynamics of the strategy operators using $\hat{n}_j(t) := e^{iHt}\hat{n}_j e^{-iHt}$, or solving the Heisenberg equations of motion for the ladder operators of $\mathcal{G}_1$ and $\mathcal{G}_2$. Finally, we will compute the mean values of $\hat{n}_1(t)$ and $\hat{n}_2(t)$ on some suitable (vector) state defined by the vector Ψ_0 in (11.2), which describes the status of the system[1] at $t = 0$. Of course, $\mathcal{S}_\mathcal{G}$ is open, since $\mathcal{G}_1$ and $\mathcal{G}_2$ are assumed to interact with their environments $\mathcal{R}_1$ and $\mathcal{R}_2$ (whatever they are!) in order to take their decisions. For this reason, H is the Hamiltonian of an open system. Contrary to what happens for the players, whose situation can be described in the simple four-dimensional Hilbert space $\mathcal{H}_\mathcal{G}$, these environments are naturally defined in an infinite-dimensional Hilbert space, see Section 2.7 for the general setting. For this reason, they can be thought to describe some sub system with infinite (or very many) degrees of freedom, as the neurons in the brain, for instance. If we adopt this interpretation, $\mathcal{R}_j$ can be seen as the neural system of $\mathcal{G}_j$,

---

[1] This is true for closed systems. But the system we are dealing with is not closed, and Ψ_0 is not sufficient to uniquely define the state. We will come back to this aspect in (11.7).

$j = 1, 2$, which is the essential tool needed to make any decision: in this case the decision is driven by the interaction of the agent and its own brain.

The full Hamiltonian H is the following:

$$\begin{cases} H = H_0 + H_I + H_{int}, \\[2mm] H_0 = \displaystyle\sum_{j=1}^{2} \omega_j b_j^\dagger b_j + \sum_{j=1}^{2} \int_{\mathbb{R}} \Omega_j(k) B_j^\dagger(k) B_j(k) \, dk, \\[4mm] H_I = \displaystyle\sum_{j=1}^{2} \lambda_j \int_{\mathbb{R}} \left(b_j B_j^\dagger(k) + B_j(k) b_j^\dagger \right) dk \\[4mm] H_{int} = \mu_{ex} \left(b_1^\dagger b_2 + b_2^\dagger b_1 \right) + \mu_{coop} \left(b_1^\dagger b_2^\dagger + b_2 b_1 \right). \end{cases} \tag{11.3}$$

Here ω_j, λ_j, μ_{ex} and μ_{coop} are real quantities, and $\Omega_j(k)$ are real functions. Hence $H = H^\dagger$.

> *Remark:* The Hamiltonian in Equation (11.3) looks like a two-agents, two-reservoirs version of that in (3.3), but with a rather different interpretation. In particular, $\mathcal{R}_1$ and $\mathcal{R}_2$ are not *external* to the agents, as in Chapter 3, since they are really *parts of the agents.*

In analogy with the b_j's, we use fermionic operators $B_j(k)$ and $B_j^\dagger(k)$ to describe the environment:

$$\left\{ B_i(k), B_l(q)^\dagger \right\} = \delta_{i,l} \delta(k - q) \, \mathbb{1}, \qquad \left\{ B_i(k), B_j(k) \right\} = 0, \tag{11.4}$$

which have to be added to those in Equation (11.1). Moreover each b_j and $b_j^\dagger$ anti-commutes with each $B_j(k)$ and $B_j^\dagger(k)$: $\{b_j, B_l(k)\} = \{b_j, B_l^\dagger(k)\} = 0$ and so on, for all $j, l = 1, 2$ and $k \in \mathbb{R}$. The various terms of H can be understood as follows: (1) H_0 is the *free* Hamiltonian, which, alone, produces no time evolution for the strategy operators $\hat{n}_j$. This is because $[H_0, \hat{n}_j] = 0$, and because H reduces to H_0 in the absence of interactions (i.e., when $\lambda_j = \mu_{ex} = \mu_{coop} = 0$). This is in agreement with our idea that any sharp strategy of $\mathcal{G}_1$ and $\mathcal{G}_2$ can be modified only in the presence of interactions.[2]

(2) H_I describes the interaction between the players and their neural systems. Of course, the one proposed here is a special kind of interaction, which is useful since it produces an analytical solution for the time evolution of (the mean values of) the strategy operators. Other choices could be considered, but these would, quite likely, break down this nice aspect of the model: the more complicated we make H_I, the lower the probability of obtaining analytical results for our model.

---

[2] We recall that "sharp strategy" means for us that the players are described, at $t = 0$, by a single eigenstate of the strategy operators, and not by some (nontrivial) liner combination of these vectors.

(3) H_{int} describes two different interactions between $\mathcal{G}_1$ and $\mathcal{G}_2$. When $\mu_{coop} = 0$, then the two players tend to act differently, since a lowering operator for one agent is coupled to a raising operator for the other, while they behave in the same way when $\mu_{ex} = 0$, since raising and lowering operators appear in pairs. Of course, when both μ_{coop} and μ_{ex} are not zero, the dynamics is includes both possibilities.

Remarks: (1) It can be useful to observe that the Hamiltonian in (11.3) was originally introduced in connection with a well-known game theory problem: the two-prisoners dilemma [15]. In the original version of this game, the two prisoners have to make a choice without the possibility of communicating between them. They cannot even communicate with the outer world. They should decide alone what is best for them, knowing only a few rules. But, of course, neither $\mathcal{G}_1$ nor $\mathcal{G}_2$ are really alone: they interact, at least, with their brains. As we have seen, this is, in fact, a possible interpretation of the interaction $\mathcal{G}_j \leftrightarrow \mathcal{R}_j$ in H_I, see (11.3). The presence of H_{int} implies that we are also giving the prisoners the possibility of communicating with each other. But this communication can be cooperative or not, depending on the values of the parameters μ_{ex} and μ_{coop}. Also, if we take $\mu_{coop} = \mu_{ex} = 0$, we restore the original impossibility of $\mathcal{G}_1$ to interact with $\mathcal{G}_2$.

(2) From Equation (11.3) it is easy to see that, in general, no integral of motion exists for our H. The reason is that, when $\mu_{coop} \neq 0$, we have *pure creation* or *pure annihilation* terms in H_{int}, and therefore in H. On the contrary, if $\mu_{coop} = 0$, each other term in the Hamiltonian (see again Equation [11.3]), contains an annihilation and a creation operator together. In this case it is possible to check that the global operator

$$\hat{N} = \sum_{j=1}^{2} \left(\hat{n}_j + \int_{\mathbb{R}} B_j^\dagger(k) B_j(k) \, dk \right)$$

commutes with H, and therefore does not change in time: $\hat{N}$ is an integral of motion.

The Heisenberg equations of motion $\dot{X}(t) = i[H, X(t)]$ can now be deduced by using the CAR (11.1) and (11.4) and by the Hamiltonian H in (11.3):

$$\begin{cases} \dot{b}_1(t) = -i\omega_1 b_1(t) + i\lambda_1 \int_{\mathbb{R}} B_1(k,t) \, dk - i\mu_{ex} b_2(t) - i\mu_{coop} b_2^\dagger(t), \\ \dot{b}_2(t) = -i\omega_2 b_2(t) + i\lambda_2 \int_{\mathbb{R}} B_2(k,t) \, dk - i\mu_{ex} b_1(t) + i\mu_{coop} b_1^\dagger(t), \\ \dot{B}_j(k,t) = -i\Omega_j(k) B_j(k,t) + i\lambda_j b_j(t), \end{cases} \quad (11.5)$$

$j = 1, 2$. The solution of this system of equations can be found similarly to what we have done before, see Section 2.7 for instance, and it looks like

$$b(t) = e^{i\,U\,t} b(0) + i \int_0^t e^{i\,U\,(t-t_1)} \beta(t_1)\, dt_1, \tag{11.6}$$

where we have introduced the following quantities:

$$b(t) = \begin{pmatrix} b_1(t) \\ b_2(t) \\ b_1^\dagger(t) \\ b_2^\dagger(t) \end{pmatrix}, \quad \beta(t) = \begin{pmatrix} \lambda_1 \beta_1(t) \\ \lambda_2 \beta_2(t) \\ -\lambda_1 \beta_1^\dagger(t) \\ -\lambda_2 \beta_2^\dagger(t) \end{pmatrix}, \quad U = \begin{pmatrix} i v_1 & -\mu_{ex} & 0 & -\mu_{coop} \\ -\mu_{ex} & i v_2 & \mu_{coop} & 0 \\ 0 & \mu_{coop} & i \overline{v_1} & \mu_{ex} \\ -\mu_{coop} & 0 & \mu_{ex} & i \overline{v_2} \end{pmatrix},$$

and where $\Omega_j(k) = \Omega_j k$, $\Omega_j > 0$, $v_j = i\omega_j + \pi \frac{\lambda_j^2}{\Omega_j}$ and $\beta_j(t) = \int_{\mathbb{R}} B_j(k) e^{-i\Omega_j k t}\, dk$, $j = 1, 2$.

We are now ready to take the average of the time evolution of the strategy operators, $\hat{n}_j(t) = b_j^\dagger(t) b_j(t)$, on a state over the full system $\mathcal{S} = \mathcal{S}_G \otimes \mathcal{R}$, where $\mathcal{S}_G = \{\mathcal{G}_1, \mathcal{G}_2\}$ and $\mathcal{R} = \{\mathcal{R}_1, \mathcal{R}_2\}$. These states are assumed to be of the usual kind, see Section 2.7: they are tensor products of vector states for $\mathcal{S}_G$ and states of the environment satisfying the usual standard requirements. More explicitly, for each operator of the form $X_S \otimes Y_R$, X_S being an operator of $\mathcal{S}_G$ and Y_R an operator of the environment, we assume that

$$\langle X_S \otimes Y_R \rangle := \langle \Psi_0, X_S \Psi_0 \rangle\, \omega_R(Y_R). \tag{11.7}$$

Here Ψ_0 is the vector introduced in (11.2), while $\omega_R(.)$ is a state satisfying the following conditions:

$$\omega_R(\mathbb{1}_R) = 1, \quad \omega_R(B_j(k)) = \omega_R(B_j^\dagger(k)) = 0, \quad \omega_R(B_j^\dagger(k) B_l(q)) = N_j\, \delta_{j,l}\delta(k-q), \tag{11.8}$$

for some constant N_j. Also, $\omega_R(B_j(k) B_l(q)) = 0$, for all j and l.

> *Remark:* Equation (11.7) suggests that some asymmetry does exist between $\mathcal{S}_G$ and $\mathcal{R}$, since their states are of an apparently different nature. This was already observed in other applications, see Chapters 3 and 10, for instance. In fact, this is not so surprising, since, among other differences, $\mathcal{S}_G$ lives in a four-dimensional Hilbert space, while $\mathcal{R}$ can be defined only in an infinite-dimensional space $\mathcal{H}_R$. However, using the so-called Gelfand-Naimark-Segal (GNS) construction, see [43], it is possible to represent ω_R as a vector state. In other words, ω_R can also be written as a vector state, but the vector that represents ω_R belongs to the Hilbert space defined by the GNS construction.

After a few computations, calling $V(t) = e^{i\,U\,t}$ and $V_{k,l}(t)$ its (k,l)-matrix element, we deduce the following general formulas for the decision functions of $\mathcal{G}_1$ and $\mathcal{G}_2$, which extend those found in [15] in absence of interference:

$$\begin{cases} n_1(t) = \left\langle b_1^\dagger(t)b_1(t) \right\rangle = \mu_1^{(\mathcal{G})}(t) + \delta\mu_1^{(\mathcal{G})}(t) + n_1^{(B)}(t), \\ n_2(t) = \left\langle b_2^\dagger(t)b_2(t) \right\rangle = \mu_2^{(\mathcal{G})}(t) + \delta\mu_2^{(\mathcal{G})}(t) + n_2^{(B)}(t). \end{cases} \tag{11.9}$$

Here, we have introduced

$$\begin{cases} \mu_1^{(\mathcal{G})}(t) = |V_{1,1}(t)|^2 \left(|\alpha_{1,0}|^2 + |\alpha_{1,1}|^2\right) + |V_{1,2}(t)|^2 \left(|\alpha_{0,1}|^2 + |\alpha_{1,1}|^2\right) \\ \qquad\quad + |V_{1,3}(t)|^2 \left(|\alpha_{0,0}|^2 + |\alpha_{0,1}|^2\right) + |V_{1,4}(t)|^2 \left(|\alpha_{0,0}|^2 + |\alpha_{1,0}|^2\right) \\ \mu_2^{(\mathcal{G})}(t) = |V_{2,1}(t)|^2 \left(|\alpha_{1,0}|^2 + |\alpha_{1,1}|^2\right) + |V_{2,2}(t)|^2 \left(|\alpha_{0,1}|^2 + |\alpha_{1,1}|^2\right) \\ \qquad\quad + |V_{2,3}(t)|^2 \left(|\alpha_{0,0}|^2 + |\alpha_{0,1}|^2\right) + |V_{2,4}(t)|^2 \left(|\alpha_{0,0}|^2 + |\alpha_{1,0}|^2\right), \end{cases} \tag{11.10}$$

$$\begin{cases} \delta\mu_1^{(\mathcal{G})}(t) = 2\Re\left[\overline{V_{1,1}(t)}\, V_{1,2}(t)\overline{\alpha_{1,0}}\, \alpha_{0,1} + \overline{V_{1,1}(t)}\, V_{1,4}(t)\overline{\alpha_{1,1}}\, \alpha_{0,0} \right] \\ \qquad\quad - 2\Re\left[\overline{V_{1,2}(t)}\, V_{1,3}(t)\overline{\alpha_{1,1}}\, \alpha_{0,0} + \overline{V_{1,3}(t)}\, V_{1,4}(t)\overline{\alpha_{0,1}}\, \alpha_{1,0} \right], \\ \delta\mu_2^{(\mathcal{G})}(t) = 2\Re\left[\overline{V_{2,1}(t)}\, V_{2,2}(t)\overline{\alpha_{1,0}}\, \alpha_{0,1} + \overline{V_{2,1}(t)}\, V_{2,4}(t)\overline{\alpha_{1,1}}\, \alpha_{0,0} \right] \\ \qquad\quad - 2\Re\left[\overline{V_{2,2}(t)}\, V_{2,3}(t)\overline{\alpha_{1,1}}\, \alpha_{0,0} + \overline{V_{2,3}(t)}\, V_{2,4}(t)\overline{\alpha_{0,1}}\, \alpha_{1,0} \right], \end{cases} \tag{11.11}$$

where $\Re[z]$ is the real part of the complex quantity z, and

$$\begin{cases} n_1^{(B)}(t) = 2\pi \int_0^t dt_1 \left[\dfrac{\lambda_1^2}{\Omega_1} \left(|V_{1,1}(t-t_1)|^2 N_1 + |V_{1,3}(t-t_1)|^2(1-N_1) \right) \right] \\ \qquad\quad + 2\pi \int_0^t dt_1 \left[\dfrac{\lambda_2^2}{\Omega_2} \left(|V_{1,2}(t-t_1)|^2 N_2 + |V_{1,4}(t-t_1)|^2(1-N_4) \right) \right], \\ n_2^{(B)}(t) = 2\pi \int_0^t dt_1 \left[\dfrac{\lambda_1^2}{\Omega_1} \left(|V_{2,1}(t-t_1)|^2 N_1 + |V_{2,3}(t-t_1)|^2(1-N_1) \right) \right] \\ \qquad\quad + 2\pi \int_0^t dt_1 \left[\dfrac{\lambda_2^2}{\Omega_2} \left(|V_{2,2}(t-t_1)|^2 N_2 + |V_{2,4}(t-t_1)|^2(1-N_4) \right) \right]. \end{cases} \tag{11.12}$$

The analytic expressions for $n_1(t)$ and $n_2(t)$ in Equation (11.9) show some similarities with those for the decision functions in Chapter 3. This is in agreement with our previous comment on the similarities between the Hamiltonians in Equations (3.3) and (11.3). However, in Equations (11.10) and (11.11), the coefficients $\alpha_{k,l}$ introduced in Equation (11.2) explicitly appear, and this is interesting for us, since we can discuss initial conditions for the agents that are more elaborated than those considered in Chapter 3, but not only.

In Equation (11.9) we have clearly separated contributions of three different natures: $\mu_j^{(\mathcal{G})}(t)$ contains contributions only due to the players $\mathcal{G}_1$ and $\mathcal{G}_2$. Their analytic expressions become particularly simple if the initial state Ψ_0 is just one of the vectors $\varphi_{j,k}$, i.e., if all the coefficients $\alpha_{k,l}$ in (11.2) are zero, except one. Moreover, in this particular situation, all the contributions in $\delta\mu_j^{(\mathcal{G})}(t)$, which also refer only to $\mathcal{G}_1$ and $\mathcal{G}_2$, are zero: they just do not contribute to Equation (11.9). For this reason, we call $\delta\mu_1^{(\mathcal{G})}(t)$ and $\delta\mu_2^{(\mathcal{G})}(t)$ *interference terms*: they are present only if Ψ_0 is some nontrivial linear superposition of eigenvectors of the (time zero) strategy operators. Otherwise, they simply disappear. Finally, $n_1^{(B)}(t)$ and $n_2^{(B)}(t)$ arise because of the interaction of the players with the environments: as we see from Equation (11.12), they are both zero if $\lambda_1 = \lambda_2 = 0$, and they do not depend on the explicit form of Ψ_0. They are, as expected, terms with memory, defined in terms of certain time integrals.

11.3 Asymptotic Values: Absence of Interferences

In this section we concentrate on what happens for very large values of t and we show, in particular, that in this limit the role of Ψ_0 is not essential. To show this, we start considering the simplest situation: $\mathcal{G}_1$ and $\mathcal{G}_2$ do not interact. This means that $\mu_{ex} = \mu_{coop} = 0$, so that $H_{int} = 0$, see (11.3). In this case the matrix U is diagonal, and the computation of $b(t)$ in (11.6) is particularly simple. We get

$$b_j(t) = e^{-\nu_j t} b_j(0) + i \int_0^t e^{-\nu_j(t-t_1)} \beta_j(t) \, dt_1,$$

$j = 1, 2$, and $b_j^\dagger(t)$ and $\hat{n}_j(t) = b_j^\dagger(t) b_j(t)$ can be deduced easily. Then, taking the mean value of $\hat{n}_j(t)$ on a state like the one in Equation (11.7), we get

$$n_j(t) = \left\langle \hat{n}_j(t) \right\rangle = e^{-\frac{2\pi\lambda_j^2}{\Omega_j} t} \|b_j \Psi_0\|^2 + N_j \left(1 - e^{-\frac{2\pi\lambda_j^2}{\Omega_j} t} \right). \tag{11.13}$$

What is interesting here is that, if $\lambda_j \neq 0$,

$$n_j(\infty) := \lim_{t\to\infty} n_j(t) = N_j \tag{11.14}$$

does not depend on the original state of mind of the two players, but only on what the reservoir suggests. This means that, when $\mathcal{G}_1$ and $\mathcal{G}_2$ do not interact, their final choices are dictated only by their environments. Stated in different words, the final decisions of the agents are completely independent of what was their original propensity. We will show that this is not so *while the decision is being processed*, i.e., for intermediate values of t.

Remarks: (1) Notice that, if $\lambda_j = 0$, Equation (11.13) reduces to $n_j(t) = \|b_j\Psi_0\|^2 = n_j(0)$, $\forall t$. This is not surprising, since in this particular situation $H = H_0$: $\mathcal{G}_1$ and $\mathcal{G}_2$ do not interact with the environments or among them. In other terms, there is no time evolution of the operator $\hat{n}_j$ at all, since $\hat{n}_j$ commutes with H and their mean values stay also constant in time.

(2) More generally, Equation (11.13) suggests the introduction of a sort of *characteristic time* for $\mathcal{G}_j$, $\tau_j = \frac{\Omega_j}{2\pi\lambda_j^2}$. The more t approaches τ_j, the bigger the influence of $\mathcal{R}_j$ on $\mathcal{G}_j$ is. In particular, if $\lambda_j \to 0$, τ_j diverges. Hence we recover our previous conclusions: $\mathcal{G}_j$ is not influenced at all by $\mathcal{R}_j$, even after a long time. A similar behavior is deduced also when Ω_j increases: the larger its value, the larger the value of τ_j. In other words, for large Ω_j the influence of the environment is effective only after a sufficiently long time interval. Of course, τ_j can be considered as a sort of *decision time*, i.e., of time needed to end up with some decision. In the next section we will discuss what happens for relatively small times, $0 < t \ll \tau_j$, and which kind of differences arises if Ψ_0 is a single eigenstate of the strategy operators or when it is a nontrivial linear combination of these.

When $\mathcal{G}_1$ and $\mathcal{G}_2$ interact, μ_{ex} or μ_{coop}, or both, are different from zero. In this case it is possible to see that the final decision of, say, $\mathcal{G}_1$, is driven by $\mathcal{R}_1$ and by $\mathcal{R}_2$. And the effect of $\mathcal{R}_2$ can be strong, depending on the numerical values of the interaction parameters. For concreteness, we fix here $\omega_1 = 1$, $\omega_2 = 2$, $\lambda_1 = \lambda_2 = 0.5$, $\Omega_1 = \Omega_2 = 0.1$, and then we see what happens to the asymptotic values of the decision functions $n_j(t)$ if we consider different choices of μ_{ex} and μ_{coop}. Their full-time behavior, also for small t, will be discussed in the next section, and interesting features will appear.

In the following we will focus on the following choices of μ_{ex} and μ_{coop}: Case (a). $\mu_{ex} = 0.01$ and $\mu_{coop} = 100$; Case (b). $\mu_{ex} = 0.01$ and $\mu_{coop} = 1$; Case (c). $\mu_{ex} = \mu_{coop} = 0.5$; Case (d). $\mu_{ex} = 1$ and $\mu_{coop} = 0.01$; Case (e). $\mu_{ex} = 100$ and $\mu_{coop} = 0.01$.

It is convenient to rewrite formulas in (11.9) as follows:

$$
\begin{aligned}
n_1(t) = &|V_{1,1}(t)|^2 \|b_1\Psi_0\|^2 + |V_{1,2}(t)|^2 \|b_2\Psi_0\|^2 + |V_{1,3}(t)|^2 \left(1 - \|b_1\Psi_0\|^2\right) \\
&+ |V_{1,4}(t)|^2 \left(1 - \|b_2\Psi_0\|^2\right) \\
&+ 2\pi \int_0^t dt_1 \frac{\lambda_1^2}{\Omega_1} \left[|V_{1,1}(t-t_1)|^2 N_1 + |V_{1,3}(t-t_1)|^2 (1-N_1)\right] \\
&+ 2\pi \int_0^t dt_1 \frac{\lambda_2^2}{\Omega_2} \left[|V_{1,2}(t-t_1)|^2 N_2 + |V_{1,4}(t-t_1)|^2 (1-N_2)\right],
\end{aligned}
$$

$$(11.15)$$

and

$$n_2(t) = |V_{2,1}(t)|^2 \, \|b_1\Psi_0\|^2 + |V_{2,2}(t)|^2 \, \|b_2\Psi_0\|^2 + |V_{2,3}(t)|^2 \left(1 - \|b_1\Psi_0\|^2\right)$$

$$+ |V_{2,4}(t)|^2 \left(1 - \|b_2\Psi_0\|^2\right)$$

$$+ 2\pi \int_0^t dt_1 \frac{\lambda_1^2}{\Omega_1} \left[|V_{2,1}(t-t_1)|^2 N_1 + |V_{2,3}(t-t_1)|^2 (1 - N_1) \right]$$

$$+ 2\pi \int_0^t dt_1 \frac{\lambda_2^2}{\Omega_2} \left[|V_{2,2}(t-t_1)|^2 N_2 + |V_{2,4}(t-t_1)|^2 (1 - N_2) \right].$$

$$(11.16)$$

The asymptotic values of these functions are the following:

Case (a), $\mu_{ex} = 0.01$ and $\mu_{coop} = 100$:

$$\begin{cases} n_1(\infty) \simeq 0.50317 N_1 + 0.49682(1 - N_2), \\ n_2(\infty) \simeq 0.49682(1 - N_1) + 0.50317 N_2. \end{cases} \qquad (11.17)$$

Case (b), $\mu_{ex} = 0.01$ and $\mu_{coop} = 1$:

$$\begin{cases} n_1(\infty) \simeq 0.91914 N_1 + 0.08075(1 - N_2), \\ n_2(\infty) \simeq 0.08075(1 - N_1) + 0.91914 N_2. \end{cases} \qquad (11.18)$$

Case (c), $\mu_{ex} = 0.5$ and $\mu_{coop} = 0.5$:

$$\begin{cases} n_1(\infty) \simeq 0.84802 N_1 + 0.10057 N_2 + 0.02543, \\ n_2(\infty) \simeq 0.10057 N_1 + 0.84802 N_2 + 0.02543. \end{cases} \qquad (11.19)$$

Case (d), $\mu_{ex} = 1$ and $\mu_{coop} = 0.01$:

$$\begin{cases} n_1(\infty) \simeq 0.99210 N_1 + 0.00795 N_2, \\ n_2(\infty) \simeq 0.00795 N_1 + 0.99210 N_2. \end{cases} \qquad (11.20)$$

Case (e), $\mu_{ex} = 100$ and $\mu_{coop} = 0.01$:

$$\begin{cases} n_1(\infty) \simeq 0.50308 N_1 + 0.49692 N_2, \\ n_2(\infty) \simeq 0.49692 N_1 + 0.50308 N_2. \end{cases} \qquad (11.21)$$

It is evident that, once again, the role of the initial state of mind of the agents is irrelevant here: Ψ_0 just disappears from the final results, which refer only to $\mathcal{R}_1$ and $\mathcal{R}_2$ (and, see Equation [11.19], to some numbers also independent of Ψ_0). Then the choice of the parameters $\alpha_{k,l}$ in Equation (11.2) seems to be irrelevant. This is the reason why we talk here of *absence of interferences*, even if some linear combinations of vectors do exist, at least in Ψ_0, at $t = 0$. We will see in the next section that this is not so if we consider the time evolution of the decision functions and not just their asymptotic values. An interesting aspect of the results above is that the presence of H_{int} in Equation (11.3) mixes the effect of the reservoirs of the two players: in all the results listed here we see that $\mathcal{R}_2$ affects the final choice of $\mathcal{G}_1$, and that $\mathcal{R}_1$ appears in the expression of $n_2(\infty)$. And these mixed contributions can be sufficiently strong for particular values of μ_{ex} or μ_{coop}. For instance, Equation (11.17) shows that the effect of $\mathcal{R}_2$ in $n_1(\infty)$ is of the same order as that of $\mathcal{R}_1$, while it is almost negligible in Case (d) (see [11.20]).

Finally, even if this is not evident from Equations (11.15) and (11.16), we expect that the value of the characteristic time for $\mathcal{G}_j$ is somehow affected by the explicit values of μ_{ex} and μ_{coop}.

> *Remarks:* (1) The fermionic nature of $\mathcal{R}_1$ and $\mathcal{R}_2$ forces N_1 and N_2 to be either zero or one. Hence $n_j(\infty)$ is always in the closed interval $[0, 1]$, $j = 1, 2$. As in Chapter 3, we interpret a value $n_1(\infty) \simeq 1$ as a strong attitude of $\mathcal{G}_1$ to choose one, for instance. If $n_j(\infty) \simeq 0.5$, $\mathcal{G}_j$ has no clear idea of what to do. Some different input is needed, in this case.
>
> (2) A different possibility for reading the results above, which might appear closer to the usual interpretation of what is *quantum* in decision-making, is to look at the $n_j(\infty)$ in a probabilistic way. For instance, rather than considering $n_j(\infty)$ as the decision made by $\mathcal{G}_j$, we could consider it as a sort of probability that $\mathcal{G}_j$ chooses zero or one. But we will not insist on this alternative here. However, we refer to [14] for more results in this direction.

11.4 Finite Time Behavior: The Role of Interference

Rather than looking at the asymptotic values of the decision functions, we want to discuss now what happens during the entire time evolution and, in particular, when the value of t is not high. This is interesting, since we will observe some peculiarities related to the coefficients $\alpha_{k,l}$ in Equation (11.2). In order to fix the ideas, we will consider here two different choices of the parameters of the Hamiltonian H, C_1 and C_2, and two different choices of initial conditions, $\mathcal{E}_1$ and $\mathcal{E}_2$. We take

C_1: $\omega_1 = 1$, $\omega_2 = 2$, $\Omega_1 = \Omega_2 = 0.1$, $\lambda_1 = \lambda_2 = 0.5$;
C_2: $\omega_1 = 0.1$, $\omega_2 = 0.2$, $\Omega_1 = \Omega_2 = 1$, $\lambda_1 = 1$ and $\lambda_2 = 0.7$.
Moreover,
$\mathcal{E}_1 = \left\{ \alpha_{k,l} = \frac{1}{2}, \forall\, k, l \right\}$, and $\mathcal{E}_2 = \left\{ \alpha_{0,1} = \frac{1}{2} = -\alpha_{1,1},\ \alpha_{0,0} = \frac{i}{2} = -\alpha_{1,0} \right\}$.

It is obvious that several other choices could also be considered. However, the ones given here already cover different interesting situations. In particular, while in C_1 the interaction parameters λ_1 and λ_2 are equal, they are different in C_2. Also, while in C_1 each ω_j is bigger than each Ω_k, the opposite inequalities are satisfied by the choice C_2. This is useful, since it is known that the ω_j's are related to the inertia of the player $\mathcal{G}_j$, so that C_1 and C_2 describe two opposite situations, from this point of view.

(A)

(a)

(b)

(B)

(c)

(d)

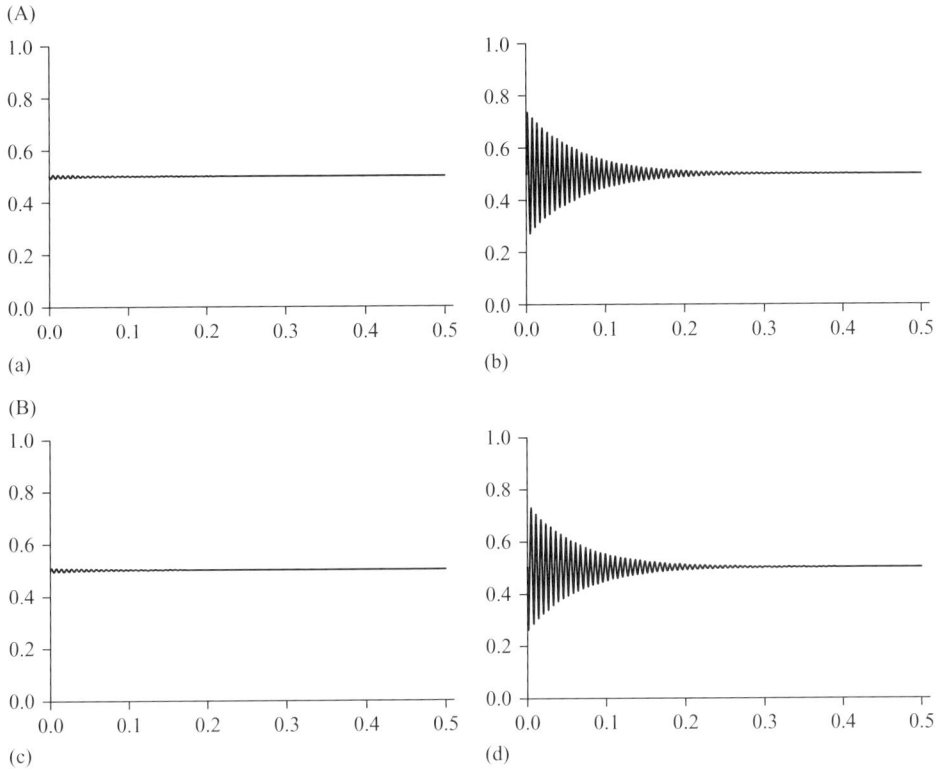

Figure 11.1 $n_1(t)$ (A) and $n_2(t)$ (B) for parameters C_1, $N_1 = 0$, $N_2 = 1$, $\mu_{ex} = 500$, $\mu_{coop} = 0$ and for $\mathcal{E}_1$ (a,c) and $\mathcal{E}_2$ (b,d) (Source: Bagarello et al. [14]).

The difference between $\mathcal{E}_1$ and $\mathcal{E}_2$ is the following: while in $\mathcal{E}_1$ all the coefficients in Equation (11.2) are equal, and in particular there is no relative phase between the various $\varphi_{k,l}$'s in Ψ_0, this is not so when adopting the choice $\mathcal{E}_2$. And, as we will see, this indeed makes a big difference for small t: we will see that, even if the asymptotic values of the two decision functions turn out to be independent of the choice of $\mathcal{E}_1$ or $\mathcal{E}_2$, there exists a certain time window, not necessarily small, in which the choice of the $\alpha_{k,l}$'s really changes the behaviors of the functions. More concretely, if we add a phase in the coefficients defining the vector Ψ_0, we may observe quite large oscillations, which are absent when there is not such a phase. Then, *interference terms in Ψ_0 make it, in general, quite difficult to get a decision in a small time*. In particular, this is the effect of the relative phases in the interference coefficients (see Figure 11.1[b],[d]), while if these coefficients have all the same phases, a decision can be reached quite soon (see Figure 11.1[a],[c]): the effect of these phases is quite evident. However, if we wait for a sufficiently long time, in both cases we reach the same final values of the decision functions: this is, in

(A)

(a)

(b)

(B)

(c)

(d)

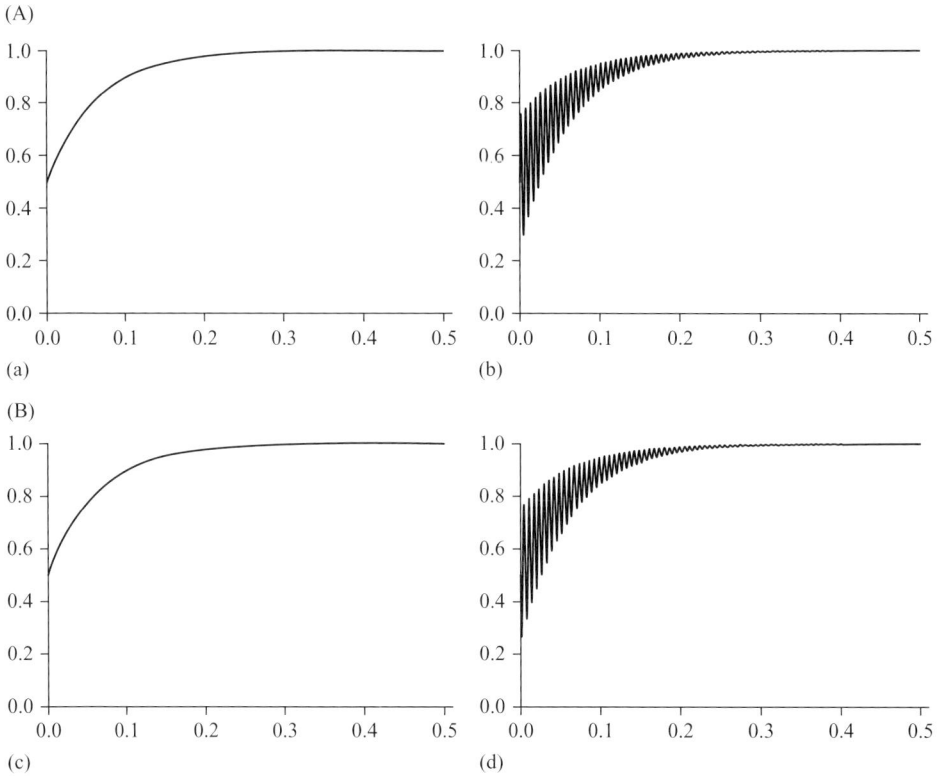

Figure 11.2 $n_1(t)$ (A) and $n_2(t)$ (B) for parameters C_1, $N_1 = 1$, $N_2 = 1$, $\mu_{ex} = 500$, $\mu_{coop} = 0$ and for $\mathcal{E}_1$ (a,c) and E_2 (b,d) (Source: Bagarello et al. [14]).

fact, what we have deduced in Section 11.3: the asymptotic values of the decision functions are independent of Ψ_0.

These conclusions are confirmed by other choices of the state on the environment. For instance, in Figure 11.2 we plot again the decision functions for the choices $\mathcal{E}_1$ (a,c) and $\mathcal{E}_2$ (b,d) and for the same choice of the parameters as in Figure 11.1, while the state of the environment is different from that of Figure 11.1, since we take now $N_1 = N_2 = 1$, rather than $N_1 = 0$ and $N_2 = 1$. We see that there is no particular difference between $n_1(t)$ and $n_2(t)$, and this is possibly due to the fact that $N_1 = N_2$. However, adding the phases to the $\alpha_{k,l}$ creates, again, a lot of noise in the decision-making process, noise that disappears, but only after a sufficiently long time.

It is useful to stress that, changing further the parameters of the Hamiltonian, we do not really modify our conclusions. Indeed, Figure 11.3 shows that even with different choices of the parameters in H, the choice C_2 with $\mu_{ex} = 100$ (rather than

(A)

(a)

(b)

(B)

(c)

(d)

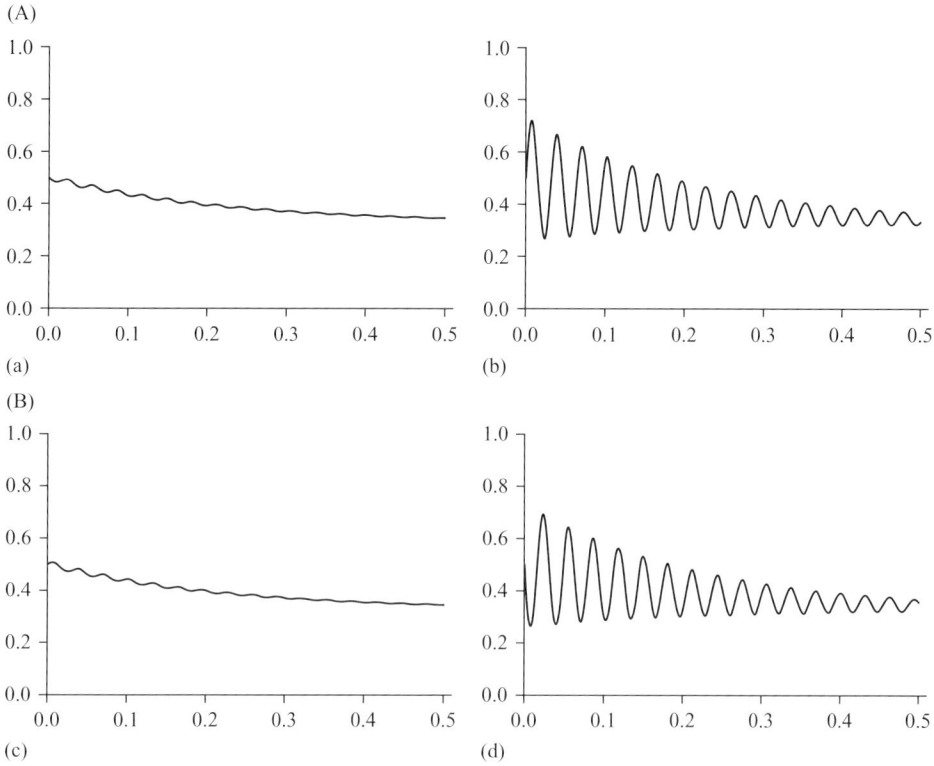

Figure 11.3 $n_1(t)$ (A) and $n_2(t)$ (B) for parameters C_2, $N_1 = 0$, $N_2 = 1$, $\mu_{ex} = 100$, $\mu_{coop} = 0$ and for $\mathcal{E}_1$ (a,c) and $\mathcal{E}_2$ (b,d) (Source: Bagarello et al. [14]).

$\mu_{ex} = 500$, as in the previous figures) phases create noise, and, again, this noise becomes smaller and smaller after some time.

In Figure 11.3, the existence of an asymptotic limit for the decision functions is not so evident, but this is only due to the small time interval considered. It is not hard to imagine that we still recover some asymptotic value for $n_j(t)$ if we consider a time interval larger than the one considered here.

We should also mention that the same behavior is observed if we take $\mu_{coop} \neq 0$: the choice $\mathcal{E}_2$ is again much more noisy than $\mathcal{E}_1$.

An interesting effect is observed when we consider both μ_{ex} and μ_{coop} different from zero, but close to each other, see Figure 11.4. It seems that when the two terms in H_{int} (see [11.3]), have the same strength, together they become capable of *filtering the noise*, making the effect of the phases in $\alpha_{k,l}$ not so strong. This is an interesting feature, which is surely worth deeper analysis.

(A)

(a)

(B)

(c)   (d)

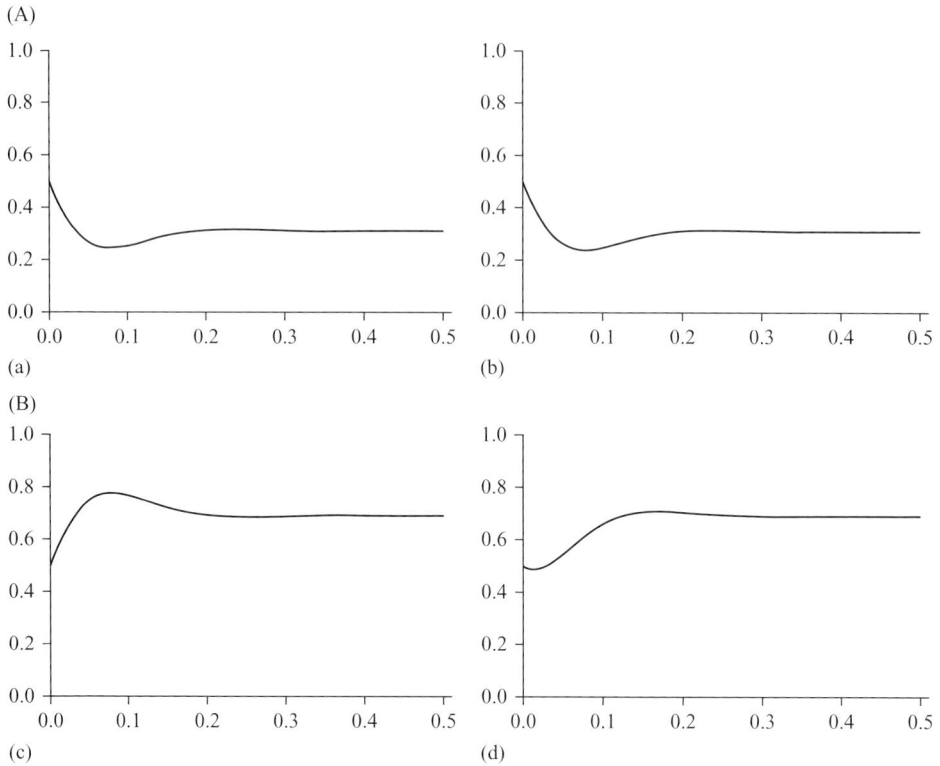

Figure 11.4 $n_1(t)$ (A) and $n_2(t)$ (B) for parameters C_1, $N_1 = 0$, $N_2 = 1$, $\mu_{ex} = 10$, $\mu_{coop} = 10$ and for $\mathcal{E}_1$ (a,c) and $\mathcal{E}_2$ (b,d) (Source: Bagarello et al. [14]).

11.5 Conclusions

We have discussed here what can be seen as a modified quantum version of the two-prisoners dilemma, based on ladder operators and open quantum systems. We have used the reservoirs to model the brains of the players, and we have shown that, in some intermediate range of time, the initial conditions can seriously affect their decisions. However, in the large time limit, it turns out that this is not so, and the final decisions of the players are independent of their initial feelings, while they are driven only by their brains. We have also deduced that when $\mathcal{G}_1$ and $\mathcal{G}_2$ can interact, the brain of $\mathcal{G}_2$ can affect the value of $n_1(\infty)$, and that of $\mathcal{G}_1$ can affect $n_2(\infty)$. Of course, this is impossible in the standard two-prisoners dilemma, since the prisoners have no way to communicate.

We refer to [14] for more details on this model, mainly for some probabilistic aspects of its interpretation. We also refer to [15] for more remarks on the relevance of our Hamiltonian in connection with an extended version of the two-prisoner dilemma.

12

Compatible and Incompatible Questions

12.1 Introduction

From what we have discussed so far in this book, it is clear that our main interest is in the dynamical behavior of some specific system $\mathcal{S}$. This was also our main concern in [1], where more applications were discussed. In many different branches of research, and in different contexts, we have always tried to analyze the time evolution of $\mathcal{S}$ by specifying first its observables, and then its Hamiltonian. All these operators are written in terms of ladder operators, which could be fermionic, bosonic or could obey some deformed version of the CCR or CAR. The time evolution of the observables, as driven by the Hamiltonian of the system, is deduced by the Heisenberg or by the Schrödinger equation.

In this chapter our interest will be focused on a different application of quantum mechanics to decision-making. In particular, we will describe an application of the Heisenberg uncertainty principle in the description of *compatible* and *incompatible* questions, which is a hot topic in psychology. We will see that, in this particular application, the time evolution of the system is not really what interests us. What is more relevant is the possibility of using other ideas coming from quantum mechanics to fit experimental data collected by psychologists.

Suppose Alice is asked a first question, Q_1, and then a second, Q_2. It may be that the two questions are somehow related, or that they are not. In the first case Q_1 and Q_2 are said to be incompatible, while they are compatible when Q_1 and Q_2 are not connected to each other. It should be clarified that this notion of compatibility is not really univocally defined. For instance, Q_1 and Q_2 could be *Are you happy?* and *Are you rich?* These questions could be considered compatible or not, depending on Alice. She can be the kind of girl who is happy only if she has enough money. So the answers to Q_1 and Q_2 are strongly related, and, according to what we said, the two questions are incompatible. But, on the opposite side, Alice could be the kind of girl who just does not care about money, at all: she is happy only when she is in love with somebody, or at peace with herself, for instance. In this case Q_1 is completely unrelated to Q_2, and the two questions can be considered compatible.

In some literature on this topic, the two classes of questions are treated with a slightly different approach, see [67] for instance. What we will propose in this chapter is a general framework in which compatible and incompatible questions can be treated together and in which a certain parameter can be introduced that measures the *degree of compatibility* of the questions. To simplify the treatment, we will consider just two binary questions (yes/no), but it is not hard to imagine how the same approach can be extended to more questions, admitting more answers.

12.2 The Mathematical Framework

We will now construct a simple mathematical structure associated with the following two binary questions asked to our agent, Alice: Q_1: *Are you happy?* and Q_2: *Are you employed?* We have already discussed that, sometimes, two different Hilbert spaces are used to consider the two possibilities (Q_1 and Q_2 are compatible or they are not), see [67] and references therein. Here we will show that a single Hilbert space can be used in both cases. Later, we will discuss how to use the same settings in other problems in decision-making, when experimental results show some lack of commutativity. In particular, we will consider the well-known problem of Gore vs. Clinton's honesty question [118].

12.2.1 The Starting Point

Let us suppose that Alice can be happy or not, and employed or not. We first consider these aspects separately, i.e., we assume Q_i is completely unrelated to Q_j, $i \neq j$. For this reason, we consider the *happiness status* as a two-state system living in a two-dimensional Hilbert space $\mathcal{H}_H = \mathbb{C}^2$. We introduce an o.n. basis $\mathcal{F}_H = \{\hat{h}_+, \hat{h}_-\}$ of $\mathcal{H}_H$, and a Hermitian operator $H = H^\dagger$, the *happiness operator*, having $\hat{h}_\pm$ as eigenstates:

$$H\hat{h}_\pm = h_\pm \, \hat{h}_\pm.$$

Here $h_\pm = \pm 1$ are the corresponding eigenvalues of H. Of course, we have $\langle \hat{h}_j, \hat{h}_k \rangle = \delta_{j,k}$, $j, k = +, -$. The interpretation is clear: if Alice's state is $\Psi = \hat{h}_+$, then she is definitively happy. But she is unhappy if $\Psi = \hat{h}_-$, while Alice is in a mixed state if $\Psi = \alpha_+ \hat{h}_+ + \alpha_- \hat{h}_-$, with $|\alpha_+|^2 + |\alpha_-|^2 = 1$. In this case, $|\alpha_+|^2$ can be thought as the probability that Alice is happy, while $|\alpha_-|^2$ is the probability she is not. We see that the answer to Q_1 is sure only if Ψ is an eigenstate of H. This is exactly what happens in ordinary quantum mechanics, when the state of a system is known exactly only when the system is in an eigenstate of a suitable set of observables on the system.

Despite the simplicity of the situation, it is possible to set up a rather rich (and well-known!) mathematical framework for H and $\mathcal{H}_H$. First we can introduce the following (rank-one) operators:

$$P_{j,k}^{(h)} f = \left\langle \hat{h}_k, f \right\rangle \hat{h}_j, \tag{12.1}$$

for each $f \in \mathcal{H}_H$. Here, again, $j, k = +, -$. It is easy to see that $P_{j,j}^{(h)}$ are orthogonal projectors: $\left(P_{j,j}^{(h)} \right)^2 = P_{j,j}^{(h)}$ and $\left(P_{j,j}^{(h)} \right)^\dagger = P_{j,j}^{(h)}$. On the other hand, if $j \neq k$, $\left(P_{j,k}^{(h)} \right)^2 = 0$ and $\left(P_{j,k}^{(h)} \right)^\dagger = P_{k,j}^{(h)}$. The operator H, which has spectrum[1] $\sigma(H) = \{-1, +1\}$, can be written in terms of the projectors $P_{j,j}^{(h)}$ as

$$H = P_{+,+}^{(h)} - P_{-,-}^{(h)}.$$

It is possible to introduce a *raising operator* σ defined by $\sigma \hat{h}_- = \hat{h}_+$ and $\sigma \hat{h}_+ = 0$. This implies that the adjoint of σ, $\sigma^\dagger$, acts as a *lowering operator*: $\sigma^\dagger \hat{h}_+ = \hat{h}_-$ and $\sigma^\dagger \hat{h}_- = 0$.

> *Remark:* The operators σ and $\sigma^\dagger$ behave like fermionic operators, and in fact they satisfy the CAR $\{\sigma^\dagger, \sigma\} = \mathbb{1}$, with $\sigma^2 = 0$. This means that $\hat{h}_-$ is the vacuum of $\sigma^\dagger$, while $\hat{h}_+$ is the excited state, see Section 2.2.

One can check that $\sigma \sigma^\dagger = P_{+,+}^{(h)}$, while $\sigma^\dagger \sigma = P_{-,-}^{(h)}$, so that the commutator between σ and $\sigma^\dagger$ is easily computed: $[\sigma, \sigma^\dagger] := \sigma \sigma^\dagger - \sigma^\dagger \sigma = H$. Moreover, we have

$$P_{j,k}^{(h)} P_{k,j}^{(h)} = P_{j,j}^{(h)}, \quad [H, \sigma] = 2H\sigma = 2P_{+,-}^{(h)}, \quad [H, \sigma^\dagger] = -2\sigma^\dagger H = -2P_{-,+}^{(h)}$$

and, probably more interesting,

$$\{H, \sigma\} = 0, \tag{12.2}$$

which also implies that $\{H, \sigma^\dagger\} = 0$: the ladder operators σ and $\sigma^\dagger$ anticommute with H.

What we have done for H can be repeated for the *employment operator* E. As for H, we assume that $\sigma(E) = \{-1, +1\}$, and we call $\hat{e}_\pm$ its eigenstates, respectively related to the eigenvalues $e_\pm = \pm 1$: $E \hat{e}_\pm = e_\pm \hat{e}_\pm$. $\mathcal{F}_E = \{\hat{e}_+, \hat{e}_-\}$ is now an o.n. basis of $\mathcal{H}_E = \mathbb{C}^2$, in general different from $\mathcal{F}_H$. In fact, in [67], this difference was a measure of the degree of compatibility between Q_1 and Q_2. Incidentally, we see that the two Hilbert spaces $\mathcal{H}_H$ and $\mathcal{H}_E$ coincide, at least as mathematical objects: $\mathcal{H}_H = \mathcal{H}_E = \mathbb{C}^2$. Of course, it is easy to introduce the analogous of the operators $P_{j,k}^{(h)}$

---

[1] In this simple case the spectrum of H, $\sigma(H)$, is just the set of its eigenvalues. The definition of the spectrum of an operator, in general, can be found in [27].

and σ, $P_{j,k}^{(e)}$ and $\sigma^{(e)}$, starting from $\mathcal{F}_E$, and to prove for them properties similar to those deduced above for H. For instance,

$$E = P_{+,+}^{(e)} - P_{-,-}^{(e)}, \qquad \{E, \sigma^{(e)}\} = 0,$$

and so on.

Once we have decided to use a quantum mechanical setting in the analysis of Q_1 and Q_2, and we have identified the relevant observables for the system, H and E, it is natural to expect that Alice's time evolution is driven by a third Hermitian operator, the Hamiltonian $\mathfrak{H}$ of the system, which, in our particular situation, should be a two-by-two matrix, at least if we consider the two questions separately. This is, of course, because the Hilbert space of each question is two-dimensional. But this has a strong implication, as we have already pointed out several times in this book. Suppose we are interested in question Q_1, and let Ψ_0 be the related vector describing the initial state of Alice and let $\mathfrak{H}$ be the related self-adjoint Hamiltonian. Hence $\Psi(t) = e^{i\mathfrak{H}t}\Psi_0$, and the mean value of each observable $\hat{X}$ at time t is $X(t) = \langle \Psi(t), \hat{X}\Psi(t)\rangle$, which is necessarily periodic or quasi-periodic in time, depending on the relation between the eigenvalues of $\mathfrak{H}$ which, of course, are not known a priori, in general. And this is true also for Hilbert spaces of finite dimensions but larger than two. In particular, this would imply that Alice oscillates (in some possibly nontrivial way) between being happy or not, or being employed or not.[2] In some articles, see [67], this problem is solved by simply fixing a *decision time* τ, which, independently of the oscillations intrinsically connected to the model, picks up particular values of the solutions. As we have already discussed many times in this book, we prefer to adopt a different strategy. In particular, we use the *environment of the system*, i.e., what is surrounding Alice, in our particular situation. In this case we can still write a self-adjoint Hamiltonian $\mathfrak{H}_{ext}$ that takes into account also the interactions between Alice and the outer world, or between Alice and her brain, see Chapter 11, and, with this refinement, the function $X(t)$ is no longer necessarily periodic or quasi-periodic. We refer to Chapters 3, 10 and 11 for several applications of this simple idea to different situations.

Other possible strategies also exist, as discussed in Chapter 2. For instance, one could use non-Hermitian effective Hamiltonians to describe the physical system, especially to describe the existence of some equilibrium state, or replace the Hamiltonian dynamics with some master equation. In both these cases, however, one loses the details of the interactions between the agents of the system, looking for some efficient description based on only a few parameters, which one tries to understand and to fix phenomenologically. This is the point of view, for instance,

---

[2] Of course being employed or not is more an objective than a subjective aspect of life. However, it may happen than Alice *feels* she is employed, since she is almost sure her part-time work is going to become a full-time job. Or something like this.

in [108, 117]. However, since in this chapter dynamics will not be an issue for us, we will not say more on this particular aspect of the problem, referring the reader to the previous chapters for more comments.

12.2.2 A Unified Scheme for Compatible and Incompatible Questions

As clearly stated in [67], the questions Q_1 and Q_2 can be thought to be compatible or incompatible: suppose Alice is unemployed and she is asked first the question Q_2. By replying to it she modifies her original mental state. In this new (*post-measurement*) state her answer to the next question, Q_1, can be different from her possible answer to Q_1 as the first question, i.e., in the initial mental state. This is because a first question can *activate thoughts*.

In this section we propose a unified framework that can be used both for compatible and for incompatible questions. Also, rather than measuring the degree of compatibility of the questions by means of the angle between, say, h_+ and e_+, as in [67], we propose a suitable deformation of two operators connected to H and E, which allow including in the same settings the two different situations, by means of some relevant commutators.

For that we adopt as our Hilbert space $\mathcal{H} = \mathbb{C}^4$, and an o.n. basis for $\mathcal{H}$ can be constructed out of $\mathcal{F}_E$ and $\mathcal{F}_H$ using the following vectors:

$$\varphi_{+,+} = \hat{e}_+ \otimes \hat{h}_+, \quad \varphi_{+,-} = \hat{e}_+ \otimes \hat{h}_-, \quad \varphi_{-,+} = \hat{e}_- \otimes \hat{h}_+, \quad \varphi_{-,-} = \hat{e}_- \otimes \hat{h}_-.$$

Then we put $\mathcal{F}_\varphi = \{\varphi_{\alpha,\beta}, \alpha, \beta = \pm\}$. The first index refers to the eigenvalues of E, while the second to those of H. Of course we have

$$\left\langle \varphi_{\alpha,\beta}, \varphi_{\alpha',\beta'} \right\rangle = \delta_{\alpha,\alpha'} \delta_{\beta,\beta'}.$$

Here the scalar product in $\mathcal{H}$ is related to those in $\mathcal{H}_E$, $\langle .,. \rangle_E$, and in $\mathcal{H}_H$, $\langle .,. \rangle_H$, as follows:

$$\langle f_E \otimes f_H, g_E \otimes g_H \rangle = \langle f_E, g_E \rangle_E \langle f_H, g_H \rangle_H,$$

for all $f_E, g_E \in \mathcal{H}_E$ and $f_H, g_H \in \mathcal{H}_H$. Of course, $\langle .,. \rangle_E$ and $\langle .,. \rangle_H$ both coicide with the standard scalar product in $\mathbb{C}^2$, $\langle f, g \rangle = \bar{f}_1 \, g_1 + \bar{f}_2 \, g_2$, where $f = (f_1, f_2)$ and $g = (g_1, g_2)$.

The various $\varphi_{\alpha,\beta}$ are common eigenstates of $\tilde{E} := E \otimes \mathbb{1}_H$ and $\tilde{H} = \mathbb{1}_E \otimes H$. Here $\mathbb{1}_E$ and $\mathbb{1}_H$ are the identity operators on $\mathcal{H}_E$ and $\mathcal{H}_H$ respectively, and the tensor product of two operators X and Y acting respectively on $\mathcal{H}_E$ and $\mathcal{H}_H$ is defined as follows:

$$(X \otimes Y)(f_E \otimes f_H) = (Xf_E) \otimes (Yf_H),$$

for all $f_E \in \mathcal{H}_E$ and $f_H \in \mathcal{H}_H$. We have

$$\tilde{E} \, \varphi_{\alpha,\beta} = e_\alpha \varphi_{\alpha,\beta}; \qquad \tilde{H} \, \varphi_{\alpha,\beta} = h_\beta \varphi_{\alpha,\beta} \tag{12.3}$$

for all α, β. We recall that $e_\alpha, h_\beta = \pm 1$. Of course $\tilde{H}$ and $\tilde{E}$ commute: $[\tilde{E}, \tilde{H}] = 0$. This is because they admit a common set of eigenstates, $\mathcal{F}_\varphi$. In fact, it is well known that two (Hermitian) operators commute if and only if they have a common set of eigenstates [2, 3].

Two relevant quantities in our analysis are the following uncertainties, defined as in (2.44):

$$(\Delta\tilde{H})^2 = \left\langle \Psi, \left(\tilde{H} - \langle\tilde{H}\rangle\right)^2 \Psi \right\rangle = \left\langle \tilde{H}^2 \right\rangle - \left\langle \tilde{H} \right\rangle^2 = \left\| \left(\tilde{H} - \langle\tilde{H}\rangle\right) \Psi \right\|^2, \qquad (12.4)$$

with a similar formula for $(\Delta\tilde{E})^2$. Incidentally, we observe that $\langle\tilde{H}\rangle$ and $(\Delta\tilde{H})$ depend on the normalized vector Ψ, but we are not making explicit this dependence here, to simplify the notation.

As we have seen in Section 2.5, whenever two Hermitian operators A and B satisfy $[A, B] = i\, C^\dagger$, for some $C = C^\dagger$, it follows that $\Delta A\, \Delta B \geq \frac{|\langle C\rangle|}{2}$, see (2.45). In particular, since $\tilde{E}$ and $\tilde{H}$ commute, $C = 0$, and this inequality becomes $\Delta\tilde{E}\,\Delta\tilde{H} \geq 0$. This is not particularly useful for us, since we already know from (12.4) that both $\Delta\tilde{E}$ and $\Delta\tilde{H}$ are nonnegative quantities, for any normalized vector $\Psi \in \mathcal{H}$. If, in particular, Ψ is an eigenstate of $\tilde{E}$, then $\Delta\tilde{E} = 0$. Analogously, if Ψ is an eigenstate of $\tilde{H}$, then $\Delta\tilde{H} = 0$. Of course, it is further possible that Ψ is a common eigenstate of $\tilde{E}$ and $\tilde{H}$, since $[\tilde{E}, \tilde{H}] = 0$; then $\Delta\tilde{E}$ and $\Delta\tilde{H}$ are both zero, and the inequality in (2.45) would become, in this particular situation, $0 \geq 0$. This implies that, since $\tilde{E}$ and $\tilde{H}$ commute, we are able to answer, with no uncertainty, Q_1 and Q_2. Notice, however, that this is true if Alice's mental state Ψ is a common eigenstate of $\tilde{E}$ and $\tilde{H}$. If Ψ is not such an eigenstate of $\tilde{E}$ or $\tilde{H}$, but is some superposition of the vectors in $\mathcal{F}_\varphi$, we are not really able to know with certainty the status of Alice. In other words, Q_1 and Q_2 are still compatible, but the state of Alice does not allow us to have any certain answer for our questions.

> *Remark:* Of course, this aspect of the framework is also time-dependent. It might easily happen that, at $t = 0$, Ψ is an eigenstate of $\tilde{E}$ and $\tilde{H}$, and that, after some time t_0, and after some thoughts, Alice's state changes to $\Psi(t_0)$, which is no longer such an eigenstate. This depends, of course, on the Hamiltonian for Alice. But we will not consider further this aspect here.

The key point in our analysis, now, is quite elementary from a mathematical point of view: we deform continuously $\tilde{E}$ and $\tilde{H}$, producing two new operators which, in general, do not commute anymore, and we use this non-commutativity as a measure of the compatibility of the questions.

Using the standard definitions of the tensor products, and observing that both H and E are diagonal two-by-two matrices with diagonal entries 1 and -1, we have

$$\tilde{H} = 1\!\!1_E \otimes H = \begin{pmatrix} 1 & 0 \\ 0 & 1 \end{pmatrix} \otimes \begin{pmatrix} 1 & 0 \\ 0 & -1 \end{pmatrix} = \begin{pmatrix} 1 & 0 & 0 & 0 \\ 0 & -1 & 0 & 0 \\ 0 & 0 & 1 & 0 \\ 0 & 0 & 0 & -1 \end{pmatrix},$$

$$\tilde{E} = E \otimes 1\!\!1_H = \begin{pmatrix} 1 & 0 \\ 0 & -1 \end{pmatrix} \otimes \begin{pmatrix} 1 & 0 \\ 0 & 1 \end{pmatrix} = \begin{pmatrix} 1 & 0 & 0 & 0 \\ 0 & 1 & 0 & 0 \\ 0 & 0 & -1 & 0 \\ 0 & 0 & 0 & -1 \end{pmatrix},$$

while $\varphi_{+,+}$, $\varphi_{+,-}$, $\varphi_{-,+}$ and $\varphi_{-,-}$ coincide with the canonical o.n. basis of $\mathbb{C}^4$, $\mathcal{F}_e = \{\hat{e}_j, j = 1, 2, 3, 4\}$. Here $\hat{e}_j$ is the four-dimensional vector with all zero components except the j-th one, which is one. Incidentally, these explicit expressions of $\tilde{H}$ and $\tilde{E}$ make it clear why these operators commute: they are both diagonal, but with different orders of ± 1 in the main diagonals.

It is well known that one can use unitary (or just invertible) operators to modify a given operator, preserving its eigenvalues and producing eigenvectors related to those of the original operator: if X is a given Hermitian operator, α and f one eigenvalue and its related eigenvector, $Xf = \alpha f$, and if we consider an invertible operator S, we can define $X_S = SXS^{-1}$ and $f_S = Sf$, and we have $X_S f_S = \alpha f_S$. Then α is also an eigenvalue of X_S, with eigenvector f_S. Notice that $X_S = X_S^\dagger$ if S is unitary, but not in general. So our idea is to deform $\tilde{H}$ and $\tilde{E}$ using this easy scheme. However, this is not enough, since the main aim of our procedure is to produce operators that do not commute. This is impossible if we use only a single invertible operator S to deform both $\tilde{E}$ and $\tilde{H}$, since $S\tilde{E}S^{-1}$ and $S\tilde{H}S^{-1}$ commute in the same way as $\tilde{E}$ and $\tilde{H}$ do:

$$\left[S\tilde{E}S^{-1}, S\tilde{H}S^{-1} \right] = S\tilde{E}S^{-1}S\tilde{H}S^{-1} - S\tilde{H}S^{-1}S\tilde{E}S^{-1} = S\left[\tilde{E}, \tilde{H} \right] S^{-1} = 0.$$

The natural way out is to use two different invertible operators, one for $\tilde{H}$ and a different one for $\tilde{E}$. Of course, the choice of these operators is quite arbitrary, and highly non-unique. However, it can be driven by some simple continuity requirements: in particular we want these deformed versions of $\tilde{H}$ and $\tilde{E}$ to depend on one single parameter (this is the simplest option, of course), which, when sent to, say, zero, returns the undeformed operators. This parameter, θ, will be called the *compatibility parameter*, for reasons that will be clarified soon.

Our explicit choice, here, is given by the following unitary matrices:

$$R_\theta = \begin{pmatrix} \cos\theta & -\sin\theta & 0 & 0 \\ \sin\theta & \cos\theta & 0 & 0 \\ 0 & 0 & \cos\theta & -\sin\theta \\ 0 & 0 & \sin\theta & \cos\theta \end{pmatrix}, \quad V_\theta = \begin{pmatrix} 1 & 0 & 0 & 0 \\ 0 & \cos\theta & -\sin\theta & 0 \\ 0 & \sin\theta & \cos\theta & 0 \\ 0 & 0 & 0 & 1 \end{pmatrix},$$

$$(12.5)$$

where θ is a parameter that we take in $[0, 2\pi]$. It is clear that $R_\theta^\dagger = R_\theta^{-1} = R_{-\theta}$, $V_\theta^\dagger = V_\theta^{-1} = V_{-\theta}$ and that $R_\theta \to 1\!\!1$ and $V_\theta \to 1\!\!1$ when $\theta \to 0$. Here $1\!\!1$ is the identity matrix in $\mathcal{H}$. We see that R_θ rotates two orthogonal subspaces of $\mathcal{H}$, the one generated by the first two vectors in $\mathcal{F}_\varphi$, $\varphi_{+,+}$ and $\varphi_{+,-}$, and the other one, generated by $\varphi_{-,+}$ and $\varphi_{-,-}$. These spaces are stable under the action of R_θ. As for V_θ, this is also a double rotation,[3] but on different subspaces of $\mathcal{H}$.

Now we use R_θ and V_θ to define two new Hermitian operators

$$H_\theta = R_\theta \tilde{H} R_\theta^{-1} = \begin{pmatrix} \cos 2\theta & \sin 2\theta & 0 & 0 \\ \sin 2\theta & -\cos 2\theta & 0 & 0 \\ 0 & 0 & \cos 2\theta & \sin 2\theta \\ 0 & 0 & \sin 2\theta & -\cos 2\theta \end{pmatrix} \qquad (12.6)$$

and

$$E_\theta = V_\theta \tilde{E} V_\theta^{-1} = \begin{pmatrix} 1 & 0 & 0 & 0 \\ 0 & \cos 2\theta & \sin 2\theta & 0 \\ 0 & \sin 2\theta & -\cos 2\theta & 0 \\ 0 & 0 & 0 & -1 \end{pmatrix}. \qquad (12.7)$$

We also define two new sets of vectors, $\mathcal{F}_\varphi^H = \{\varphi_{\alpha,\beta}^H = R_\theta \varphi_{\alpha,\beta}, \ \alpha, \beta = \pm\}$ and $\mathcal{F}_\varphi^E = \{\varphi_{\alpha,\beta}^E = V_\theta \varphi_{\alpha,\beta}, \ \alpha, \beta = \pm\}$. The explicitly expressions of these vectors are

$$\varphi_{+,+}^H = \begin{pmatrix} \cos\theta \\ \sin\theta \\ 0 \\ 0 \end{pmatrix}, \quad \varphi_{+,-}^H = \begin{pmatrix} -\sin\theta \\ \cos\theta \\ 0 \\ 0 \end{pmatrix}, \quad \varphi_{-,+}^H = \begin{pmatrix} 0 \\ 0 \\ \cos\theta \\ \sin\theta \end{pmatrix}, \quad \varphi_{-,-}^H = \begin{pmatrix} 0 \\ 0 \\ -\sin\theta \\ \cos\theta \end{pmatrix},$$

and

$$\varphi_{+,+}^E = \begin{pmatrix} 1 \\ 0 \\ 0 \\ 0 \end{pmatrix}, \quad \varphi_{+,-}^E = \begin{pmatrix} 0 \\ \cos\theta \\ \sin\theta \\ 0 \end{pmatrix}, \quad \varphi_{-,+}^E = \begin{pmatrix} 0 \\ -\sin\theta \\ \cos\theta \\ 0 \end{pmatrix}, \quad \varphi_{-,-}^E = \begin{pmatrix} 0 \\ 0 \\ 0 \\ 1 \end{pmatrix}.$$

It is evident that both $\mathcal{F}_\varphi^H$ and $\mathcal{F}_\varphi^E$ are (different) o.n. bases for $\mathcal{H}$, and this was expected since R_θ and V_θ are unitary operators. Also, each $\varphi_{\alpha,\beta}^H$ is an eigenstate of H_θ, while the $\varphi_{\alpha,\beta}^E$'s are eigenstates of E_θ. The eigenvalues are ± 1, each with multiplicity two. Also, $\mathcal{F}_\varphi^H$ and $\mathcal{F}_\varphi^E$ are mutually related by the unitary operator $T_\theta = R_\theta V_\theta^{-1}$: $\varphi_{\alpha,\beta}^H = T_\theta \varphi_{\alpha,\beta}^E$, for all α and β. Moreover, when $\theta \to 0$, both $\mathcal{F}_\varphi^H$ and $\mathcal{F}_\varphi^E$ converge[4] to the original set $\mathcal{F}_\varphi$ and, in the same limit, $H_\theta \to \tilde{H}$ and $E_\theta \to \tilde{E}$.

---

[3] One of these rotations is trivial.

[4] Here convergence is meant in a very naive sense, i.e., as the convergence of each component of the matrices and vectors considered. This is sufficient for our present aims.

The essential aspect of our deformed operators is that, while $[\tilde{E}, \tilde{H}] = 0$, E_θ and H_θ do not commute anymore, in general, in fact they satisfy the following:

$$[E_\theta, H_\theta] = iU_\theta, \tag{12.8}$$

where

$$U_\theta = \begin{pmatrix} 0 & -4i\cos\theta\sin^3\theta & i\sin^2 2\theta & 0 \\ 4i\cos\theta\sin^3\theta & 0 & -i\sin 4\theta & -i\sin^2 2\theta \\ -i\sin^2 2\theta & i\sin 4\theta & 0 & -4i\cos\theta\sin^3\theta \\ 0 & i\sin^2 2\theta & 4i\cos\theta\sin^3\theta & 0 \end{pmatrix}, \tag{12.9}$$

which is, in a sense, what we were looking for: two non-commuting operators related to Q_1 and Q_2. We see that $U_\theta = U_\theta^\dagger$ and that $U_\theta \to 0$ when $\theta \to 0$. This reflects the fact that, in this limit, the two operators E_θ and H_θ converge to the commuting matrices $\tilde{E}$ and $\tilde{H}$. This is the reason why θ is called the *compatibility parameter*. For the same reason, it is also natural to call the pair (R_θ, V_θ) the *compatibility operators*.

> *Remark:* Once again, we stress that the choice of these operators is highly not unique. This is, for us, a bonus of our strategy, since this extra freedom could be useful to adapt our framework to fit other aspects of the system we are considering, or of the questions we are dealing with. Moreover, it is not hard to imagine how the same idea behind our proposal can be generalized to a larger number of compatible/incompatible questions, independently of whether these questions are binary or not.

Now, Equation (12.8) implies that

$$\Delta E_\theta \, \Delta H_\theta \geq \frac{1}{2} \, |\langle U_\theta \rangle|, \tag{12.10}$$

where we recall that $\langle X \rangle = \langle \Psi, X\Psi \rangle$, $\Delta X = \langle X^2 \rangle - \langle X \rangle^2$, for $X = E_\theta, H_\theta, U_\theta$, and Ψ is a fixed normalized vector of $\mathcal{H}$. This inequality is sometimes called the Heisenberg-Robertson inequality.

Let us consider first the case when Alice's state is an eigenstate of H: $\Psi = \varphi^H_{\alpha,\beta}$, for some α and β. Then $\Delta H_\theta = \langle U_\theta \rangle = 0$, while $\Delta E_\theta \geq 0$, in general. This means that we know if Alice is happy or not, but we do not know exactly if she is employed. In fact, if we take for concreteness $\Psi = \varphi^H_{+,+}$, then

$$(\Delta E_\theta)^2 = 2\cos^2(\theta)(3 - \cos(2\theta))\sin^4(\theta),$$

which is clearly not identically zero. The plot of $(\Delta E_\theta)^2$ is given in Figure 12.1, which shows that, for almost all values of $\theta \in [0, 2\pi]$, ΔE_θ is not zero. This reflects the impossibility of having an exact idea of the employment of Alice. Moreover it

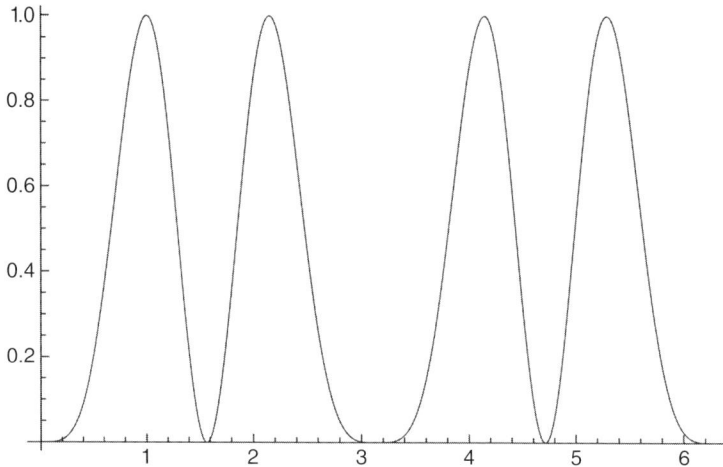

Figure 12.1 $(\Delta E_\theta)^2$ as a function of θ.

is clear that this lack of knowledge increases in some range of θ, then decreases to increase again, and so on.

However, $\Delta E_\theta = 0$, as expected, if $\theta = 0$. But it is also zero if $\theta = \pi, 2\pi$ and if $\theta = \frac{\pi}{2}, \frac{3\pi}{2}$. To understand these results it is enough to observe that $H_\theta = \tilde{H}$ and $E_\theta = \tilde{E}$ if $\theta = 0, \pi, 2\pi$. For these values of θ, because of our choice of R_θ and V_θ we go back to the original commuting operators, so that $[H_\theta, E_\theta] = 0$. Analogously, for $\theta = \frac{\pi}{2}$, we find that $H_\theta = -\tilde{H}$ and $E_\theta = \tilde{E}$, so that, again, $[H_\theta, E_\theta] = 0$. A similar conclusion can be deduced for $\theta = \frac{3\pi}{2}$. This commutativity can also be easily recovered by looking directly at U_θ in (12.9), which is zero for all these values of θ. Then also the deformed versions of $\tilde{E}$ and $\tilde{H}$ may correspond to compatible questions, at least for these particular values of θ, and Equation (12.10) becomes a strict equality, $0 = 0$. The situation is completely analogous if we assume that, at $t = 0$, Alice's state is some eigenstate of E_θ, $\Psi = \varphi^E_{\alpha,\beta}$. In this case ΔE_θ is zero, for each θ, while ΔH_θ exhibits an oscillating behavior similar to the one plotted in Figure 12.1. Moreover, even if U_θ is not always zero, its mean value is zero $\langle U_\theta \rangle = 0$, independently of the value of θ.

Let us now see what happens if Alice is described by a generic normalized vector

$$\Psi = \begin{pmatrix} c_{+,+} \\ c_{+,-} \\ c_{-,+} \\ c_{-,-} \end{pmatrix}, \tag{12.11}$$

with $\sum_{\alpha,\beta=\pm} |c_{\alpha,\beta}|^2 = 1$. Since $E_\theta^2 = H_\theta^2 = 1\!\!1$ for all values of θ, we have $\langle E_\theta^2 \rangle = \langle H_\theta^2 \rangle = 1$. Moreover, simple computations give

$$\langle E_\theta \rangle = |c_{+,+}|^2 - |c_{-,-}|^2 + \left(|c_{+,-}|^2 - |c_{-,+}|^2 \right) \cos(2\theta)$$
$$+ (c_{+,-} \overline{c_{-,+}} + c_{-,+} \overline{c_{+,-}}) \sin(2\theta),$$

while

$$\langle H_\theta \rangle = \left(|c_{+,+}|^2 + |c_{-,+}|^2 - |c_{+,-}|^2 - |c_{-,-}|^2 \right) \cos(2\theta)$$
$$+ 2\Re \; (c_{+,+} \overline{c_{+,-}} + c_{-,-} \overline{c_{-,+}}) \sin(2\theta).$$

Therefore $(\Delta E_\theta)^2 = 1 - \langle E_\theta \rangle^2$ while $(\Delta H_\theta)^2 = 1 - \langle H_\theta \rangle^2$.

As for $\langle \Psi, U_\theta \Psi \rangle$, using the explicit expression of U_θ we deduce the following interesting result: if $c_{\alpha,\beta}$, $\alpha, \beta = \pm$, are all real-valued, then $\langle \Psi, U_\theta \Psi \rangle = 0$. This is because

$$\langle \Psi, U_\theta \Psi \rangle = 2i \left[u_{12} \Im \left(\overline{c_{+,+}} c_{+,-} + \overline{c_{-,+}} c_{+,+} \right) + u_{13} \Im \left(\overline{c_{+,+}} c_{-,+} + \overline{c_{-,-}} c_{+,-} \right) \right.$$
$$\left. + u_{23} \Im \left(\overline{c_{+,-}} c_{-,+} \right) \right],$$

where $\Im(z)$ is the imaginary part of z, and u_{jk} are the entries of U_θ: $u_{12} = -4i \cos(\theta) \sin^3(\theta)$, $u_{13} = i \sin^2(2\theta)$ and $u_{23} = -i \sin(4\theta)$. This means that, under this assumption on $c_{\alpha,\beta}$, inequality (12.10) is trivially satisfied. Notice that neither ΔE_θ nor ΔH_θ are also zero, in general.

Let us now see what happens for a few explicit choices of Ψ.

The first concrete situation we want to consider now is produced by the following vector $\Psi_a^T = \frac{1}{2}(1, 1, 1, 1)$. Here Ψ_a^T is the transpose of Ψ_a. In Figure 12.2 we plot $(\Delta E_\theta)^2$ (a), $(\Delta H_\theta)^2$ (b), and the product of these two (c).

We see that, in this case, both uncertainties oscillates with θ, but while the status of the happiness can be known exactly for some particular values of θ (those for which $\Delta H_\theta = 0$), it is impossible to have a net answer for Q_2, since ΔE_θ is never zero: we always have some uncertainty about Alice's employment, with this choice of Ψ_a.

In we rather consider the vector $\Psi_b^T = \frac{1}{3}(1, \sqrt{2}, \sqrt{3}, \sqrt{3})$, Figure 12.3 shows that neither ΔH_θ nor ΔE_θ are zero, even if ΔH_θ approaches zero for some θ. This means that there exists a certain uncertainty on both Q_1 and Q_2 for all values of θ, even for those for which $[E_\theta, H_\theta] = 0$, but while ΔH_θ can become small for certain values of θ, ΔE_θ cannot. So we can be almost sure about Q_1, but we cannot say much about Q_2.

Another interesting situation is when the state of Alice depends explicitly on θ. For instance, if we consider the third normalized vector $\Psi_c^T = (\cos(\theta), \frac{\sin(\theta)}{\sqrt{2}},$ $\frac{\sin(\theta)}{\sqrt{2}}, 0)$, the results are given in Figure 12.4.

In this situation we have three values of θ, $\theta = 0, \pi, 2\pi$, for which $\Delta E_\theta = \Delta H_\theta = 0$ simultaneously,[5] but for all the other values of θ, the uncertainty ΔH_θ is different

---

[5] This in not surprising, since when $\theta = 0, \pi, 2\pi$ the vector Ψ_c is equal, or proportional, to $\varphi_{+,+}$. Hence Ψ_c is a common eigenstate of $\tilde{H}$ and $\tilde{E}$, in these cases.

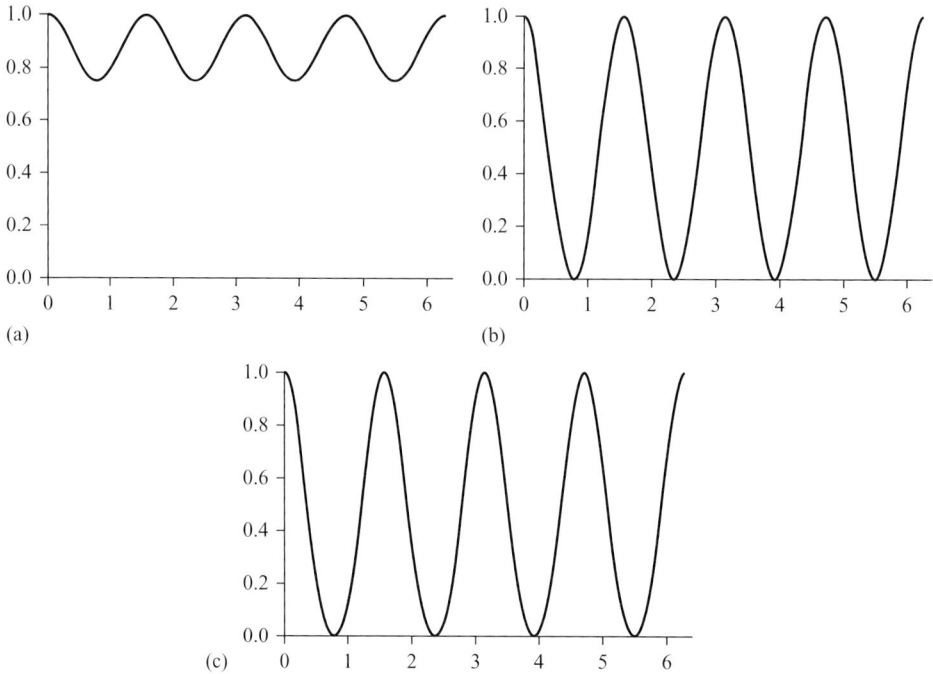

Figure 12.2 $(\Delta E_\theta)^2$ (a), $(\Delta H_\theta)^2$ (b), and the product of these two (c), as functions of θ for Ψ_a.

from zero (we cannot know if Alice is happy or not). However, more values of θ exist for which $\Delta E_\theta = 0$, so that we could still know if she is employed or not.

Let us now see what happens when the right-hand side of Robertson inequality in Equation (12.10) is also nonzero, and, in particular, let us compare the two sides of Equation (12.10) as functions of θ. As we have discussed before, not all the coefficients of Ψ in Equation (12.11) can be taken to be real. Otherwise $\langle \Psi, U_\theta \Psi \rangle = 0$, which is not what we want, now. We slightly modify our choices above, considering the following vectors:

$$\tilde{\Psi}_a = \frac{1}{2}\begin{pmatrix} i \\ 1 \\ 1 \\ 1 \end{pmatrix}, \qquad \tilde{\Psi}_b = \frac{1}{3}\begin{pmatrix} i \\ \sqrt{2} \\ \sqrt{3} \\ \sqrt{3} \end{pmatrix}, \qquad \tilde{\Psi}_c = \frac{1}{3}\begin{pmatrix} i\cos(\theta) \\ \frac{\sin(\theta)}{\sqrt{2}} \\ \frac{\sin(\theta)}{\sqrt{2}} \\ 0 \end{pmatrix}.$$

These vectors differ from those considered before only for the first component, which is taken as purely imaginary. Figures 12.5–12.7 show our results, see also [16]: the bold solid lines refer to the product of standard deviations in (12.10), while the dashed, gray line, to the term $\frac{1}{2}|\langle U_\theta \rangle|$.

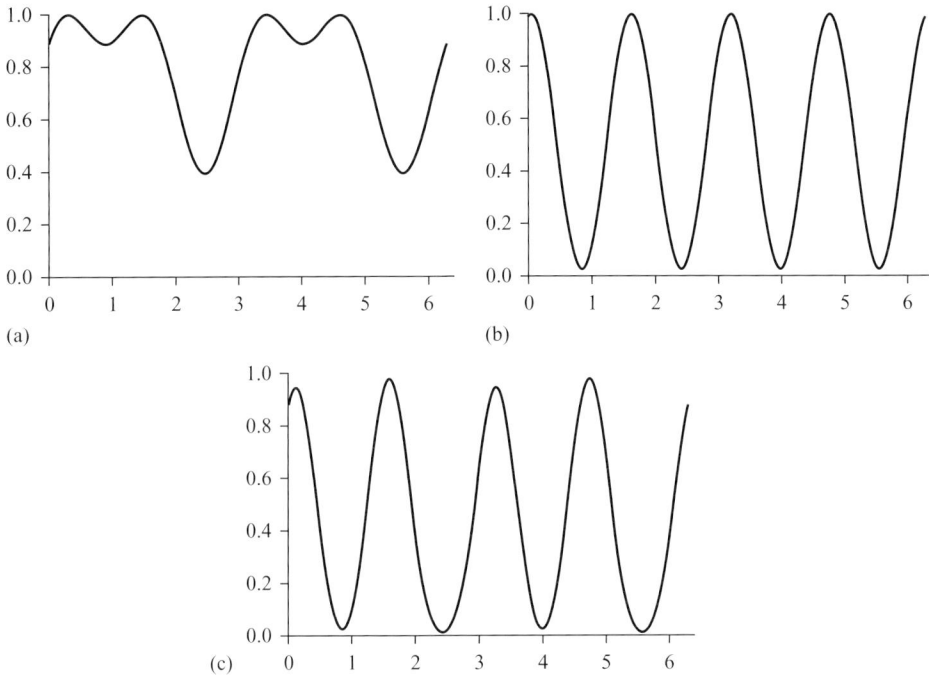

Figure 12.3 $(\Delta E_\theta)^2$ (a), $(\Delta H_\theta)^2$ (b), and the product of these two (c), as functions of θ for Ψ_b.

We see that both sides of the Robertson inequality fluctuate with respect to the deformation parameter θ. In complete accordance with quantum theory, the graph of the right-hand side is majorized by the graph of the left-hand side. In particular, the product of uncertainties can vanish only for those values of θ for which the average of commutator vanishes as well, see Figure 12.7. Of course, the opposite relation is not necessarily true, as the same figures clearly show.

Remarks: (1) We observe that, for all those values of θ for which $U_\theta = 0$, we go back to the case of compatible questions: in this case (consider, in particular, $\theta = 0$), we recover the o.n. basis considered in [67], and E_θ and H_θ become those considered in that article. So everything discussed in [67] can be restated in our framework.

(2) It is useful to notice that the idea described here can be easily extended to other questions that can be compatible or not, not necessarily living in a four-dimensional Hilbert space. This is the case, for instance, of the following two questions: Q_1: Is Alice employed or not? and Q_2: Is Alice somehow connected with some politician of the right wing of the parliament? Or of the left wing? Or of the center? Or, maybe, is Alice not connected to any politician at all? Of course, the idea behind these questions is that the politician can help Alice, or not, to find a job. In this case we have an eight-dimensional Hilbert space, obtained as the tensor product of $\mathcal{H}_E$ and $\mathcal{H}_P$,

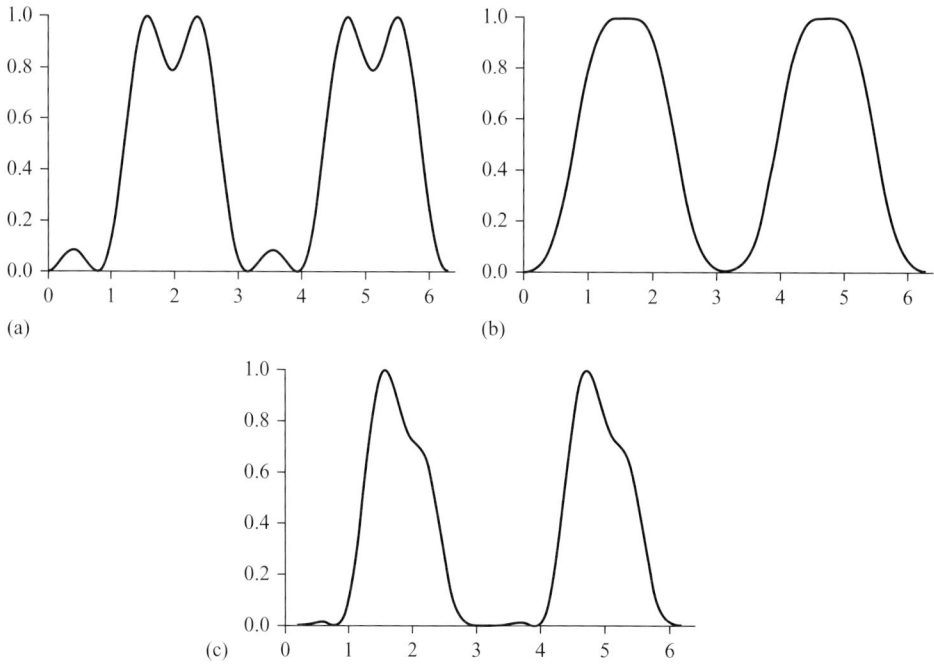

Figure 12.4 $(\Delta E_\theta)^2$ (a), $(\Delta H_\theta)^2$ (b), and the product of these two (c), as functions of θ for Ψ_c.

the four-dimensional Hilbert space associated with Q_2. Of course, in situations like this, it is not easy (or not even possible) to compare orthonormal bases as proposed in [67], since Q_1 and Q_2 are naturally *attached* to different Hilbert spaces, with different dimensions. On the contrary, the deformation parameter θ can always be defined, independently of the dimension of the Hilbert space where the model is defined and of the different Hilbert spaces related to Q_1 and Q_2.

(3) It is not hard to extend our framework to more than two questions, which could be (again) compatible or not. In this case, more than just two operators should be deformed, giving rise to a new set of, in general, non-commuting Hermitian matrices for which the same analysis proposed here can be repeated. In this case we would obtain a set of inequalities like the one in Equation (12.10), one for each pair of (non-commuting) observables.

(4) If we compare Figure 12.5 with the plot in Figure 12.2(c), we see a different behavior, even if the difference between the two different situations is just a complex value (i) as the first component of the two vectors Ψ_a and $\tilde{\Psi}_a$. More differences can also be observed by comparing Figures 12.6 and 12.3, and Figures 12.7 and 12.4. Then we conclude as in Section 11.4: phases (and relative phases) are important! They can produce serious differences both at a dynamical level, and in what happens even before the time evolution begins.

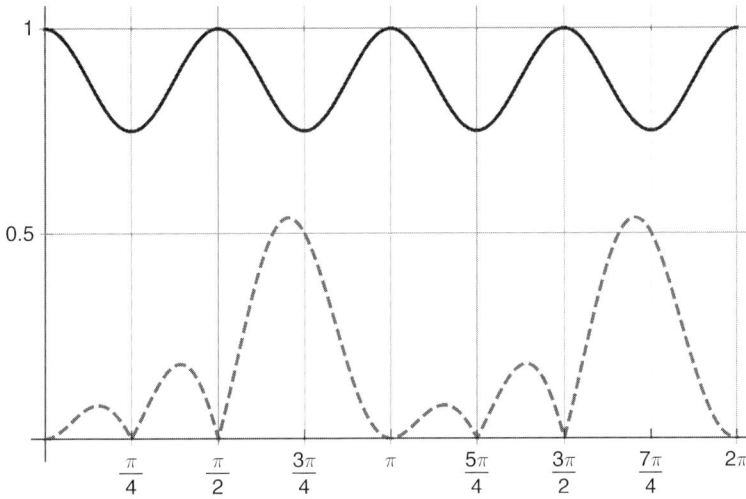

Figure 12.5 Dashed line: RHS of the Robertson inequality and the solid line: LHS of the Robertson inequality as functions of θ, for $\Psi = \tilde{\Psi}_a$ (Source: Bagarello et al. [16]).

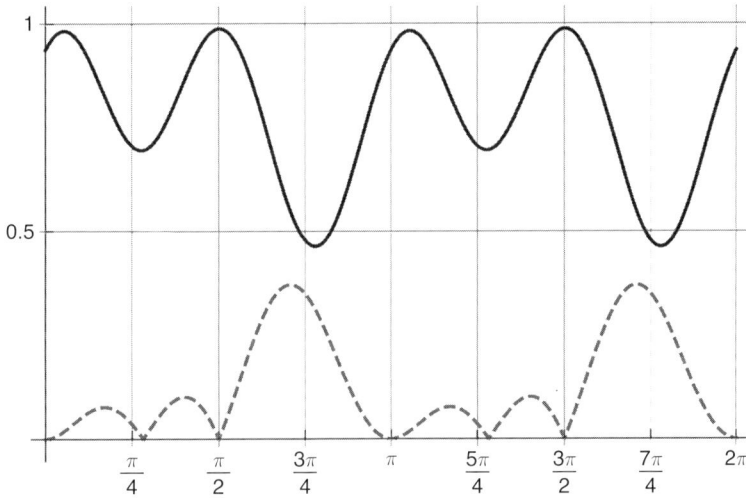

Figure 12.6 Dashed line: RHS of the Robertson inequality and the solid line: LHS the Robertson inequality as functions of θ, for $\Psi = \tilde{\Psi}_b$ (Source: Bagarello et al. [16]).

The analysis considered here, and in particular the role of commuting and non-commuting operators, or of operators with zero mean values, suggests to recall here the classification proposed by Ozawa in a slightly different context, in a series of articles [119–121]: let us consider two Hermitian operators A and B, acting on some

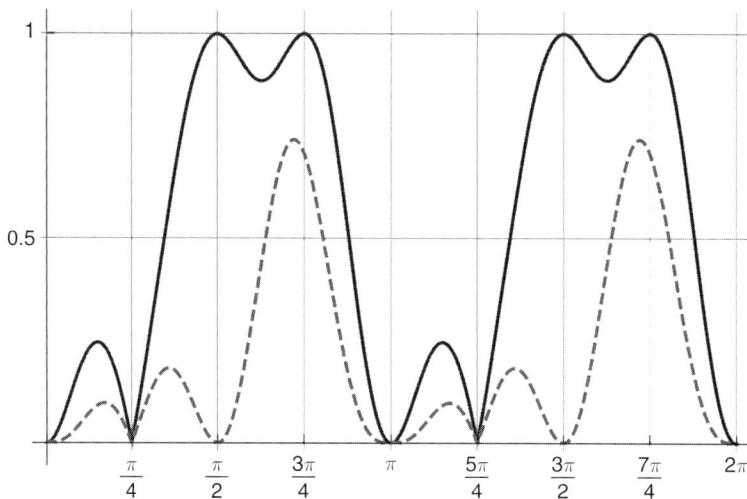

Figure 12.7 Dashed line: RHS of the Robertson inequality and the solid line: LHS the Robertson inequality as functions of θ, for $\Psi = \tilde{\Psi}_c$ (Source: Bagarello et al. [16]).

finite-dimensional Hilbert space, $\mathcal{H}$, and their spectral families $E^A(x), E^B(y)$. In our case these are just families of projectors on the eigen-subspaces of $\mathcal{H}$ of these operators, like the projectors $P_{+,+}^{(h)}$ and $P_{-,-}^{(h)}$ introduced in (12.1). The following possibilities are interesting:

1. A, B **commuting:** $[A, B] = 0$: there exists an orthonormal basis consisting of common eigenvectors.
2. A, B **non-commuting:** $[A, B] \neq 0$: an orthonormal basis consisting of common eigenvectors does not exist. Still, in principle, some common eigenvectors can exist.
3. A, B **nowhere-commuting: for any state** ψ, $[A, B]\psi \neq 0$: there exists no common eigenvector at all.
4. A, B **spectrally commuting in** ψ: $[E^A(x), E^B(y)]\psi = 0$: this vector ψ is a superposition of common eigenstates of A, B.
5. A, B **weakly commuting in** ψ: $\langle \psi, [A, B]\psi \rangle = 0$: this is the weakest version of commutativity.

Regarding our original questions Q_1 and Q_2, we are in the following position: the original operators $\tilde{H}$ and $\tilde{E}$ are commuting. Their deformed versions, H_θ and E_θ, are non-commuting, in general, but they can be commuting, or weakly commuting, for particular values of θ, or for some particular choice of Ψ. In our treatment, spectral commutativity has no role, as we have seen so far. We refer to [16] for some discussion on connections with probability theory of these commuting or weakly commuting operators.

12.3 An Application to Decision-Making and Order Effect

The problem we are going to consider in this section is quite simple, but rather interesting: let us suppose we ask a set of people, $\mathcal{P}$, two questions Q_C and Q_G in two different orders, $O_{C \to G}$ or $O_{G \to C}$. Here Q_C and Q_G stand respectively for *Is Clinton honest?* and *Is Gore honest?* Order $O_{C \to G}$ means that Q_C is asked first, and Q_G after, while $O_{G \to C}$ describes the reversed temporal order. The results of this experiment show that, if we adopt $O_{C \to G}$, then 50% of $\mathcal{P}$ answer yes to Q_C, while 68% of $\mathcal{P}$ answer yes to Q_G. On the other hand, if we consider the reverse order, $O_{G \to C}$, interestingly enough the output changes: in this case the above percentages change respectively to 57% and 60%. It is clear that the order is important and the explanation proposed in the literature is that *the first question activates thoughts*, see [67]. This is an example of what, in the literature, is called *order effect*, but is slightly different from a similar effect already considered in the previous section, in connection with the two questions Q_1 and Q_2. One evident difference is that, here, we are no longer dealing with a single agent, Alice, but with an entire group of people.

In this section we want to show how the general settings proposed before can be efficiently used in the description of this Gore–Clinton order effect.

We start introducing two Hermitian operators C and G, two-by-two matrices, with eigenvalues ± 1 and eigenvectors $\hat{c}_{\pm}$ and $\hat{g}_{\pm}$:

$$C\hat{c}_{\pm} = c_{\pm}\hat{c}_{\pm}, \qquad G\hat{g}_{\pm} = g_{\pm}\hat{g}_{\pm},$$

where $c_{\pm} = g_{\pm} = \pm 1$. Of course we have $\langle \hat{c}_{\alpha}, \hat{c}_{\beta} \rangle = \langle \hat{g}_{\alpha}, \hat{g}_{\beta} \rangle = \delta_{\alpha,\beta}$, while, in general $\langle \hat{c}_{\alpha}, \hat{g}_{\beta} \rangle > \neq \delta_{\alpha,\beta}$. The interpretation reflects the one for Q_1 and Q_2 of the previous section: the state Ψ of $\mathcal{P}$ is $\hat{c}_+$ if $\mathcal{P}$ thinks that Clinton is honest, while $\Psi = \hat{c}_-$ otherwise. Also, $\Psi = \hat{g}_+$ if $\mathcal{P}$ thinks that Gore is honest, while $\Psi = \hat{g}_-$ otherwise. In general, $\mathcal{P}$ is not expected to be in any such state, but in some suitable superposition. In complete analogy with what we have done for H and E, we introduce now the four-dimensional representations of C and G, and an o.n. basis $\mathcal{F}_{\Phi} = \{\Phi_{\alpha,\beta} = \hat{c}_{\alpha} \otimes \hat{g}_{\beta}\}$ in the Hilbert space $\mathcal{H} = \mathbb{C}^4$: $\tilde{C} = C \otimes \mathbb{1}$ and $\tilde{G} = \mathbb{1} \otimes G$. It is clear that $\tilde{C}$ and $\tilde{G}$ can be identified, as mathematical objects, with the matrices $\tilde{E}$ and $\tilde{H}$ of the previous section, while the set $\mathcal{F}_{\Phi}$ is analogous to $\mathcal{F}_{\varphi}$. For this reason we do not repeat the full procedure, but we list just a few relevant facts:

$$[\tilde{C}, \tilde{G}] = 0, \qquad \tilde{C}\Phi_{\alpha,\beta} = c_{\alpha}\Phi_{\alpha,\beta}, \qquad \tilde{G}\Phi_{\alpha,\beta} = g_{\beta}\Phi_{\alpha,\beta}.$$

Also, we can deform $\tilde{C}$ and $\tilde{G}$ by means of the same operators V_{θ} and R_{θ} which we have used in Section 12.2.2 respectively:

$$C_\theta = V_\theta \tilde{C} V_\theta^{-1}, \qquad G_\theta = R_\theta \tilde{G} R_\theta^{-1}, \qquad (12.12)$$

and we can introduce two new o.n. bases of $\mathcal{H}$ as

$$\mathcal{F}_\Phi^{\theta,C} = \left\{ \Phi_{\alpha,\beta}^{\theta,C} = V_\theta \Phi_{\alpha,\beta} \right\}, \qquad \mathcal{F}_\Phi^{\theta,G} = \left\{ \Phi_{\alpha,\beta}^{\theta,G} = R_\theta \Phi_{\alpha,\beta} \right\},$$

which are eigenstates of C_θ and G_θ, respectively. Of course, we get $[C_\theta, G_\theta] = iU_\theta$, where U_θ is given in Equation (12.9).

> *Remark:* As in Section 12.2, the choice of R_θ and V_θ as particular compatibility operators is rather arbitrary. However, as we will show soon, the results they produce are quite in agreement with the experimental data. This fully justifies our choice.

Let us now define the linear operators $P_{\alpha,\alpha}^{\theta,G}$ and $P_{\alpha,\alpha}^{\theta,C}$ as follows:

$$P_{\alpha,\alpha}^{\theta,G} f = \left\langle \Phi_{\alpha,\alpha}^{\theta,G}, f \right\rangle \Phi_{\alpha,\alpha}^{\theta,G}, \qquad P_{\alpha,\alpha}^{\theta,C} f = \left\langle \Phi_{\alpha,\alpha}^{\theta,C}, f \right\rangle \Phi_{\alpha,\alpha}^{\theta,C},$$

for all $f \in \mathcal{H}$, and for $\alpha = \pm$. Then $C_\theta = P_{+,+}^{\theta,C} - P_{-,-}^{\theta,C}$, and $G_\theta = P_{+,+}^{\theta,G} - P_{-,-}^{\theta,G}$.

Our idea is to use this framework in the analysis of the experimental data related to Q_C and Q_G. For that, we first list the following results:

$$p_{C \to G}(+,+) = 0.4899; \quad p_{C \to G}(+,-) = 0.0447;$$
$$p_{C \to G}(-,+) = 0.1767; \quad p_{C \to G}(+,+) = 0.2886;$$

and

$$p_{G \to C}(+,+) = 0.5625; \quad p_{G \to C}(+,-) = 0.1991;$$
$$p_{G \to C}(-,+) = 0.0255; \quad p_{G \to C}(+,+) = 0.2130.$$

The meaning of these numbers is the following: $p_{C \to G}(+,+)$ gives the percentage of people who, when asked first Q_C and then Q_G, answer that both Clinton and Gore are honest. Analogously, $p_{G \to C}(+,-)$ is the percentage of $\mathcal{P}$ who, when asked first Q_G and then Q_C, answer that Clinton is honest while Gore is not. In particular we see that, while $p_{C \to G}(+,+) = 0.4899$, when asked the same questions in the reverse order the result changes: $p_{G \to C}(+,+) = 0.5625$. These data can be found in [118, 122].

Suppose now that $\mathcal{P}$ is in a normalized state Ψ. When people are asked Q_C, the new state of $\mathcal{P}$ is $P_{+,+}^{\theta,C} \Psi$ or $P_{-,-}^{\theta,C} \Psi$, depending on the answers of the people (+, that is, yes; or −, that is, no). Then, if we now ask Q_G, we repeat the same operation on the two different vectors and we get

$$P_{+,+}^{\theta,G} \left(P_{+,+}^{\theta,C} \Psi \right), \qquad P_{+,+}^{\theta,G} \left(P_{-,-}^{\theta,C} \Psi \right),$$

or

$$P_{-,-}^{\theta,G} \left(P_{+,+}^{\theta,C} \Psi \right), \qquad P_{-,-}^{\theta,G} \left(P_{-,-}^{\theta,C} \Psi \right).$$

If we ask $\mathcal{P}$ the same questions in the reverse order, the G-projectors and the C-projectors should be exchanged. Summarizing, we have the following matching conditions:

$$\left\| P_{\beta,\beta}^{\theta,G} P_{\alpha,\alpha}^{\theta,C} \Psi \right\|^2 = p_{C\to G}(\alpha,\beta), \qquad \left\| P_{\alpha,\alpha}^{\theta,C} P_{\beta,\beta}^{\theta,G} \Psi \right\|^2 = p_{G\to C}(\beta,\alpha), \qquad (12.13)$$

for $\alpha, \beta = \pm$.

Since the sum of probabilities for each of the experiments $Q_C \to Q_G$ and $Q_G \to Q_C$ equals one, we have a system of 6 equations. Each state in the four-dimensional complex Hilbert space can be encoded by $3 + 3$ real parameters, $\Psi = x\Phi_{+,+} + y\Phi_{+,-} + z\Phi_{-,+} + w\Phi_{-,-}$, where $\Phi_{\alpha,\beta}$ are those constructed as above. A numerical optimization can be easily carried out, and the vector that fits conditions (12.13) is the following:

$$\Psi = ae^{i\phi_1}\Phi_{+,+} + be^{i\phi_2}\Phi_{+,-} + ce^{i\phi_3}\Phi_{-,+} + \sqrt{1 - a^2 - b^2 - c^2}\,\Phi_{-,-}, \qquad (12.14)$$

with the following values of the parameters: $a = 0.538$; $b = 0.424$; $c = 0.342$; $\phi_1 = 0.217$; $\phi_2 = 0.593$; $\phi_3 = -0.188$. Also, the best value of θ is found in the same way: $\theta = 0.427$. The order of matching between our theoretical results and the experimental data is very high: $\epsilon = 3 \times 10^{-6}$. This suggests that our approach based on non-commuting operators and degrees of compatibility between observables works very well, at least in this concrete problem. More details can be found in [16].

> *Remark:* The uncertainties ΔC_θ and ΔG_θ could also be used in connection with the order effect problem discussed in this section. For instance, they can be used as a further measure of the validity of our approach, by checking whether the errors we make in the fitting analysis producing the vector in (12.14) stands within the window defined by ΔC_θ and ΔG_θ. This analysis is not really necessary, here, due to the fact that the error is already extremely small. However, it could be relevant in absence of such an (evident) good numerical result.

12.4 Conclusions

What we have discussed in this section is a possible role of the Heisenberg inequality, in its Robertson form, in the analysis of some classical problems in decision-making. The two problems we have considered in some details are the following: decision under uncertainty and order effects. In both cases, the role of non-commuting operators is quite evident, and this is reflected in the nontriviality of the Heisenberg inequality, and in the consequence of this.

As we have already mentioned, several possible extensions of our framework could be analyzed since they can be relevant in more, and more refined, problems in decision-making. Also, it would be interesting to introduce some dynamics to the framework. This is relevant since the process of decision-making is, in fact, a dynamical process, so that time is surely an important variable of the model, as we have already seen in Chapters 3, 10 and 11. In this case, of course, the deformed operators H_θ and E_θ in Equations (12.6) and (12.7), and C_θ and G_θ in Equation (12.12), should be endowed with an extra time dependence. This would surely improve the model, since happiness is not really a static feature of the human being, and the opinion on Gore's honesty could change with time as well, also because of some external input, for instance. Of course, a similar extension is not really automatic. For instance, we should first understand how the Hamiltonian must be defined. Then, we should clarify if $H_\theta(t) = (H(t))_\theta$, or not. In other words: Does the deformations of the observables commute with the time evolution? And why? But these are clearly only preliminary questions: many other surely will follow, making this already fascinating problem even more interesting.

13

This Is Not the End...

What we have shown in this book is that some of the essential tools normally adopted in quantum mechanics, like the Heisenberg and Schrödinger equations of motion, linear (bounded or unbounded) operators, Heisenberg uncertainty relation and CCR and CAR and their ladder operators, have the right to be considered outside a microscopic context. This is a pragmatic point of view. In fact, adopting these tools, we can discuss the dynamical behavior of several interesting (and macroscopic) systems, getting, most of the time, interesting (if not useful) results. Of course, since these tools are more complicated than those typically used in classical mechanics (functions, rather than operators), the *degree of difficulty* is often higher than in a standard approach. However, in several situations we have been able to deduce analytical solutions for our problems, exact or approximated, and we have easily (or *almost easily*) found interesting consequences of these.

The fact that the applications discussed in this book are so different should not be seen as something strange. In fact, except for Chapter 12, all the systems we have analyzed are examples of dynamical systems, and, as such, they share some similarities and have some differences: although most often different, in all of them, *time is running*, and then the system changes as a consequence of this evolution. The essential differences between a classical approach and the one considered here are the way in which the dynamics is described and the nature of the relevant variables needed to describe the evolution of the system.

It is possibly worth stressing that we have not really considered here alternative (with respect to our) quantum approaches to macroscopic systems, as those discussed in [17–21], and in [22], for instance. The main reason is that in none of these books is the role of ladder operators as emphasized as in this book. But, of course, overlaps exist and more can be created, hopefully.

13.1 What Next?

We have discussed several aspects of the general framework that we have introduced already some years ago, and that have been refined and refined again, thanks
also to the people involved along the years in this line of research. It will be possibly refined even more in the future. This is part of our work in progress. But
there are many other aspects which are interesting for us. Some of them are briefly
mentioned in what follows.

One of the main limits of our strategy, in our experience, is the possibility of
using a rigorous easy strategy to describe phenomena that are intrinsically *one-
directional*. This is because the Hamiltonian we use is (except for few exceptions,
in this book) Hermitian: $H = H^\dagger$. This means that, if we have an annihilation term
αa, $\alpha \in \mathbb{R}$, H must also contain a creation contribution $\alpha a^\dagger$. But, if for instance H
is the Hamiltonian of some biological system, while it makes sense to imagine that
some organism of the system dies, it makes far less sense to imagine that it can
also be created out of nothing. The approach we have considered often here and
in other applications, to avoid this *strange effect*, is to replace H with an *effective*
Hamiltonian, which is not necessarily Hermitian, or to add a reservoir to the system. In both cases, the system is driven in some specific direction, and *going back*
is almost impossible. However, these strategies have different drawbacks: the use
of an effective Hamiltonian breaks down many of the properties of the quantum
dynamics: this is no longer unitary and, more dangerous, it is no longer an auto-
morphism. This means that $A(t)B(t) \neq (AB)(t)$, for two generic observables A and
B. Then, the analysis of the Heisenberg equations of motion becomes very hard.
The use of a reservoir does not cause these kind of problems, but transforms a possibly very simple quantum system into a system with infinitely many degrees of
freedom, for which some analytic solution can still be computed, but only under
special (but, we should add, natural) assumptions on the full Hamiltonian and on
its *components*.

The alternative we find more exciting and promising, also because of its mathematical elegance, is what we have called the (H, ρ)-induced dynamics. In our
opinion, this is both interesting as a purely mathematical tool, and in fact it fits
well into the so-called *algebraic quantum mechanics* [43], which is used mainly
for systems with infinite degrees of freedom, and for applications. We have considered several such applications in this book, but many others are possible, also
in view of its interpretation as a step-wise version of a (much more complicated)
time-dependent Hamiltonian. This will be a possible line of research in the next
future.

Another possible extension of our framework that we plan to construct soon
merges the ideas discussed in Chapter 12 with the dynamical approach adopted
in the various applications considered in this book, and in [1], where the time

evolution was always deduced for commuting observables, described in terms of suitable number-like operators. It is surely interesting to extend this approach to non-commuting observables or, better to say, to observables which could commute or not, depending on some particular factor. This is exactly what happens for compatible and incompatible questions, as we have widely discussed in the previous chapter. More in detail, let $\hat{N}$ and $\hat{M}$ be two Hermitian observables of a system $\mathcal{S}$, and let $H = H^\dagger$ be its Hamiltonian. $\hat{N}$ and $\hat{M}$ can be *compatible*, and then $[\hat{N}, \hat{M}] = 0$, or incompatible, so that $[\hat{N}, \hat{M}] \neq 0$. If we let the (undeformed, see below) system to evolve according to the Heisenberg rule, $\hat{N}(t) = e^{iHt}\hat{N}e^{-iHt}$ and $\hat{M}(t) = e^{iHt}\hat{M}e^{-iHt}$, it is clear that similar commutators are found between $\hat{N}(t)$ and $\hat{M}(t)$ at each time t. More explicitly,

$$\left[\hat{N}(t), \hat{M}(t)\right] = e^{iHt}\left[\hat{N}, \hat{M}\right]e^{-iHt},$$

and therefore, in particular, if $\hat{N}$ and $\hat{M}$ commute at $t = 0$, they commute at each time. This is exactly the case in all the examples considered in this book, from Chapter 3 to Chapter 11, where all the observables are assumed to commute at $t = 0$ and, in fact, it was always possible to find a common set of eigenvectors. Now, suppose we want to introduce some possible *connection* between $\hat{N}$ and $\hat{M}$, and suppose this connection should exist already at $t = 0$. Then, we could follow the idea described in Chapter 12, and deform $\hat{N}$ and $\hat{M}$ using a suitable choice of compatibility operators, introduced in the same spirit as R_θ and V_θ in Equation (12.5). Then we get the deformed operators $\hat{N}_\theta$ and $\hat{M}_\theta$, with $[\hat{N}_\theta, \hat{M}_\theta] \neq 0$, in general (i.e., for some θ), even when $[\hat{N}, \hat{M}] = 0$. The time evolution of these operators is, with obvious notation, $\hat{N}_\theta(t)$ and $\hat{M}_\theta(t)$. But this result should be compared with $(\hat{N}(t))_\theta$ and $(\hat{M}(t))_\theta$, which are the result of the deformation on the time-evoluted operators. Of course, $\hat{N}_\theta(t) = (\hat{N}(t))_\theta$ and $\hat{M}_\theta(t) = (\hat{M}(t))_\theta$ if the Heisenberg dynamics commute with the θ-deformation, but not otherwise. This poses a lot of questions: When do we have this commutativity? And what does it mean? Which kind of systems can be usefully analyzed using this strategy? Many other questions are expected to emerge in concrete situations, of course, which look relevant both from their purely applicative potentialities and for the mathematical difficulties that should be considered.

We have left for the end the possibly most relevant problem to consider as soon as possible: after so many applications, in so many different topics, we really need to test the general strategy not only at a *qualitative level*, but also checking its *quantitative aspects*. It must be fully clarified that our strategy is not only able to produce reasonable results but that it is also able to reproduce data! We are optimistic, in fact, since when we have compared our theoretical results with data, they were very much in agreement (see Chapters 9 and 12, or references [59] and [60]). So, hopefully, this will not be the last book on these methods...

References

[1] F. Bagarello, *Quantum dynamics for classical systems: With applications of the number operator*, John Wiley & Sons, Hoboken, NJ, 2012.

[2] E. Merzbacher, *Quantum mechanics*, Wiley, New York, 1970.

[3] A. Messiah, *Quantum mechanics*, vol. 2, North Holland Publishing Company, Amsterdam, the Netherlands, 1962.

[4] F. Bagarello, G. Bravo, F. Gargano, L. Tamburino, *Large-scale effects of migration and conflict in pre-agricultural human groups: insights from a dynamic model*, PLoS ONE, 12, e0172262, 2017.

[5] F. Bagarello, R. Di Salvo, F. Gargano, F. Oliveri, (H, ρ)-*induced dynamics and the quantum game of life*, Applied Mathematical Modelling, **43**, 15–32, 2017.

[6] F. Bagarello, R. Di Salvo, F. Gargano, F. Oliveri, (H, ρ)-*induced dynamics and large time behaviors*, Physica A, **505**, 355–373 (2018).

[7] F. Bagarello, F. Oliveri, *An operator description of interactions between populations with applications to migration*, Mathematical Models & Methods in Applied Sciences, **23**, 471–492, 2013.

[8] F. Bagarello, F. Gargano, F. Oliveri, *A phenomenological operator description of dynamics of crowds: Escape strategies*, Applied Mathematical Modelling, **39**, 2276–2294, 2015.

[9] F. Bagarello, F. Oliveri, *An operator-like description of love affairs*, SIAM Journal of Applied Mathematics, **70**, 3235–3251, 2011.

[10] F. Bagarello, *Damping in quantum love affairs*, Physica A, **390**, 2803–2811, 2011.

[11] F. Bagarello, I. Basieva, A. Khrennikov, *Quantum field inspired model of decision making: Asymptotic stabilization of the belief state via interaction with surrounding mental environment*, Journal of Mathematical Psychology, **82**, 159–168, 2018.

[12] F. Bagarello, *An operator view on alliances in politics*, SIAM Journal of Applied Mathematics, **75**, No. 2, 564–584, 2015.

[13] F. Bagarello, F. Gargano, *Modeling interactions between political parties and electors*, Physica A, **481**, 243–264, 2017.

[14] F. Bagarello, E. Haven, A. Khrennikov, *A model of adaptive decision making from representation of information environment by quantum fields*, Philosophical Transactions A, **375**, 20170162, 2017.

[15] F. Bagarello, *A quantum-like view to a generalized two players game*, International Journal of Theoretical Physics, **54**, No. 10, 3612–3627, 2015.

[16] F. Bagarello, I. Basieva, A. Khrennikov, E. Pothos, *Quantum like modeling of decision making: Quantifying uncertainty with the aid of Heisenberg-Robinson inequality*, Journal of Mathematical Psychology, **84**, 49–56, 2018.

[17] D. Abbott, P.C.W. Davies, A.K. Pati, *Quantum aspects of life*, Imperial College Press, London, UK, 2008.

[18] B.E. Baaquie, *Quantum finance, path integrals and Hamiltonians for options and interest rates*, Cambridge University Press, New York, 2004.

[19] B.E. Baaquie, *Interest rates and coupon bonds in quantum finance*, Cambridge University Press, Cambridge, UK, 2009.

[20] J.R. Busemeyer, P.D. Bruza, *Quantum models of cognition and decision*, Cambridge University Press, Cambridge, UK, 2012.

[21] E. Haven, A. Khrennikov, *Quantum social science*, Cambridge University Press, 2013.

[22] A. Khrennikov, *Ubiquitous quantum structure: From psychology to finance*, Springer, Berlin, Germany, 2010.

[23] F. Bagarello, J.P. Gazeau, F.H. Szafraniec, M. Znojil, Eds., *Non-selfadjoint operators in quantum physics: Mathematical aspects*, John Wiley & Sons, Hoboken, NJ, 2015.

[24] C. Bender, *Making sense of non-hermitian hamiltonians*, Reports on Progress in Physics, **70**, 947–1018, 2007.

[25] A. Mostafazadeh, *Pseudo-Hermitian representation of quantum mechanics*, International Journal of Geometric Methods in Modern Physics, **7**, 1191–1306, 2010.

[26] F. Bagarello, F. Gargano, *Non-hermitian operator modelling of basic cancer cell dynamics*, Entropy, 2018. doi:10.3390/e20040270.

[27] M. Reed, B. Simon, *Methods of modern mathematical physics*, I, Academic Press, New York, 1980.

[28] J.-P. Antoine, A. Inoue, C. Trapani, *Partial*-algebras and their operator realizations*, Kluwer, Dordrecht, the Netherlands, 2002.

[29] F. Bagarello, *Algebras of unbounded operators and physical applications: A survey*, Reviews in Mathematical Physics, **19**, No. 3, 231–272, 2007.

[30] K. Schmüdgen, *Unbounded operator algebras and representation theory*, Birkhäuser, Basel, Switzerland, 1990.

[31] P. Roman, *Advanced quantum mechanics*, Addison-Wesley, New York, 1965.

[32] F. Bagarello, *Quons, coherent states and intertwining operators*, Physics Letters A, **373**, 2637–2642, 2009.

[33] F. Bagarello, *Deformed quons and bi-coherent states*, Proceedings of the Royal Society A, **473**, 20170049, 2017.

[34] V.V. Eremin, A.A. Meldianov, *The q-deformed harmonic oscillator, coherent states and the uncertainty relation*, Theoretical and Mathematical Physics, **147**, No. 2, 709–715, 2006.

[35] D.I. Fivel, *Interpolation between Fermi and Bose statistics using generalized commutators*, Physical Review Letters, **65**, 3361–3364, 1990; Erratum, Physical Review Letters, **69**, 2020, 1992.

[36] O.W. Greenberg, *Particles with small violations of Fermi or Bose statistics*, Physical Review D, **43**, 4111–4120, 1991.

[37] R.N. Mohapatra, *Infinite statistics and a possible small violation of the Pauli principle*, Physics Letters B, **242**, 407–411, 1990.

[38] B. Bagchi, S.N. Biswas, A. Khare, P.K. Roy, *Truncated harmonic oscillator and parasupersymmetric quantum mechanics*, Pramana, **49**, No. 2, 199–204, 1997.

[39] H.A. Buchdahl, *Concerning a kind of truncated quantized linear harmonic oscillator*, American Journal of Physics, **35**, 210–218, 1967.

[40] F. Bagarello, *Deformed canonical (anti-)commutation relations and non hermitian hamiltonians*, in *Non-selfadjoint operators in quantum physics: Mathematical aspects*, F. Bagarello, J.P. Gazeau, F.H. Szafraniec and M. Znojil, Eds., John Wiley & Sons, Hoboken, 2015.

[41] F. Bagarello, *Finite-dimensional pseudo-bosons: A non-Hermitian version of the truncated harmonic oscillator*, Physics Letters A. doi:10.1016/j.physleta.2018.06.044.

[42] J.J. Sakurai, *Modern quantum mechanics*, Addison-Wesley, Reading, MA, 1994.

[43] O. Bratteli and D.W. Robinson, *Operator algebras and quantum statistical mechanics 1*, Springer-Verlag, New York, 1987.

[44] G.L. Sewell, *Quantum theory of collective phenomena*, Oxford University Press, Oxford, UK, 1989.

[45] S.M. Barnett, P.M. Radmore, *Methods in theoretical quantum optics*, Clarendon Press, Oxford, UK, 1997.

[46] Y. Ben-Aryeh, A. Mann, I. Yaakov, *Rabi oscillations in a two-level atomic system with a pseudo-hermitian hamiltonian*, Journal of Physics A, **37**, 12059–12066, 2004.

[47] O. Cherbal, M. Drir, M. Maamache, D.A. Trifonov, *Fermionic coherent states for pseudo-Hermitian two-level systems*, Journal of Physics A, **40**, 1835–1844, 2007.

[48] F. Bagarello, *Linear pseudo-fermions*, Journal of Physics A, **45**, 444002, 2012.

[49] S. Pascazio, *All you ever wanted to know about quantum Zeno effect in 70 minutes*, Open Systems & Information Dynamics, **21**, 1440007, 2014.

[50] S. Galam, *Sociophysics, A physicist's modeling of psycho-political phenomena*, Springer, Berlin, Germany, 2012.

[51] B. Buonomo, A. d'Onofrio, *Modeling the influence of publics memory on the corruption popularity dilemma in politics*, Journal of Optimization Theory and Applications, **158**, 554–575, 2013.

[52] O.A. Davis, M.J. Hinich, P. Ordeshook, *An expository development of a mathematical model of the electoral process*, The American Political Science Review, **64**, No. 2, 426–448, 1970.

[53] G. Feichtinger, F. Wirl, *On the stability and potential cyclicity of corruption in governments subject to popularity constraints*, Mathematical Social Sciences, **28**, 113–131, 1994.

[54] S. Galam, *Majority rule, hierarchical structures, and democratic totalitarianism: A statistical approach*, Journal of Mathematical Psychology, **30**, 426–434, 1986.

[55] S. Galam, *The drastic outcomes from voting alliances in three-party democratic voting (1990→2013)*, Journal of Statistical Physics, **151**, 46–68, 2013.

[56] S.E. Page, L.M. Sander, C.M. Schneider-Mizell, *Conformity and dissonance in generalized voter models*, Journal of Statistical Physics, **128**, 1279–1287, 2007.

[57] F. Palombi, S. Toti, *Stochastic dynamics of the multi-state voter model over a network based on interacting cliques and zealot candidates*, Journal of Statistical Physics, **156**, 336–367, 2014.

[58] G. Raffaelli, M. Marsili, *Statistical mechanics model for the emergence of consensus*, Physical Review E, **72**, 016114, 2005.

[59] R. Di Salvo, M. Gorgone, F. Oliveri, *Political dynamics affected by turncoats*, International Journal of Theoretical Physics, **56**, No. 11, 3604–3614, 2017.

[60] R. Di Salvo, M. Gorgone, F. Oliveri, *(H, ρ)-induced political dynamics: Facets of the disloyal attitudes into the public opinion*, International Journal of Theoretical Physics, **56**, No. 12, 3912–3922, 2017.

[61] P. Khrennikova, E. Haven, A. Khrennikov, *An application of the theory of open quantum systems to model the dynamics of party governance in the US Political System*, International Journal of Theoretical Physics, **53**, No. 4, 1346–1360, 2014.

[62] P. Khrennikova, *Order effect in a study on voters preferences: Quantum framework representation of the observables*, Physica Scripta, **163**, 18, 2014.

[63] P. Khrennikova, *Quantum like modelling of the nonseparability of voters preferences in the US political system*, In Atmanspacher et al. (Eds.), Lecture notes in computer science: Vol. 8951. Quantum Interactions, 8th International Conference, QI 2014, Filzbach, Switzerland, June 30–July 3, 2014 (pp. 196–209).

[64] P. Khrennikova, *Quantum dynamical modeling of competition and cooperation between political parties: The coalition and non-coalition equilibrium model*, Journal of Mathematical Psychology, **71**, 3950, 2016.

[65] M. Makowski, E.W. Piotrowski, *Decisions in elections transitive or intransitive quantum preferences*, Journal of Physics A, **44**, No. 21, 215303, 2011.

[66] F. Bagarello, A.M. Cherubini, F. Oliveri, *An operatorial description of desertification*, SIAM Journal of Applied Mathematics, **76**, No. 2, 479–499, 2016.

[67] E.M. Pothos, J.R. Busemeyer, *Can quantum probability provide a new direction for cognitive modeling?* Behavioral and Brain Sciences, **36**, 255327, 2013.

[68] F. Bagarello, *An improved model of alliances between political parties*, Ricerche di Matematica, **65**, No. 2, 399–412, 2016.

[69] J.F. Reynolds, D.M. Stafford Smith, E.F. Lambin, B.L. Turner II, M. Mortimore, S.P.J. Batterbury, T.E. Downing et al., *Global desertification: Building a science for dryland development*, Science, **316**, 847–851, 2007.

[70] V. Dakos, S.R. Carpenter, W. Brock, A.M. Ellison, V. Guttal, A.R. Ives, S. Kéfi, V. Livina, D.A. Seekell, E.H. van Nes, M. Scheffer, *Methods for detecting early warnings of critical transitions in time series illustrated using simulated ecological data*, PLoS ONE, **7**, e41010, 2012.

[71] S. Kéfi, V. Guttal, W.A. Brock, S.R. Carpenter, A.M. Ellison, V.N. Livina, D.A. Seekell, M. Scheffer, E.H. van Nes, V. Dakos, *Early warning signals of ecological transitions: Methods for spatial patterns*, PLoS One, **9**, E92097, 2014.

[72] R. Corrado, A.M. Cherubini, C. Pennetta, *Critical desertification transition in semi-arid ecosystems: The role of local facilitation and colonization rate*, Communications in Nonlinear Science and Numerical Simulation, **22**, 3–12, 2015.

[73] E.W. Dijkstra, *A note on two problems in connexion with graphs*, Numerische Mathematik, **1**, 269–271, 1959.

[74] D.-M. Shi, B.-H. Wang, *Evacuation of pedestrians from a single room by using snowdrift game theories*. Physical Review E, **87**, 022802, 2013.

[75] R.-Y. Guo, H.J. Huang, *Route choice in pedestrian evacuation: Formulated using a potential field*. Journal of Statistical Mechanics: Theory and Experiment, **4**, P04012, 2011.

[76] D. Helbing, I.J. Farkas, T. Vicsek. *Simulating dynamical features of escape panic*, Nature, **407**, 487–490, 2000.

[77] A. Kirchner, A. Schadschneider, *Simulation of evacuation processes using a bionics-inspired cellular automaton model for Pedestrian dynamics*, Physica A, **312**, 260–276, 2002.

[78] R. Nagai, M. Fukamachi, T. Nagatani, *Evacuation of crawlers and walkers from corridor through an exit*, Physica A, **367**, 449–460, 2006.

[79] A. Varas, M.D. Cornejo, D. Mainemer, B. Toledo, J. Rogan, V. Muñoz, J.A. Valdivia, *Cellular automaton model for evacuation process with obstacles*, Physica A, **2**, 631–642, 2007.

[80] W. Yuan, K.H. Tan, *A model for simulation of crowd behaviour in the evacuation from a smoke-filled compartment*, Physica A, **390**, 4210–4218, 2011.

[81] Z. Xiaoping, Z. Tingkuan, L. Mengting, *Modeling crowd evacuation of a building based on seven methodological approaches*, Building and Environment, **44**, 437–445, 2009.

[82] Z. Xiaoping, T.K. Zhong, M.T. Liu, *Study on numeral simulation approaches of crowd evacuation*, Journal of System Simulation, **21**, 3503–3508, 2009.

[83] A.V. Brilkov, V.V. Ganusov, E.V. Morozova, N.S. Pechurkin, *Computer modeling of the biotic cycle formation in a closed ecological system*, Advances in Space Research, **27**, 1587–1592, 2001.

[84] D.L. De Angelis, *Mathematical modeling relevant to closed artificial ecosystems*, Advances in Space Research, **31**, 1657–1665, 2003.

[85] Y.M. Svirezhev, V.P. Krysanova, A.A. Voinov, *Mathematical modelling of a fish pond ecosystem*, Ecological Modelling, **21**, 315–337, 1984.

[86] R.M. May, *Stability and complexity in model ecosystems*, Princeton University Press, Princeton, NJ, 1973.

[87] F. Bagarello, F. Oliveri, *Dynamics of closed ecosystems described by operators*, Ecological Modeling, **275**, 89–99, 2014.

[88] R. Di Salvo, F. Oliveri, *On fermionic models of a closed ecosystem with application to bacterial populations*, Atti della Accademia Peloritana dei Pericolanti, **94**, No. 2, A5, 2016.

[89] R. Di Salvo, F. Oliveri, *An operatorial model for long-term survival of bacterial populations*, Ricerche di Matematica, **65**, 435–447, 2016.

[90] T.R. Robinson, A.M. Fry, E. Haven, *Quantum counting: Operator methods for discrete population dynamics with applications to cell division*, Progress in Biophysics and Molecular Biology, **130**, 106–119, 2017.

[91] F. Bagarello, R. Passante, C. Trapani, *Non-Hermitian Hamiltonians in Quantum Physics*, Selected Contributions from the 15th International Conference on Non-Hermitian Hamiltonians in Quantum Physics, Palermo, Italy, 18–23 May 2015, Springer, Cham, Switzerland, 2016.

[92] T.M. Walker, C.J. Burger, K.D. Elgert, *Tumor growth alters T cell and macrophage production of and responsiveness to granulocyte-macrophage colony-stimulating factor: Partial dysregulation through interleukin-10*, Cellular Immunology, **154**, No. 2, 342–357, 1994.

[93] M. Gardner, *Mathematical games: The fantastic combinations of John Conways new solitiare game of Life*, Scientific American, **223**, No. 10, 120, 1970.

[94] M. Gardner, *Mathematical games: On cellular automata, self-reproduction, the Garden of Eden and the game of Life*, Scientific American, **224**, No. 2, 116, 1971.

[95] J. Lee, S. Adachi, F. Peper, K. Morita, *Asynchronous game of life*, Physica D, **194**, 369–384, 2004.

[96] P. Arrighi, J. Grattage, *A quantum game of life*, TUSC. Journées Automates Cellulaires 2010, Finland, 31–42, 2010.

[97] D. Bleh, T. Calarco, S. Montagero, *Quantum game of life*, Europhysics Letters A, **97**, 20012, 2012.

[98] A.P. Flitney, D. Abbott, *Towards a quantum game of life*, Game of Life Cellular Automata, Springer, London, UK, 465–486, 2010.

[99] A. Golberg, A.M. Mychajliw, E.A. Hadly, *Post-invasion demography of prehistoric humans in South America*, Nature, **532**, 232–245, 2016.

[100] E. Boserup, *The conditions of agricultural growth: The economics of agrarian change under population pressure*, Transaction Publishers, New Brunswick, NJ, 2005.

[101] G. Bravo, L. Tamburino, *Are two resources really better than one? Some unexpected results of the availability of substitutes*, Journal of Environmental Management, **92** No. 11, 2865–2874, 2011.

[102] M.L. Sentís, *Quantum theory of open systems*, Thesis, ETH Zuerich, 2002.

[103] A.S. Kyle, *Continuous auctions an insider trading*, Econometrica, **53**, 1315–1335, 1985.

[104] E. Haven, *The variation of financial arbitrage via the use of an information wave function*, International Journal of Theoretical Physics, **51**, 193–199, 2008.

[105] F. Bagarello, E. Haven, *The role of information in a two-traders market*, Physica A, **404**, 224–233, 2014.

[106] F. Bagarello, E. Haven, *Towards a formalization of a two traders market with information exchange*, Physica Scripta, **90**, 015203, 2015.

[107] P.M. Agrawal, R. Sharda, *OR Forum – Quantum mechanics and human decision making*, Operations Research, **61** No. 1, 1–16, 2013.

[108] M. Asano, M. Ohya, Y. Tanaka, I. Basieva, A. Khrennikov, *Quantum-like dynamics of decision-making*, Physica A, **391**, 2083–2099, 2012.

[109] J.R. Busemeyer, E.M. Pothos, *A quantum probability explanation for violations of "rational" decision theory*, Proeedings of the Royal Society B, **276** No. 1665, 2171–2178, 2009.

[110] J.R. Busemeyer, Z. Wang, J.T. Townsend, *Quantum dynamics of human decision-making*, Journal Mathematical Psychology, **50**, 220–241, 2006.

[111] E. Conte, O. Todarello, A. Federici, F. Vitiello, M. Lopane, A. Khrennikov, J.P. Zbilut, *Some remarks on an experiment suggesting quantum-like behavior of cognitive entities and formulation of an abstract quantum mechanical formalism to describe cognitive entity and its dynamics*, Chaos Solitons Fractals, **31**, 1076–1088, 2006.

[112] E. Manousakis, *Quantum formalism to describe binocular rivalry*, Biosystems, **98**, No. 2, 5766, 2009.

[113] I. Martínez-Martínez, *A connection between quantum decision theory and quantum games: The Hamiltonian of Strategic Interaction*, Journal of Mathematical Psychology, **58**, 3344, 2014.

[114] A. Bevilacqua Leoneti, *Utility function for modeling group multicriteria Decision Making problems as games*, Operations Research Perspectives, **3**, 21–26, 2016.

[115] M. Machina, *Expected utility analysis without the independence axiom*, Econometrica, **50**, No. 2, 277–323, 1982.

[116] C.L. Sheng, *A general utility functions for decision making*, Mathematical Modeling, **5**, 265–274, 1984.

[117] M. Asano, M. Ohya, Y. Tanaka, I. Basieva, A. Khrennikov, *Quantum-like model of brains functioning: Decision making from decoherence*, Journal of Theoretical Biology, **281**, No. 1, 56–64, 2011.

[118] D.W. Moore, *Measuring new types of question-order effects*, Public Opinion Quarterly, **66**, 8091, 2002.

[119] M. Ozawa, *Quantum perfect correlations*, Annals of Physics, **321**, 744769, 2006.

[120] M. Ozawa, *Quantum reality and measurement: A quantum logical approach*, Foundations of Physics, **41**, 592607, 2011.

[121] M. Ozawa, *Probabilistic interpretation of quantum theory*, New Generation Computing, **34**, 125–152, 2016.

[122] Z. Wang, J.R. Busemeyer, *A quantum question order model supported by empirical tests of an a priori and precise prediction*, Topics in Cognitive Sciences, **5**, 689–710, 2013.

Index